Population growth, economic development, and urbanisation are putting pressure on the world's water resources. The need to develop a more sustainable approach to the management of water resources, and the need to protect ecosystems where these resources are found, has become increasingly important.

This comprehensive volume describes how ecosystem services-based approaches can assist in addressing major global and regional water challenges, such as climate change, biodiversity loss, and water security in the developing world, by integrating scientific knowledge from different disciplines, such as hydrological modelling and environmental economics. As well as consolidating current thinking, the book also takes a more innovative approach to these challenges, involving disciplines such as psychology and international law.

Empirical assessments at the national, catchment, and regional levels are used to critically appraise this systemic approach, and the merits and potential limitations are presented. The practicalities of this approach with regard to water resources management, nature conservation, and sustainable business practices are discussed, and the role of society in underpinning the concept of ecosystem services is explored.

Presenting new insights and perspectives on how to shape future strategies, this contributory volume is a valuable reference for researchers, academics, students, and policy makers, in environmental studies, hydrology, water resource management, ecology, environmental law, policy and economics, and conservation biology.

DR. JULIA MARTIN-ORTEGA is Senior Environmental Economist at the James Hutton Institute, Scotland. With ten years of research experience, she has an interdisciplinary background in environmental sciences and economics, with an emphasis on environmental valuation, and an interest in qualitative approaches to social science. Her research, focusing on the understanding of the relationships between people and water ecosystem services, has informed policy making in relation to issues such as the implementation of the European Water Framework Directive and the development of the Water Resources (Scotland) Act. She currently leads the James Hutton Institute – UNESCO Global Initiatives Directive on water ecosystem services.

PROFESSOR ROBERT C. FERRIER is Director of Research Impact at the James Hutton Institute. His extensive research interests centre on the role of policy and land use change on water resources and the development of systems-based approaches for sustainable management. He has 30 years' experience in water resources management, particularly the hydrological, hydrochemical, and ecological consequences of environmental change, in both freshwater and coastal systems. He is also Director of the Centre of Expertise for Waters (CREW), and is involved in delivering the Scottish Government's Hydro Nation agenda, both domestically and internationally.

PROFESSOR IAIN J. GORDON is Chief Executive and Director of the James Hutton Institute, Scotland. He has played an active role in promoting the value of biodiversity and its importance in the provision of ecosystem services and human wellbeing, and has managed projects in Africa, Asia, South America, and Australia, including major research portfolios on land management to improve the water quality of the catchments flowing into the lagoon of the Great Barrier Reef.

PROFESSOR SHAHBAZ KHAN is Deputy Director of Asia and the Pacific Regional Science Bureau of UNESCO, where he advises UN member states on environmental policies, review of curricula, and securing multilateral support for research and education projects, especially in the Asia-Pacific and African regions. He has been involved in a range of innovative water management programmes through his work with the Australian government. Professor Khan was previously Chief of Water and Sustainable Development at UNESCO, Paris, and prior to this was Professor of Hydrology and Director of the International Centre of Water at the Charles Sturt University, Australia.

INTERNATIONAL HYDROLOGY SERIES

The **International Hydrological Programme** (IHP) was established by the United Nations Educational, Scientific and Cultural Organization (UNESCO) in 1975 as the successor to the International Hydrological Decade. The long-term goal of the IHP is to advance our understanding of processes occurring in the water cycle and to integrate this knowledge into water resources management. The IHP is the only UN science and educational programme in the field of water resources, and one of its outputs has been a steady stream of technical and information documents aimed at water specialists and decision-makers.

The **International Hydrology Series** has been developed by the IHP in collaboration with Cambridge University Press as a major collection of research monographs, synthesis volumes, and graduate texts on the subject of water. Authoritative and international in scope, the various books within the series all contribute to the aims of the IHP in improving scientific and technical knowledge of fresh-water processes, in providing research know-how and in stimulating the responsible management of water resources.

TITLES IN PRINT IN THIS SERIES
M. Bonell, M. M. Hufschmidt, and J. S. Gladwell *Hydrology and Water Management in the Humid Tropics: Hydrological Research Issues and Strategies for Water Management*
Z. W. Kundzewicz *New Uncertainty Concepts in Hydrology and Water Resources*
R. A. Feddes *Space and Time Scale Variability and Interdependencies in Hydrological Processes*
J. Gibert, J. Mathieu, and F. Fournier *Groundwater/Surface Water Ecotones: Biological and Hydrological Interactions and Management Options*
G. Dagan and S. Neuman *Subsurface Flow and Transport: A Stochastic Approach*
J. C. van Dam *Impacts of Climate Change and Climate Variability on Hydrological Regimes*
D. P. Loucks and J. S. Gladwell *Sustainability Criteria for Water Resource Systems*
J. J. Bogardi and Z. W. Kundzewicz *Risk, Reliability, Uncertainty, and Robustness of Water Resource Systems*
G. Kaser and H. Osmaston *Tropical Glaciers*
I. A. Shiklomanov and J. C. Rodda *World Water Resources at the Beginning of the Twenty-First Century*
A. S. Issar *Climate Changes during the Holocene and their Impact on Hydrological Systems*
M. Bonell and L. A. Bruijnzeel *Forests, Water and People in the Humid Tropics: Past, Present and Future Hydrological Research for Integrated Land and Water Management*
F. Ghassemi and I. White *Inter-Basin Water Transfer: Case Studies from Australia, United States, Canada, China and India*
K. D. W. Nandalal and J. J. Bogardi *Dynamic Programming Based Operation of Reservoirs: Applicability and Limits*
H. S. Wheater, S. Sorooshian, and K. D. Sharma *Hydrological Modelling in Arid and Semi-Arid Areas*
J. Delli Priscoli and A. T. Wolf *Managing and Transforming Water Conflicts*
H. S. Wheater, S. A. Mathias, and X. Li *Groundwater Modelling in Arid and Semi-Arid Areas*
L. A. Bruijnzeel, F. N. Scatena, and L. S. Hamilton *Tropical Montane Cloud Forests*
S. Mithen and E. Black *Water, Life and Civilization: Climate, Environment and Society in the Jordan Valley*
K. A. Daniell *Co-Engineering and Participatory Water Management*
R. Teegavarapu *Floods in a Changing Climate: Extreme Precipitation*
P. P. Mujumdar and D. Nagesh Kumar *Floods in a Changing Climate: Hydrologic Modeling*
G. Di Baldassarre *Floods in a Changing Climate: Inundation Modelling*
S. Simonović *Floods in a Changing Climate: Risk Management*
S. Hendry *Frameworks for Water Law Reform*
J. Martin-Ortega, R. C. Ferrier, I. J. Gordon, and S. Khan *Water Ecosystem Services: A Global Perspective*

Water Ecosystem Services

A Global Perspective

Edited by

Julia Martin-Ortega
The James Hutton Institute

Robert C. Ferrier
The James Hutton Institute

Iain J. Gordon
The James Hutton Institute

Shahbaz Khan
UNESCO

International Hydrology Series

CAMBRIDGE
UNIVERSITY PRESS

CAMBRIDGE
UNIVERSITY PRESS

University Printing House, Cambridge CB2 8BS, United Kingdom

One Liberty Plaza, 20th Floor, New York, NY 10006, USA

477 Williamstown Road, Port Melbourne, VIC 3207, Australia

4843/24, 2nd Floor, Ansari Road, Daryaganj, Delhi - 110002, India

79 Anson Road, #06-04/06, Singapore 079906

Cambridge University Press is part of the University of Cambridge.

It furthers the University's mission by disseminating knowledge in the pursuit of
education, learning and research at the highest international levels of excellence.

www.cambridge.org
Information on this title: www.cambridge.org/9781107496187

First published 2015
First paperback edition 2017

A catalogue record for this publication is available from the British Library

Library of Congress Cataloging in Publication data
Martin-Ortega, Julia, author.
   Water ecosystem services : a global perspective.
     pages cm. – (International hydrology series)
Authors: J. Martin-Ortega, R.C. Ferrier, I.J. Gordon, and S. Khan.
ISBN 978-1-107-10037-4 (Hardback)
1. Ecohydrology. 2. Ecosystem services. I. Title. II. Series: International hydrology series.
QH541.15.E19M37 2015
577.6–dc23   2014039488

ISBN UNESCO 978-92-3-100068-3
ISBN Cambridge 978-1-107-10037-4 Hardback
ISBN Cambridge 978-1-107-49618-7 Paperback

Contents

List of contributors   *page* vii
Preface   xi

1   Introduction   1
    *Iain J. Gordon, Julia Martin-Ortega, and Robert
    C. Ferrier*

2   What defines ecosystem services-based approaches?   3
    *Julia Martin-Ortega, Dídac Jorda-Capdevila, Klaus
    Glenk, and Kirsty L. Holstead*

Part I Addressing global challenges

3   Assessing climate change risks and prioritising
    adaptation options using a water ecosystem
    services-based approach   17
    Samantha J. Capon and Stuart E. Bunn

4   Operationalizing an ecosystem services-based approach
    for managing river biodiversity   26
    *Catherine M. Febria, Benjamin J. Koch,
    and Margaret A. Palmer*

5   Water for agriculture and energy: the African quest under
    the lenses of an ecosystem services-based approach   35
    Maher Salman and Alba Martinez

**Part II Applying frameworks for water management and
biodiversity conservation under an ecosystem services-based
approach**

6   Using ecosystem services-based approaches in Integrated
    Water Resources Management: perspectives from the
    developing world   49
    Madiodio Niasse and Jan Cherlet

7   Implementation of the European Water Framework
    Directive: what does taking an ecosystem
    services-based approach add?   57
    *Kirsty L. Blackstock, Julia Martin-Ortega,
    and Chris J. Spray*

8   How useful to biodiversity conservation are ecosystem
    services-based approaches?   65
    Craig Leisher

Part III Assessing water ecosystem services

9   The first United Kingdom's National Ecosystem
    Assessment and beyond   73
    *Marije Schaafsma, Silvia Ferrini, Amii R. Harwood,
    and Ian J. Bateman*

10 Using an ecosystem services-based approach to
    measure the benefits of reducing diversions of
    freshwater: a case study in the Murray-Darling
    basin, Australia   82
    *Neville D. Crossman, Rosalind H. Bark, Matthew
    J. Colloff, Darla Hatton MacDonald,
    and Carmel A. Pollino*

11 An ecosystem services-based approach to integrated
    regional catchment management: the South East
    Queensland experience   90
    *Simone Maynard, David James, Stuart Hoverman,
    Andrew Davidson, and Shannon Mooney*

12 Policy support systems for the development of
    benefit-sharing mechanisms for water-related
    ecosystem services   99
    *Mark Mulligan, Silvia Benítez-Ponce,
    Juan S. Lozano-V, and Jorge Leon Sarmiento*

13 Assessing biophysical and economic dimensions of
    societal value: an example for water ecosystem
    services in Madagascar   110
    *Ferdinando Villa, Rosimeiry Portela, Laura Onofri,
    Paulo A. L. D. Nunes, and Glenn-Marie Lange*

14 Rapid land use change impacts on coastal ecosystem
    services: a South Korean case study   119
    Hojeong Kang, Heejun Chang, and Min Gon Chung

Part IV Broadening the perspective

15 Ecosystem services-based approaches to water
    management: what opportunities and challenges
    for business?   129
    Joël Houdet, Andrew Johnstone,
    and Charles Germaneau

16 Key factors for successful application of ecosystem
    services-based approaches to water resources
    management: the role of stakeholder participation   138
    Jos Brils, Al Appleton, Nicolaas van Everdingen,
    and Dylan Bright

17 Cultural ecosystem services, water, and aquatic
    environments   148
    Andrew Church, Rob Fish, Neil Ravenscroft,
    and Lee Stapleton

18 The psychological dimension of water ecosystem
    services   156
    Victor Corral-Verdugo, Martha Frías-Armenta,
    César Tapia-Fonllem, and Blanca Fraijo-Sing

19 The interface between human rights and ecosystem
    services   163
    Stephen J. Turner

20 Water ecosystem services: moving forward   170
    Julia Martin-Ortega, Robert C. Ferrier,
    and Iain J. Gordon

Index   174

Colour plate section between pages 84 and 85

Contributors

Al Appleton
The Cooper Union, New York, New York 10128, USA

Rosalind H. Bark
CSIRO Ecosystem Sciences, GPO Box 2583, Brisbane, Queensland, Australia

Ian J. Bateman
Centre for Social and Economic Research on the Global Environment (CSERGE), School of Environmental Sciences, University of East Anglia, Norwich, NR4 7TJ, UK

Silvia Benítez-Ponce
The Nature Conservancy, Los Naranjos N 44 491 y Azucenas (Monteserrín), 170507 Quito, Ecuador

Kirsty L. Blackstock
Social, Economic and Geographical Sciences Group, The James Hutton Institute, Craigiebuckler, Aberdeen, AB15 8QH, Scotland, UK

Dylan Bright
South West Water, Peninsular House, Rydon Lane, Exeter, EX2 7HR, UK

Jos Brils
Deltares, Princetonlaan 6, 3584 CB Utrecht, the Netherlands

Stuart E. Bunn
Australian Rivers Institute, Griffith University, Nathan, Queensland, Australia

Samantha J. Capon
Australian Rivers Institute, Griffith University, Nathan, Queensland, Australia

Heejun Chang
Department of Geography and Institute for Sustainable Solutions, Portland State University, Portland, Oregon, USA

Jan Cherlet
International Land Coalition, Secretariat at IFAD, Via Paolo di Dono 44, 00142 Rome, Italy

Andrew Church
School of Environment and Technology, University of Brighton, Brighton, BN2 4GJ, UK

Matthew J. Colloff
CSIRO Ecosystem Sciences, GPO Box 1700, Canberra, ACT, 2601, Australia

Victor Corral-Verdugo
Department of Psychology, University of Sonora, Luis Encinas y Rosales S/N, Hermosillo, Sonora, 83000, Mexico

Neville D. Crossman
CSIRO Ecosystem Sciences, Private Bag 2, Glen Osmond, SA, 5064, Australia

Andrew Davidson
SEQ Catchments, PO Box 13204, George St, Queensland, Australia

Nicolaas van Everdingen
Watermaatwerk, Rietveldlaan 11, 6717 KZ Ede, the Netherlands

Catherine M. Febria
Freshwater Ecology Research Group, School of Biological Sciences, University of Canterbury, Christchurch, Canterbury, New Zealand

Robert C. Ferrier
Director of Research Impact, The James Hutton Institute, Craigiebuckler, Aberdeen, AB15 8QH, Scotland, UK

Silvia Ferrini
Centre for Social and Economic Research on the Global Environment (CSERGE), School of Environmental Sciences, University of East Anglia, Norwich, NR4 7TJ, UK; Department of Political Science and International, University of Siena, Italy

Rob Fish
School of Anthropology and Conservation, University of Kent, CT2 7NR, UK

Blanca Fraijo-Sing
Department of Psychology, University of Sonora, Luis Encinas y Rosales S/N, Hermosillo, Sonora, 83000, Mexico

Martha Frías-Armenta
Department of Law, University of Sonora, Luis Encinas y
Rosales S/N, Hermosillo, Sonora, 83000, Mexico

Charles Germaneau
Head of Communications, Synergiz, Paris, France

Klaus Glenk
Land Economy, Environment and Society Group, Scotland's
Rural College (SRUC), King's Buildings, West Mains Road,
Edinburgh, EH9 3JG, Scotland, UK

Min Gon Chung
School of Civil and Environmental Engineering, Yonsei
University, Seoul, South Korea

Iain J. Gordon
Chief Executive Officer, The James Hutton Institute,
Invergowrie, Dundee, DD2 5DA, Scotland, UK

Amii R. Harwood
Centre for Social and Economic Research on the Global
Environment (CSERGE), School of Environmental Sciences,
University of East Anglia, Norwich, NR4 7TJ, UK

Darla Hatton MacDonald
CSIRO Ecosystem Sciences, Private Bag 2, Glen Osmond, SA,
5064, Australia

Kirsty L. Holstead
Social, Economics and Geographical Sciences Group, The James
Hutton Institute, Craigiebuckler, Aberdeen, AB15 8QH,
Scotland, UK

Joël Houdet
Senior Research Fellow, Responsible Natural Resources
Economics Programme Leader, African Center for Technical
Studies (ACTS), Head of Applied Research, Synergiz,
Johannesburg, South Africa

Stuart Hoverman
Hoverman NRM and Associates, 22 Todd Street, Shorncliffe,
Queensland, Australia

David James
University Sunshine Coast, 90 Sippy Downs Dr, Sippy Downs,
Queensland, Australia

Andrew Johnstone
Managing Director, Groundwater Consulting Services – GCS
(Pty) Ltd, Johannesburg, South Africa

Dídac Jorda-Capdevila
Institute for Environmental Science and Technology,
Autonomous University of Barcelona, Barcelona, Bellaterra,
08193, Spain

Hojeong Kang
School of Civil and Environmental Engineering, Yonsei
University, Seoul, South Korea

Shahbaz Khan
UNESCO Regional Science Bureau for Asia and the Pacific,
DKI Jakarta 12110, Indonesia

Benjamin J. Koch
Center for Ecosystem Science and Society,
Northern Arizona University, Box 5620, Flagstaff, Arizona,
86011-5620, USA

Glenn-Marie Lange
World Bank, 818 H Street, NW Washington,
DC 20433, USA

Craig Leisher
Central Science, The Nature Conservancy, 4245 North Fairfax
Drive, Arlington, VA 22203, USA

Juan S. Lozano-V
The Nature Conservancy, Calle 67 # 7-94, Piso 3, Torre
Colfondos 11001000 Bogotá, Colombia

Alba Martinez
Land and Water Division, Food and Agriculture Organization of
the United Nations, Viale delle Terme di Caracalla, 00153
Rome, Italy

Julia Martin-Ortega
Social, Economic and Geographical Sciences Group,
The James Hutton Institute, Craigiebuckler, Aberdeen, AB15
8QH, Scotland, UK

Simone Maynard
Australian National University,
Fenner School of Environment and Society,
Canberra, ACT, Australia

Shannon Mooney
SEQ Catchments, PO Box 13204, George St,
Queensland 4003, Australia

Mark Mulligan
Department of Geography, King's College London, Strand,
London, WC2R 2LS, UK

Madiodio Niasse
International Land Coalition, Secretariat at IFAD,
Via Paolo di Dono 44, 00142 Rome, Italy

Paulo A. L. D. Nunes
Department of Agricultural and Resource Economics,
University of Padova, Via 8 Febbraio 1848, 35122 Padova,
Italy

Laura Onofri
Department of Economics, University Cà Foscari of Venice, Dorsoduro 3246, 30123 Venezia, Italy

Margaret A. Palmer
National Socio-Environmental Synthesis Center (SESYNC), University of Maryland, Maryland, 21401 USA

Carmel A. Pollino
CSIRO Land and Water, GPO Box 1666, Canberra, ACT, Australia

Rosimeiry Portela
Conservation International, 2011 Crystal Drive 500, Arlington, Virginia, USA

Neil Ravenscroft
School of Environment and Technology, University of Brighton, Brighton, BN2 4GJ, UK

Maher Salman
Land and Water Division, Food and Agriculture Organization of the United Nations, Viale delle Terme di Caracalla, 00153 Rome, Italy

Jorge Leon Sarmiento
The Nature Conservancy, Calle 67 # 7-94 Piso 3, Torre Colfondos 11001000 Bogotá, Colombia

Marije Schaafsma
Centre for Social and Economic Research on the Global Environment (CSERGE), School of Environmental Sciences, University of East Anglia, Norwich, NR4 7TJ, UK

Chris J. Spray
Centre for Water Law, Policy and Science, under the auspices of UNESCO, University of Dundee, Nethergate, Dundee, DD1 4HN, UK

Lee Stapleton
SPRU (Science & Technology Policy Research), University of Sussex, Brighton, BN1 9SL, UK

César Tapia-Fonllem
Department of Psychology, University of Sonora, Luis Encinas y Rosales S/N, Hermosillo, Sonora, 83000, Mexico

Stephen J. Turner
Lincoln Law School, Lincoln University, Brayford Wharf, Lincoln, LN5 7AT, UK

Ferdinando Villa
Basque Centre for Climate Change (BC3), IKERBASQUE, Basque Foundation for Science, Alameda Urquijo 4-4, 48008 Bilbao, Spain

Preface

Rapidly rising demand for food, raw materials, and energy is leading to intensifying human environmental footprints locally, nationally, and globally. This has consequences for the health of ecosystems and the services these provide. By considering the complex socio-economic interactions between the water, carbon-energy, food production, and climate cycles, UNESCO's Natural Science Programme (principally through its International Hydrological Programme and the Man and Biosphere Programme) promotes trans-disciplinary approaches to help restore, enhance, and protect the sustainability of land and water systems. This new book, in the International Hydrology Series, is well aligned with the aims and objectives of UNESCO's International Hydrological Programme by furthering the understanding and championing the potential of the ecosystem services-based approaches. Phase VIII (2014–2021) of the Programme is focused on 'Water security: responses to local, regional, and global challenges' and has a special theme on 'Ecohydrology, engineering harmony for sustainable world', to which this book is of direct relevance. The UNESCO Ecohydrology initiative involves the development of tools that integrate basin-wide human activities with hydrological cycles in order to sustain, improve, and restore the ecological functions of, and services in, river basins and coastal zones as a basis to support positive socio-economic development. Experiences gained through previous phases of the International Hydrological Programme, have shown that freshwater availability will become a major concern if no immediate action is taken to restore and enhance the associated ecosystems. Therefore, the knowledge presented in this book adds great value to inform global efforts on ensuring water security through UNESCO's ongoing initiatives at the river basin level, such as the Hydrology for the Environment, Life and Policy, Ecohydrology and Integrated Water Resources Management Guidelines.

The Hydrology for the Environment, Life and Policy in river basins initiative aims to deliver social, economic, and environmental benefits to stakeholders through research towards the sustainable use of water for human and environmental purposes. Ecosystem services-based approaches fit well with that aim by looking specifically at the complexity in order to improve understanding of the relationships between hydrological processes, water resources management, ecology, socio-economics, and policy making. UNESCO's Ecohydrology initiative aims to enhance water-based ecosystem services through dual regulation of flow and biota in freshwater and estuarine environments. The ecosystem services-based approach is expected to offer adaptable tools to river basin managers who are struggling to implement Integrated Water Resources Management that achieves a better balance between consumptive and environmental uses of water. In this regard, UNESCO's Integrated Water Resources Management Guidelines, at the river basin level, can benefit from the practical case studies presented in this book.

This book will also help open the paradigm lock between the research community and policy makers who struggle to manage complex interactions between biological diversity, climate change, land use change, and the limits and constraints on freshwater use. These represent four of the nine boundaries of the Earth System processes that the Stockholm Resilience Centre recommend are not crossed if environmental change that is disastrous to humanity is to be avoided. The conceptual and theoretical discussions and case studies presented in this book will help guide UNESCO's member states in devising river basin management plans for maintaining and enhancing the multifunctional productivity of water and ecosystem resources to optimise physical, economic, social, and environmental benefits without compromising the quality of these resources.

Finally, I congratulate the editorial team at the James Hutton Institute for their enthusiasm and strong commitment to lead and deliver this complex body of knowledge through close contact with the authors, who are to be praised for their valuable contributions. This collation of knowledge from esteemed academics and practitioners will help in achieving a step change towards enhanced ecosystem service delivery through sustainable land and water management for present and future generations.

Shahbaz Khan
UNESCO Regional Science Bureau for Asia
and the Pacific, Jakarta, Indonesia

1 Introduction

Iain J. Gordon, Julia Martin-Ortega, and Robert C. Ferrier

Fresh water is vital for the function of all terrestrial ecosystems – the flora and the fauna that make up those ecosystems, and, of course, for humans. Humanity relies on water not just for drinking, but also for food production, dealing with waste, providing energy and transport, to name but a few. To meet its needs humanity harnesses water through dams, irrigation networks, and pumps and pipes that supply drinking water and remove wastes. It is estimated that humanity consumes 1000–1700 m^3 of the globe's surface and groundwater resources per year; that is between 22% and 150% of the annual global supply of fresh water (Hoekstra & Wiedmann 2014). This proportion is likely to increase as the global human population increases in the next 30 years and the demands for water in developing countries catches up with that of developed countries. According to the Intergovernmental Panel on Climate Change, changes in climate will amplify existing stress on water availability and will exacerbate different forms of water pollution, with impacts on ecosystems, human health, and water system reliability in large parts of the world (Stocker et al. 2013).

For a number of years, academics have tried to understand the linkages between the water system and human needs and the impacts that anthropogenic activities have on the water system itself. In the early days, the scientific approach sat within individual domains (e.g. hydrology for the water cycle (Thompson 1999); ecology for ecological impacts of water pollution (Abel 1996)). Given the complexity of the interactions and the centrality of humans in the water environment, more recently interdisciplinary approaches have come to the fore (e.g. Ferrier & Jenkins 2010; Renaud & Kuenzer 2014). The latest of such approaches is what we define in this book as ecosystem services-based approaches. These encompass a range of ways of understanding, assessing, and managing ecosystems at which core is the notion of ecosystem services, understood as the benefits that humans obtain from ecosystems.

The water cycle intimately embraces the ecosystems services paradigm. From regulating to provisioning and cultural services, the water environment provides a unique context through which to express the state of natural capital and flows between different ecosystems and the effects they produce on human wellbeing.

Much has been written about ecosystem services, and approaches using this notion are now being applied to the practical management of ecosystems around the world. Given that ten years have passed since the publication of the Millennium Ecosystem Assessment (2005), it is time to reflect on what has been achieved, what lessons can be learnt, and how we can improve the application of ecosystem services-based approaches for managing water ecosystems in the future.

This book aims to develop a better understanding of water as a service delivered by ecosystems, by furthering the understanding and the potential of ecosystem services-based approaches. This understanding is necessary not only to identify and quantify the critical linkages that regulate the interrelationships of hydrology and biota, but also to elucidate how the control of these linkages contributes to environmental sustainability, human livelihoods, and wellbeing.

In this book, leading academic and non-academic authors, from prestigious research institutions, world global organisations, and international non-governmental organisations, describe the forefront of the application of ecosystem services-based approaches to address global water challenges. Recognising that the challenge is multi-faceted, the authors come from a range of disciplinary backgrounds (from hydrological modelling, to environmental economics, through environmental psychology, international law, and ecological sciences) and 'real world' experiences in conservation, water management, and business. The result is a balance between global and world-regional visions and national and regional case studies from across the world.

The second chapter of the book provides an in-depth history of the notion of ecosystem services and proposes a definition of ecosystem services-based approaches based on four defining core elements (i.e. (1) focusing on the status of ecosystems, and the recognition of its effects on human wellbeing; (2) understanding the biophysical underpinning of ecosystems in terms of service delivery; (3) integrating natural and social sciences and other strands of knowledge for a comprehensive understanding of the service delivery process; and (4) assessing the services provided by ecosystems for its incorporation into decision-making). These core elements articulate discussions on a range of broad issues on each of the individual chapters, which are organised in four parts:

- Part I looks at how ecosystem services-based approaches can help address major global challenges, such as climate change, food and energy supply, and biodiversity loss at regional and global scales.
- Part II reflects upon whether the notion of ecosystem services is useful in the context of frameworks for water resources management and biodiversity conservation, with a focus on the practicalities of the implementation of the approach.
- Part III provides examples of assessments of ecosystem services through a number of case studies from across the world, showing the latest advances in the integration of the biophysical quantification of water ecosystem service delivery with economic valuation techniques.
- Part IV broadens the perspective, providing innovative insights from less explored areas such as business, cultural ecosystem services, human rights, beliefs, and emotions towards water ecosystem services and the role of community partnerships.

Addressing global challenges and development goals requires a vision for water management beyond protection and restoration. It has to recognise the carrying capacity of ecosystems threatened by increasing human impact and find ways to enhance the resilience of socio-ecological systems. This book provides a global synthesis of current thinking and applications of ecosystem services-based approaches to inform future water decision-making. The book consolidates current thinking and opens up new perspectives, with contributions from top scholars and practitioners, who take a critical and forward-thinking view aimed at stimulating the debate. The book highlights the potential benefits and challenges of adopting ecosystem services-based approaches and gives an insight on how to shape future strategies for water management and ecosystems conservation.

References

Abel, P. D. (1996) *Water Pollution Biology*. Taylor & Francis, London.

Ferrier, R. C. & Jenkins, A. (2010) *Handbook of Catchment Management*. Wiley, Chichester.

Hoekstra, A. Y. & Wiedmann, T. O. (2014) Humanity's unsustainable environmental footprint. *Science* **344**: 1114–1117.

Millennium Ecosystem Assessment (2005). *Ecosystems and Human Well-being: General Synthesis*. Island Press, Washington, DC.

Renaud, F. G. & Kuenzer, C. (2014) *The Mekong Delta System: Interdisciplinary Analyses of a River Delta*. Springer, New York.

Stocker, T. F. D., Qin, G.-K., Plattner, L. V. *et al.*, (2013) Technical summary. In:, T. F. D. Stocker, G.-K. Qin, M. Plattner, *et al.* (eds), *Climate Change: The Physical Science Basis. Contribution of Working Group I to the Fifth Assessment Report of the Intergovernmental Panel on Climate Change*. Cambridge University Press, Cambridge and New York.

Thompson, S. A. (1999) *Hydrology for Water Management*. AA Balkema, Rotterdam.

2 What defines ecosystem services-based approaches?

Julia Martin-Ortega, Dídac Jorda-Capdevila, Klaus Glenk, and Kirsty L. Holstead

2.1 INTRODUCTION

It has long been held that human life depends on the existence of a finite natural resource base, and that nature contributes to the fulfilment of human needs (Malthus 1888; Meadows *et al.* 1972). This knowledge has led to different and evolving ways of understanding the relationship between humans and nature (Raymond *et al.* 2013). The notion of ecosystem services is one of these, which began to be developed in the late 1960s (King 1966; Helliwell 1969; Study of Critical Environmental Problems 1970; Odum and Odum 1972). How human needs and wellbeing interact with quantities and qualities of the finite natural resource base, and how changes to the natural environment impact on human activities and vice versa, are key questions underlying the conceptual development of ecosystem services and related concepts.

In 2000, the Secretary-General of the United Nations called for a worldwide initiative, the Millennium Ecosystem Assessment, 'to assess the consequences of ecosystem change for human wellbeing and the scientific basis for action needed to enhance the conservation and sustainable use of those systems' (Millennium Ecosystem Assessment 2003). Ecosystem services were defined as 'the benefits that people obtain from ecosystems' and the Millennium Ecosystem Assessment emphasised the need to incorporate the value of ecosystem services into decision-making to reverse increasing degradation of ecosystems. Since the publication of the Millennium Ecosystem Assessment in 2005, economic approaches to the understanding and management of natural resources based on the notion of ecosystem services have been increasingly discussed in the scientific literature (Fisher *et al.* 2009; Norgaard 2010; Ojea *et al.* 2012). The Millennium Ecosystem Assessment was followed by a number of other initiatives to assess ecosystem services, the most significant global assessment being *The Economics of Ecosystem Services and Biodiversity* (Kumar 2010). Other national-level assessments, for example, the UK National Ecosystem Assessment (2011; see Schaafsma *et al.*, this book) and the Spanish Millennium Ecosystem Assessment (EME 2011) have also been published. Incorporation of these assessments into policy making is not yet well established; however, there is clear interest in very diverse contexts across the world. For example, there are ongoing discussions about how to incorporate ecosystem services in the upcoming river basin planning cycles within the Common Implementation Strategy of the European Water Framework Directive (Martin-Ortega 2012; Blackstock *et al.*, this book). Also, in Malawi, the Decentralised Environmental Management Guidelines produced by the Ministry of Local Government and Development (2012) to guide environmental management at the district level include elements of an ecosystem services-based approach (Waylen and Martin-Ortega 2013), and the South East Queensland Ecosystem Services Framework in Australia provides an example at the catchment level (Maynard *et al.*, this book).

In parallel to the popularisation of the idea of ecosystem services, related concepts such as payments for ecosystem services have increasingly been considered as economic instruments to enhance or safeguard ecosystem service supply for the benefit of society across both developing and industrialised countries (Schomers and Matzdorf 2013). Payment for ecosystem services schemes aim to reach mutually beneficial agreements between providers and beneficiaries of ecosystem services, and entail a reward mechanism for ecosystem managers to maintain or improve provision of services valued by beneficiaries (Engel *et al.* 2008; Wunder *et al.* 2008). The number of payments for ecosystem services schemes and related applications has grown significantly in the past two decades, particularly in Latin America (Brouwer *et al.* 2011; Martin-Ortega *et al.* 2013; Mulligan *et al.*, this book).

Integration of ecosystem services and ecosystem capital into national accounts is also of growing academic and policy interest (Edens & Hein 2013). Beyond academia and the policy domain, preliminary research has been initiated to explore business opportunities in managing ecosystem services, and there is increasing recognition that enhanced understanding of how businesses depend on natural resources can lead to better decision-making and contribute to reductions in biodiversity loss (Houdet *et al.* 2012). Growing pressure on businesses to consider ecosystems was reflected in the official petition for the business community to contribute to the Convention on Biological Diversity in 2006, highlighting the need for businesses to develop best-practice guidelines to reduce the impact of their activities on biodiversity (Houdet *et al.* 2012). The need for, and the opportunities of, business engagement in sustainable ecosystem management is evident from other initiatives, including, for example, the Economics of Ecosystems and Biodiversity for Business in Brazil (Pavese *et al.* 2012), the UK Ecosystem Markets Task Force (2013), and the World Business Council for Sustainable Development (2014; Houdet *et al.* 2014; Houdet *et al.*, this book).

The concept of ecosystem services has arguably inspired collaboration and enhanced communication between scientists from different disciplines to address complex socio-ecological problems. It has certainly led to wider debate about the representation of environmental issues in decision-making processes among researchers, policy makers, practitioners, and conservation groups. However, popularisation of the concept has also resulted in a lack of clarity about the meaning of 'ecosystem services' and in confusion about terminology, for example in relation to the broader Ecosystem Approach, as defined by the Convention on Biological Diversity (2000) (see Box 2.1).

There is also concern about the gap between the conceptualisation and endorsement of ecosystem services by policy makers and the incorporation of ecosystem services-based approaches into actual natural resources management practice (Nahlik *et al.* 2012). Many initiatives are at an early stage, or remain at a conceptual level. Mechanisms to monitor the effectiveness of ecosystem services-based management approaches are not widely in place, or do not yet provide sufficient evidence. Also, it remains subject to debate whether at least some of those initiatives are being influenced and driven by a genuine ecosystem services paradigm, or whether part of the popularisation of ecosystem services can be attributed to re-framing or re-labelling existing approaches, i.e. 'old wine in new bottles'. The rapid and widespread adoption of the term 'ecosystem services' in the scientific literature (see Figure 2.1) and in the policy domain carries the risk of its use becoming detached from any specific meaning. Gómez-Baggethun *et al.* (2010) express concerns that mainstreaming ecosystem services may result in applications that diverge from the purpose of the concept. Specifically, they are concerned about the shift away from its original purpose

---

Box 2.1 The Ecosystem Approach (versus ecosystem services-based approaches)

The terms ecosystem approach and ecosystem services are often used interchangeably and it is worth discussing the differences (Waylen *et al.* 2013, 2014).

The Ecosystem Approach (capitalised) is a specific *framework for action* adopted by the Convention on Biological Diversity (2000) as 'a strategy for the integrated management of land, water and living resources that promotes conservation and sustainable use in an equitable way'. It is based on the application of the 12 Malawi Principles, which are explicit and prescriptive characteristics of this framework for action. While being different in essence, the specific Ecosystem Approach and the more generic ecosystem services-based approaches as defined in this book (i.e. as a flexible *way of understanding*), overlap in certain critical areas. Notably, the Ecosystem Approach considers humans as an integral part of ecosystems (close to core element 1 in this book's definition of ecosystem service-based approaches – see Section 2.3). It also recognises, in Malawi Principle 4, the need to understand ecosystems in an economic context (e.g. internalising the benefits), which is implicit in our core element 4. Both the Ecosystem Approach and ecosystem services-based approaches prescribe the involvement of stakeholders and various forms of knowledge in natural resource management (Malawi Principle 11; our core element 3). However, the Ecosystem Approach goes further in that it involves prescription of how ecosystems should be managed. By contrast, in our definition, ecosystem services-based approaches may or may not encompass action.

It could be said that existing management and conservation frameworks, such as the Ecosystem Approach,[a] have shaped ecosystem services-based approaches, and, conversely, ecosystem services-based approaches have influenced the general paradigm of natural resource management and the operationalisation of the Ecosystem Approach in practice. For example, the conceptual framework of phase 2 of the UK National Ecosystem Assessment has now been clearly embedded within the wider Ecosystem Approach to include aspects of governance and decision-making (Scott *et al.* 2014). Conversely, after the release of the Millennium Ecosystem Assessment reports, the Convention on Biological Diversity has suggested that the use of ecosystem services concepts and language could help support its goals (Convention on Biological Diversity 2006)

In summary, while the terms 'Ecosystem Approach' and 'ecosystem services-based approaches' are sometimes used interchangeably, it is important to note that the two are not the same and that the adoption of an ecosystem services-based approach is not a substitute for, or equal to, adopting the Ecosystem Approach. Although an ecosystem service-based approach can fit within an Ecosystem Approach, implementing an ecosystem service-based approach does not necessarily involve the range of considerations encapsulated by the 12 Malawi Principles.

[a] As well as Integrated Water Resources Management (see Niasse and Cherlet, this book).

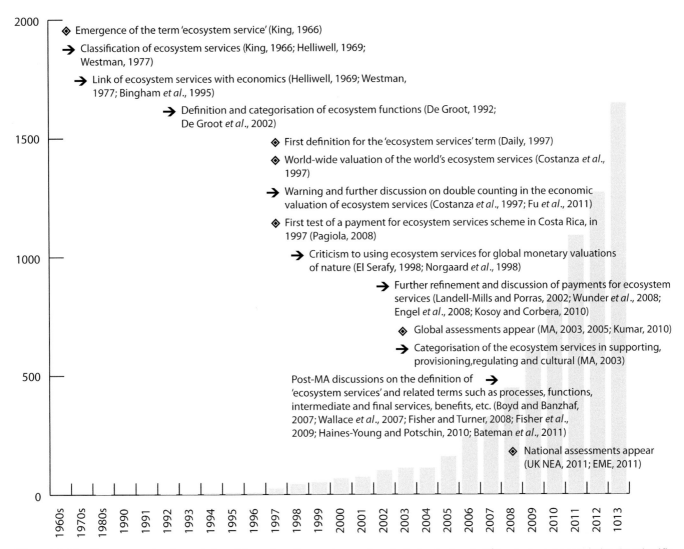

Figure 2.1 Timeline representing the evolution of the notion of ecosystem services, including landmarks (◈) and new aspects entering the scientific discussion (➔). The bar chart illustrates the increase in publications using the term 'ecosystem services' or 'ecological services' based on a computerised search of the ISI Web of Science database during the time period up to 2013 (as an update of Fisher *et al.* 2009).

as an *educational* concept to increase public interest in biodiversity conservation (Peterson *et al.* 2010), towards an emphasis on the commodification of nature for trade in potential markets (Kosoy & Corbera 2010; Corbera & Pascual 2012). Norgaard (2010) states that we might be 'blinded' by the ecosystem services 'metaphor' and thus not see the ecological, economic, and political complexities of the challenges we actually face. Some have argued that the 'economic production metaphor' does not incorporate the important moral and ethical dimensions that humans associate with nature, and which are embedded in held values, beliefs, and norms about nature (Raymond *et al.* 2013) and in the multiple and complex values that humans attribute to nature (Kosoy & Corbera 2010).

In addition, excessive, uncritical faith in the potential of management approaches based on some form of an ecosystem services

framework to address complex and conflict-laden resource management problems is likely to result in disillusion if solutions prove to be unsatisfactory. For example, great expectations are currently being placed on the potential of payments for ecosystem services schemes in mitigating water-related problems derived from forest degradation, despite the fact that robust evidence on the positive impacts of existing schemes is lacking (Porras *et al.* 2012; Martin-Ortega *et al.* 2013).

This chapter aims to disentangle the notion of what we call 'ecosystem services-based approaches'. First, we review the evolution of the term 'ecosystem services'. Then we propose a way of characterising ecosystem services-based approaches for research and decision-making. Our purpose is not to provide an ultimate definition of *the* ecosystem service approach, but rather to establish a basis for characterising its applications (in policy initiatives or research projects). Because we acknowledge that definitions and classifications of ecosystem services are case-specific and purpose-driven, we focus on common key (core) elements that constitute and characterise ways of approaching environmental problems within the ecosystem services paradigm.

The terminology adopted here has been carefully considered. We refer to approaches and not frameworks, because we refer to the *way* complex relationships between humans and the environment are understood, and not to a formalised supporting structure. We use the plural because we consider ecosystem services-based approaches to be based on a paradigm that encompasses different ways of articulating that understanding. These different articulations can take the form of conceptual theoretical frameworks, such as the ones proposed by the UK National Ecosystem Assessment (Bateman *et al.* 2011; Schaafsma *et al.*, this book), the Valuing Nature Network (UK National Ecosystem Assessment 2014) or the well-established ecosystem service's cascade from Haines-Young and Potschin (2010); frameworks of action such as the Ecosystem Approach (Box 2.1) and Integrated Water Resources Management (Niasse and Cherlet, this book); or classification or accounting frameworks (such as the Common International Classification of Ecosystem Services developed by the European Environment Agency.[1]) The term *services-based* is used to explicitly differentiate from the Ecosystem Approach. The term *core elements* is used rather than principles, to further ensure clear differentiation with the Malawi *Principles* of the Ecosystem Approach, and to reflect the idea that the elements we propose are at the *core* of what we understand is an ecosystem services-based approach.

2.2 ORIGINS AND EVOLUTION OF THE NOTION OF 'ECOSYSTEM SERVICES'

Gómez-Baggethun *et al.* (2010) link the historic development of the concept of ecosystem services to the evolution of general economic concerns about nature, and the emergence and expansion of environmental economics as a discipline. In this context, the authors describe the evolution from the original economic conception of nature's benefits as use values in Classical economics; their conceptualisation in terms of 'exchange values' in Neoclassical economics; and the expansion of monetary valuation to what they call the 'mainstreaming of the new economics of ecosystems', in which the ecosystem services notion is embedded. Here we focus on the emergence of the term 'ecosystem service' itself, and the evolution of its meaning and use (see Figure 2.1 for a graphical representation).

The term *ecosystem services* was first mentioned in the 1960s. King (1966) was concerned with the interaction between ecological and economic relationships of humans, and defined six values associated with wildlife that are 'positive' to people'.[2] Helliwell (1969) identifies recognisable benefits from wildlife and proposed the monetisation of values to incorporate them into conventional cost–benefit analysis. Westman (1977, p.961) discusses the importance of accounting for the benefits of nature's *services*, understood as the 'dynamics of ecosystems' that 'impart to society a variety of benefits', and differentiated them from ecosystems' standing stock or nature's free goods. In their article 'Extinction, substitution and ecosystem services', Ehrlich and Mooney (1983) highlight that extinctions of species would result in the loss of *services* to humanity, which could range from trivial to catastrophic. Further publications appeared in the early 1990s (e.g. Ehrlich & Wilson 1991; Costanza & Daly 1992; Ehrlich & Ehrlich 1992). Bingham *et al.* (1995) discuss the relationship between ecosystem services and economic valuation. These studies used the term ecosystem service, but none gave specific definitions.

Key milestones were the publication of Daily's book *Nature's Services: Societal Dependence on Natural Ecosystems* (1997), and Costanza *et al.*'s (1997) seminal work 'The value of the world's ecosystem services and natural capital'. Daily (1997, p.3) provides the first definition of the term 'ecosystem services', as 'the conditions and processes through which natural ecosystems, and the species that make them up, sustain and fulfil human life'. She also highlights that failure to foster delivery of ecosystem services undermines economic prosperity, forecloses options, and diminishes other aspects of human wellbeing. Costanza *et al.* (1997) set the ambitious goal of assigning a monetary value to the world's ecosystems and estimated an aggregated value of the entire biosphere. Costanza *et al.*'s work has been subject to criticism; El Serafy (1998) raises concerns about the comparison between the world's ecosystem services values and the global gross national product; Norgaard and Bode (1998) focus their criticism on the use of marginal values 'when the total collapse of some services seemed not only plausible but the driving concern'. Both highlight the fact that separate valuations of ecosystem services could result in double counting (a fact that had been acknowledged by Constanza *et al.* themselves). Despite these criticisms, this work contributed significantly to placing the valuation of ecosystem services very high on the research agenda.

From the late 1990s onwards, the literature on ecosystem services grew rapidly (e.g. Limburg & Folke 1999; Bockstael *et al.* 2000; De Groot *et al.* 2002). In particular, De Groot *et al.* (2002) made a critical contribution by emphasising the role of the *ecosystem functions* underlying the provision of services and

---

[1] www.cices.eu

[2] The six values listed by King (1966) are: commercial, recreational, biological, esthetic, scientific, and social values.

goods. They list and describe a set of ecosystem functions as 'the capacity of natural processes and components to provide goods and services that satisfy human needs, directly or indirectly' (De Groot *et al.* 2002, p.394). Based on an earlier paper (De Groot 1992), four general types of ecosystem functions were defined: regulation, habitat, production, and information functions.

These publications provided the foundation for the Millennium Ecosystem Assessment (2003 2005), which is undoubtedly the turning point in the popularisation of the ecosystem services concept. The assessment aimed to demonstrate how the decline in biodiversity (and degradation of ecosystems more generally) directly affect ecosystem functions that underpin services essential for human wellbeing. It provided a broad definition of ecosystem services as 'the benefits people obtain from ecosystems' (2003, p.49) and the most frequently quoted typology of services: provisioning (production), regulating (regulation), supporting (habitat), and cultural (information) services. The Millennium Ecosystem Assessment explicitly promoted the use of the notion of ecosystem services to inform decision-makers across the globe, and has clearly inspired the development and application of different forms of ecosystem services-based approaches.

Since publication of the Millennium Ecosystem Assessment, different interpretations and critiques of the definition and classification of ecosystem services have emerged. Ojea *et al.* (2012) reviewed the range of definitions that have been proposed, and found that interpretations differ according to the nature and types of services that are considered to have value for society. One post-Millennium Ecosystem Assessment definition is that of Boyd and Banzhaf (2007, p.619), who define final ecosystem services as 'the components of nature directly enjoyed, consumed, or used to yield human well-being'. The authors consider services as the end products of nature (and hence the term *final* ecosystem services), and distinguish them from intermediate natural components and from benefits. Boyd and Banzhaf propose to value only services as defined above, and exclude benefits in which anthropogenic inputs are involved (e.g. recreational angling would have non-natural inputs such as tackle and boats) and intermediate components, which they define as part of the process resulting in ecosystem services. Fisher *et al.* (2009) define ecosystem services as the aspects of ecosystems utilised (actively or passively) to generate human wellbeing. Based on this definition, they distinguish between (1) abiotic inputs such as rainfall; (2) intermediate services such as water regulation; (3) final services such as constant stream flow; and (4) benefits, such as water for irrigation, for hydroelectric power, or recreation. Wallace (2007) and Fisher and Turner (2008) highlight that the same service can be either intermediate or final, depending on the context (e.g. primary production to regulate water or to benefit directly as food). Fisher *et al.* (2009) also point to the importance of stakeholders' perceptions in defining whether a service is intermediate or final.

The focus on final ecosystem services is motivated by the need to avoid double counting when valuing ecosystem services. As Lele (2009) explains for the case of water services, structural changes in ecosystems (e.g. timber plantations) can influence watershed processes (e.g. increase of erosion rates). These changes can result in different kinds of human impact, which can be negative (e.g. decreased reservoir capacity due to sediment load resulting in reduced hydropower production capacity) or positive (e.g. increased fertilisation of floodplains). Lele points out that the 'process' should not be the focus of valuation. Rather, it is the outcome of the process (the final service), which has an impact on human wellbeing and, therefore, has economic value. According to Fu *et al.* (2011), the exclusion of intermediate services in economic valuation does not indicate that they have no value, but that their values are realised through the value of the final ecosystem services.

The idea of *final* ecosystem services has been incorporated into recent assessments of ecosystem services, for example, in the latest report on The Economics of Ecosystems and Biodiversity (Kumar 2010); the UK National Ecosystem Assessment (Bateman *et al.* 2011); and other literature (Haines-Young & Potschin 2010). Supporting (and in some cases even regulating) ecosystem services have been located in the intermediate ecosystem services group due to their indirect repercussions on human wellbeing (Wallace 2007; Fu *et al.* 2011) – for example, their role in preserving the delivery of provisioning services.

A further distinction is that of final services and *goods*. The UK National Ecosystem Assessment defines goods as the objects (both of use and non-use character) that people value (Bateman *et al.* 2011; Schaafsma *et al.*, this book). Goods should therefore be at the centre of any assessment, while services are the flows that originate from ecosystems and contribute to the provisioning of goods. The Common International Classification of Ecosystem Services also recognises the need to distinguish between final ecosystem services and ecosystem goods and benefits (collectively referred to as 'products') and defines ecosystem services as the contributions that ecosystems make to human wellbeing (Common International Classification of Ecosystem Services 2012). These services are final in that they are the outputs of ecosystems that most directly affect the wellbeing of people. According to the Common International Classification of Ecosystem Services, a fundamental characteristic is that final services retain a connection with the underlying ecosystem functions, processes, and structures that generate them. Ecosystems products are the goods and benefits that people create or derive from final ecosystem services. These final outputs from ecosystems have been turned into products or experiences that are not functionally connected to the systems from which they were derived.

A parallel discussion has developed around the monetisation of the value of ecosystem services. In environmental economics, the predominant paradigm for the interpretation of the notion of value of ecosystem services has been that of Neoclassical economics (Gómez-Baggethun *et al.* 2010). Within this paradigm, the value of ecosystem services is measured in terms of the welfare change associated with changes in ecosystem status in monetary units (Pearce & Turner 1989). The need for and validity of monetary assessments of ecosystem services values has been, and continues to be, heavily criticised, particularly from ecological economics perspectives (Proops 1989; Martinez-Alier *et al.* 1998; Azqueta & Delacámara 2006; Spangenbergh & Settele 2010). Even though alternative indicators of wellbeing that do not rely on monetary values have been suggested and applied (Byg 2015), they have only recently found their way into actual assessments of ecosystem services. For example, Kenter *et al.* (2013) investigated the recreational use and non-use values of UK divers and sea anglers in potential marine protected areas in the context of the UK National Ecosystem Assessment, using a combination of monetary and non-monetary valuation methods and an interactive mapping application to assess site visit numbers.

2.3 ECOSYSTEM SERVICES-BASED APPROACHES: DEFINITION AND CORE ELEMENTS

As demonstrated above, there is no clear consensus on how exactly ecosystem services should be defined and classified, and as research on ecosystem services evolves, further interpretations might emerge. Major differences between definitions arise from the *purpose* the ecosystem service concept is expected to serve (Fisher & Turner 2008; Fisher *et al.* 2009). A purely *descriptive* objective, for example, illustrating human–nature relationships, can use the most generic and broad definitions, such as those given by the Millennium Ecosystem Assessment (2005) and Daily (1997). For the specific purpose of creating an ecosystem services or 'green' inventory that can be balanced against economic national accounts – and therefore an *evaluative* use of the term – it is useful to think beyond aspects that are 'valued' and define ecosystem services more narrowly, as in the Common International Classification of Ecosystem Services developed from the work on environmental accounting by the European Environment Agency. Frameworks of identified ecosystem services will then differ depending on the specific descriptive or evaluative objectives behind the task (see Fisher *et al.* 2009).

Instead of drawing upon extensive but generic 'lists' of services such as the ones published in the Millennium Ecosystem Assessment, the selection and definition of relevant ecosystem services should be on a project-by-project basis to avoid a mismatch of purpose and underlying conceptual framework. Research papers should make clear the underlying purpose of the work and how the term ecosystem service is defined. Unlike

Nahlik *et al.* (2012), we understand that specific projects should define and operationalise frameworks to achieve their own specific targets. As stated previously, rather than trying to provide an ultimate definition of *the* Ecosystem Services *Framework*, we propose a set of common guiding core elements of *generic ecosystem services-based approaches* that underpins the characterisation of research and policy applications.

Broadly, then, an ecosystem services-based approach is *a way of understanding the complex relationships between nature and humans to support decision-making, with the aim of reversing the declining status of ecosystems and ensuring the sustainable use/management/conservation of resources.* An ecosystem services-based approach entails the following core elements:

(1) *The focus on the status of ecosystems, and the recognition of its effects on human wellbeing.* An ecosystem services-based approach takes a viewpoint of anthropocentric instrumentalism, placing the emphasis on the benefits that *humans obtain from nature*, and recognising that *humans* are the ones who assign value to aspects of ecosystems. This is in contrast to alternative ways of interpreting the relationships between humans and nature, which consider the human system to be part of a broader ecological system and reject the idea of decision-making being purely driven by anthropocentric views, including notions of intrinsic value and bio- or eco-centric viewpoints.

(2) *The understanding of the biophysical underpinning of ecosystems in terms of service delivery.* This represents a new way of understanding and describing ecosystems in terms of the biophysical structures, processes, and functions leading to the delivery of services to humans (production chain). Traditionally, ecologists and other natural scientists have not thought about ecosystems in terms of human wellbeing, but rather in terms of biogeochemical cycles, energy flows, species behaviour, population dynamics, etc. An ecosystem services-based approach implies that there should be a 're-phrasing' of science in terms of how nature delivers to humans and what roles humans play in that delivery. Moreover, it requires the description and adequate quantification of the interactions of an ecosystem's components and their effects upon a single service or a range of services (acknowledging complex interdependencies), across temporal and spatial scales.

(3) *The integration of natural and social sciences and other strands of knowledge for a comprehensive understanding of the service delivery process.* An ecosystem services-based approach is, by definition, transdisciplinary in nature; this requires the integration of different academic disciplines, for example, via jointly developed models, which inevitably trade-off precision in disciplinary approaches to achieve outcomes that are of use to

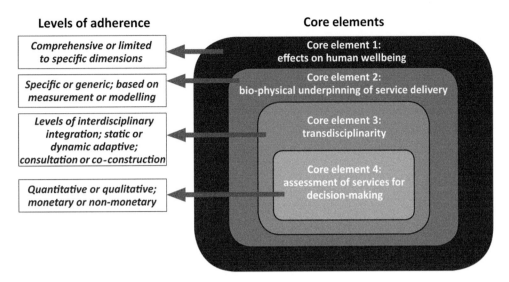

Figure 2.2 Nested core elements characterising ecosystem services-based approaches.

decision-making. An ecosystem services-based approach also requires the consideration of non-academic strands of knowledge, including the views and perceptions of stakeholders at the relevant scales. Co-construction of knowledge with stakeholders is essential to understand the variety of ways in which ecosystems generate wellbeing, and to establish the legitimacy of decisions based on the valuation of ecosystem services.

(4) *The assessment of the services provided by ecosystems for its incorporation into decision-making.* An ecosystem services-based approach inherently implies an assessment (qualitative or quantitative) of the services delivered by ecosystems, and the identification of the social/individual values of services in monetary and/or non-monetary terms. This is motivated by the need to incorporate these values into decision-making processes.

The above core elements are logically related to each other in a nested structure. Core element 1 is a necessary condition for core element 2 to apply. Similarly, core elements 1 and 2 are implied in the integrative work of core element 3, and for the assessment established in core element 4, i.e. as pre-requisites for the assessment of ecosystem services and the incorporation of their values into decision-making. Figure 2.2 illustrates this.

The nested structure of the core elements accommodates variations in the application of ecosystem services-based approaches. Our proposition is that an ecosystem services-based approach necessarily implies that the core elements are present, but that different research or policy case studies vary in how the core elements are represented.[3] According to the nested structure, any

ecosystem services-based application is necessarily grounded in the acknowledgement that ecosystem status and human wellbeing are linked (core element 1); however, the effects on human wellbeing can be perceived in a comprehensive manner, or be focused on specific dimensions of wellbeing only (for example, whether solely economic welfare effects are considered, or whether shared social values, happiness, health, security, etc., are included as well). In core element 2, variation may arise from the way the biophysical underpinning of service delivery is established. For example, biophysical analysis can be predominantly based on either measurement or modelling. Also, some applications might be based on a more complex, site-specific biophysical analysis than others that, for example, rely on transferring knowledge on biophysical effects of ecosystem changes from similar contexts. Similarly, the integration of knowledge across disciplines and domains (core element 3) can also be examined along a range of dimensions; the degree of knowledge integration can involve only a few scientific disciplines and domains, or co-generation of knowledge can involve many disciplines and domains; integration can be either static, following pre-defined paths in which knowledge flows between all the parties involved, or dynamic, allowing for feedback loops and adjustments in the conditions and assumptions underlying knowledge creation. Adherence to core element 3 can be achieved through (quantitative) surveys or (qualitative) participatory processes with stakeholders that aim to co-construct knowledge. Finally, the assessment of services (core element 4) can be quantitative or qualitative, or be conducted in monetary or non-monetary terms. The suggestions for characterising adherence to the four core elements (see Figure 2.2) are not meant to be comprehensive. Rather, we hope that the idea of the nested core elements will stimulate discussions among researchers and policy makers about plausible and useful characterising terms. Furthermore, any characterisation may be adjusted over time to

---

[3] Examples of levels of adherence can be found in the boxes on the left-hand side of Figure 2.2. Many of the pairs of terms in the figure describe extremes, while a case study may actually sit somewhere in between.

Box 2.2 An ecosystem services-based approach to the understanding of water-related forest ecosystem services

In a water context, an ecosystem services-based approach:

- recognises that structural changes to forests can influence several watershed processes (e.g. erosion rates, sediment load, water chemistry, peak flow levels, total flow, base flow, or groundwater recharge) in different ways and that, in turn, these changes result in different kinds of impact on human wellbeing (e.g. increased costs of water purification, increased fertilisation of floodplain lands, decreased reservoir capacity due to siltation, flood damage, changes in agriculture) (Lele 2009) – core element 1.
- requires the understanding of the biophysical processes that determine the way forest cover, forest structure, soil–vegetation dynamics, etc. affect the amount and quality of freshwater to the extent that it impacts on human wellbeing (through use or non-use) by the beneficiaries (core element 2).
- combines knowledge of the service delivery processes that are based on natural sciences (e.g. plant physiology, ecology, hydrology) with information from social sciences (e.g. economics, psychology, political science) and (local) stakeholder knowledge (e.g. farmers, drinking water users, floodplain residents, hydropower companies, regulators) that jointly help to understand, for example, where benefits arise in relation to where ecosystem change takes place (core element 3).
- requires at least some degree of measurement of changes in the final services delivered (e.g. increase of the flow of water associated with forest cover) coupled with a qualitative interpretation of the implications for human wellbeing, or the valuation of associated benefits through, for example, willingness to pay for increased water availability, so that these benefits can be incorporated into decision making (for example, on afforestation or the creation of protected forest areas) – core element 4.

---

accommodate novel developments in ecosystem services-based approaches methodology or application.

Box 2.2 describes the core elements of an ecosystem services-based approach using the understanding of forests' water-related ecosystem services as an example.

2.4 CONCLUSIONS

We view ecosystem services-based approaches as a particular way of understanding the complex relationships between humans and nature; that is, a particular *way of looking* at socio-oecological issues. An ecosystem services-based approach is *not* a management tool *per se*, but rather a pair of *glasses* that one (researcher, analyst, policy maker, or land manager) might wear to tackle the problem at hand. As such, it is expected to promote holistic systems thinking, identifying connections between an ecosystem's components, and to help understand how ecosystem services benefit different social groups at different locations, revealing what dis-services and trade-offs might exist.

The concept of ecosystem services has arguably inspired collaboration and enhanced communication between scientists from different disciplines to address complex socio-ecological problems. It has certainly led to wider debate about the representation of environmental issues in decision-making processes among researchers, policy makers, practitioners, and conservation groups. It has helped to incorporate into the debate often ignored benefits that people derive from ecosystems and to recognise the

many values of nature within different decision-making contexts that affect a broad range of stakeholders.

Despite this enthusiasm and popularisation, or maybe precisely because of it, we see three major risks associated with the adoption of ecosystem services concepts. The first risk relates to current confusion about terminology and the understanding of related concepts. We believe that an increasingly blind and uncritical adoption of ecosystem service terminology that is devoid of any specific meaning can over time be detrimental to the targeted application of ecosystem services-based approaches and their potential to inform decision-making processes. This is because consensus (between researchers, policy makers, stakeholders) may be based on each party's own interpretation of the terminology and associated/underlying conceptual foundations, and it may create 'fake consensus' situations where problems only surface when affected parties are probed more deeply about what they actually mean. At the extreme, the ecosystem services 'discourse' may be exploited to sell 'business as usual' in research and decision-making, and solely to create new research demands rather than to clarify existing needs. We therefore think there is a greater need for researchers and decision-makers alike to question their use of ecosystem service terminology, and also the use of ecosystem service terminology by their peers.

The second risk stems from overlooking the limitations and potential negative consequences of applying ecosystem services-based approaches. Among the limitations of moving from the conceptual level to the practical implementation of ecosystem services-based approaches are the challenges associated with the current capacity of understanding of the effects of interventions

---

Box 2.3 Key messages

• The concept of ecosystem services has inspired collaboration and enhanced communication between scientists, policy makers, practitioners, and conservation groups.

• The popularisation of the concept has also resulted in a lack of clarity about the meaning of 'ecosystem services' and in confusion about terminology.

• This chapter defines 'ecosystem services-based approaches' as a way of understanding the complex relationships between nature and humans to support decision-making, characterised by four core elements:

(1) The recognition that the status of ecosystems affects human wellbeing, from an anthropocentric point of view.

(2) The understanding of the biophysical underpinning of ecosystems in terms of service delivery, which implies that science should acknowledge what nature delivers to humans and what roles humans play in that delivery.

(3) The integration of natural and social sciences for a comprehensive understanding of the service delivery process, and consideration of non-scientific stakeholder perceptions to both understand the variety of ways in which ecosystems generate wellbeing, and establish legitimacy of decisions based on the valuation of ecosystem services.

(4) The assessment of the services provided by ecosystems for the incorporation of (monetary or non-monetary) values into decision-making.

• Ecosystem services-based approaches are neither a silver bullet nor a panacea and need to be assessed and monitored appropriately.

• They have the great virtue of having stimulated dialogue, but it is now important to make sure this dialogue remains meaningful and purpose-driven.

---

impacting on land use and water management in terms of final ecosystem services and, hence, the possibility of accurately valuing benefits. If incentive mechanisms for land and water management, such as payments for ecosystems services, are put in place based on the false assumption that the desired benefits will be delivered, then the process is likely to be counterproductive.

Finally, even if non-monetary assessments are used, the essentially anthropogenic nature of ecosystem services-based approaches might indeed lead to the 'commodification' of nature and natural assets. This could introduce unforeseeable effects on societies if the service notion clashes with their world views (e.g. according to Ibarra et al. 2011, a payment for ecosystem services scheme caused the food insecurity of an indigenous community in Mexico), and/or result in the neglect of negative impacts on aspects of ecosystems for which final services and benefits have not yet been identified.

In summary, ecosystem services-based approaches are neither a silver bullet nor a panacea and need to be assessed and monitored appropriately. They have the great virtue of having stimulated dialogue, but it is now important to make sure that this dialogue remains meaningful and purpose-driven.

ACKNOWLEDGEMENTS

This work has been funded by Scottish Government Rural Affairs and the Environment Portfolio Strategic Research Programme 2011–2016. The work of Dídac Jorda-Capdevila has been funded by the project CSO2010-21979 from the Spanish National Programme for Basic Research. The authors are grateful to Steve Albon, Rob Brooker, Kirsty Blackstock, Kerry Waylen, Lisa Norton, Iain Gordon, Bob Ferrier, Iain Brown, Beatriz Rodriguez-Labajos, and Joan Martinez-Alier for very valuable comments to different versions of this chapter.

References

Azqueta, D. & Delacámara, G. (2006). Ethics, economics and environmental management. *Ecological Economics* **56**(4), 524–533.

Bateman, I. J., Mace, G. M., Fezzi, C., et al. (2011). Economic analysis for ecosystem service assessments. *Environmental and Resource Economics* **48**(2), 177–218.

Bingham, G., Bishop, R., Brody, M., et al. (1995). Issues in ecosystem valuation: improving information for decision making. *Ecological Economics* **4**, 73–90.

Bockstael, N. E., Freeman, A. M., Koop, R. J., et al. (2000). On measuring economic values for nature. *Environmental Science & Technology* **34**(8), 1384–1389.

Boyd, J. & Banzhaf, S. (2007). What are ecosystem services? The need for standardized environmental accounting units. *Ecological Economics* **63**(2–3), 616–626.

Brouwer, R., Tesfaye, A., & Pauw, P. (2011). Meta-analysis of institutional-economic factors explaining the environmental performance of payments for watershed services. *Environmental Conservation* **38**, 380–392.

Byg, A. (2015). Non-monetary valuation of ecosystem services. Report for the Scottish Government Research Portfolio Workpackage 1.2: The value of ecosystem services. The James Hutton Institute. http://www.hutton.ac.uk/research/themes/safeguarding-natnal-capital/research-outputs.

Common International Classification of Ecosystem Service (2012). Consultation on Version 4, August–December 2012. EEA Framework Contract No EEA/IEA/09/003. http://cices.eu (last accessed 5 June 2014).

Convention on Biological Diversity. (2006). Decision adopted by the conference of the 451 parties to the convention on biological diversity at its eighth meeting; VIII/9. 452 Implications of the findings of the millennium ecosystem assessment.

Corbera, E. & Pascual, U. (2012). Ecosystem services: heed social goals. *Science* **335**(10), 355–356.

Costanza, R. & Daly, H. E. (1992). Natural capital and sustainable development. *Conservation Biology* **6**(1), 37–46.

Costanza, R., d'Arge, R., De Groot, R. S., *et al.* (1997). The value of the world's ecosystem services and natural capital. *Nature* **387**(6630), 253–260.

Daily, G. C. (1997). *Nature's Services: Societal Dependence on Natural Ecosystems.* Island Press, Washington, DC.

De Groot, R. S. (1992). *Functions of Nature: Evaluation of Nature in Environmental Planning, Management and Decision Making.* Wolters-Noordhoff, Groningen.

De Groot, R. S., Wilson, M. A., & Boumans, R. M. J. (2002). A typology for the classification, description and valuation of ecosystem functions, goods and services. *Ecological Economics* **41**(3), 393–408.

Ecosystem Markets Task Force. (2013). *Realising Nature's Value: Final Report.* Available at www.defra.gov.uk/ecosystem-markets/files/Ecosystem-Markets-Task-Force-Final-Report-.pdf (last accessed 8 July 2013).

Edens, B. & Hein, L. (2013). Towards a consistent approach for ecosystem accounting. *Ecological Economics* **90**, 41–52.

Ehrlich, P. R. & Ehrlich, A. H. (1992). The value of biodiversity. *Ambio* **21**(3), 219–226.

Ehrlich, P. R. & Mooney, H. A. (1983). Extinction, substitution, and ecosystem services. *BioScience* **33**(4), 248–254.

Ehrlich, P. R. & Wilson, E. O. (1991). Biodiversity studies: science and policy. *Science* **253**, 758–762.

El Serafy, S. (1998). Pricing the invaluable: the value of the world's ecosystem services and natural capital. *Ecological Economics* **25**(1), 25–27.

EME (Spanish Millennium Ecosystem Assessment) (2011). *Ecosistemas y biodiversidad para el bienestar humano. Evaluación de los Ecosistemas del Milenio de España.* Fundación Biodiversidad, Ministerio de Medio Ambiente y Medio Rural y Marino, Madrid.

Engel, S., Pagiola, S., & Wunder, S. (2008). Designing payments for environmental services in theory and practice: an overview of the issues. *Ecological Economics* **65**(4), 663–674.

Fisher, B. & Turner, K. R. (2008). Ecosystem services: classification for valuation. *Biological Conservation* **141**(5), 1167–1169.

Fisher, B., Turner, R. K., & Morling, P. (2009). Defining and classifying ecosystem services for decision making. *Ecological Economics* **68**(3), 643–653.

Fu, B. J., Su, C. H., Wei, Y. P., *et al.* (2011). Double counting in ecosystem services valuation: causes and countermeasures. *Ecological Research* **26**(1), 1–14.

Gómez-Baggethun, E., De Groot, R. S., Lomas, P. L., *et al.* (2010). The history of ecosystem services in economic theory and practice: from early notions to markets and payment schemes. *Ecological Economics* **69**(6), 1209–1218.

Haines-Young, R. & Potschin, M. (2010). The links between biodiversity, ecosystem services and human well-being. In *Ecosystem Ecology: A New Synthesis*, ed. D. Raffaelli & C. Frid. Cambridge University Press, Cambridge, pp. 1–31.

Helliwell, D. R. (1969). Valuation of wildlife resources. *Regional Studies* **3**, 41–49.

Houdet, J., Trommetter, M., & Weber, J. (2012). Understanding changes in business strategies regarding biodiversity and ecosystem services. *Ecological Economics* **73**, 37–46.

Houdet, J., Burritt, R., Farrell, K., *et al.* (2014). What natural capital disclosure for integrated reporting? Designing & modelling an Integrated Financial–Natural Capital Accounting and Reporting Framework. Synergiz–ACTS, Working Paper 2014-01.

Ibarra, J. T., Barreau, A., del Campo, C., *et al.* (2011). When formal and market-based conservation mechanisms disrupt food sovereignty: impacts of community conservation and payments for environmental services on an indigenous community of Oaxaca, Mexico. *International Forestry Review* **13**(3), 318–337.

Kenter, J. O., Bryce, R., Davies, A., *et al.* (2013). *The Value of Potential Marine Protected Areas in the UK to Divers and Sea Anglers.* UNEP-WCMC, Cambridge. http://uknea.unep-wcmc.org/LinkClick.aspx?fileticket=Mb8nUAphh%2BY%3D&tabid=82 (last accessed 21 October 2014).

King, R. T. (1966). Wildlife and man. *NY Conservationist* **20**(6), 8–11.

Kosoy, N. & Corbera, E. (2010). Payments for ecosystem services as commodity fetishism. *Ecological Economics* **69**, 1228–1236.

Kumar, P. (ed.) (2010). *The Economics of Ecosystems and Biodiversity (TEEB): Ecological and Economic Foundations.* Earthscan, London and Washington, DC.

Landell-Mills, N. & Porras, I. T. (2002). *Silver Bullet or Fools' Gold? A global Review of Markets for Forest Environmental Services and their Impact on the Poor.* International Institute for Environment and Development, London.

Lele, S. (2009). Watershed services of tropical forests: from hydrology to economic valuation to integrated analysis. *Current Opinion in Environmental Sustainability* **1**(2), 148–155.

Limburg, K. E. & Folke, C. (1999). The ecology of ecosystem services: introduction to the special issue. *Ecological Economics* **29**(2), 179–182.

Malthus, T. R. (1888). *An Essay on the Principle of Population: Or, A View of its Past and Present Effects on Human Happiness.* Reeves and Turner, London.

Martinez-Alier, J., Munda, G., & O'Neill, J. (1998). Weak comparability of values as a foundation for ecological economics. *Ecological Economics* **26**(3), 277–286.

Martin-Ortega, J. (2012). Economic prescriptions and policy applications in the implementation of the European Water Framework Directive. *Environmental Sciences and Policy* **24**, 83–91.

Martin-Ortega, J., Ojea, E., & Roux, C. (2013). Payments for water ecosystem services in Latin America: a literature review and conceptual framework. *Ecosystem Services* (in press).

Meadows, D. H., Meadows, D. L., Randers, J., *et al.* (1972). *Limits to Growth.* Universe Books, New York.

Millennium Ecosystem Assessment (2003). *Ecosystems and Human Wellbeing: A Framework for Assessment.* Island Press, Washington, DC.

Millennium Ecosystem Assessment (2005). *Ecosystems and Human Wellbeing: General Synthesis.* Island Press, Washington, DC.

Ministry of Local Government and Rural Development of Malawi (2012). Revised decentralized environmental guidelines.

Nahlik, A. M., Kentula, M. E., Fennessy, M. S., *et al.* (2012). Where is the consensus? A proposed foundation for moving ecosystem service concepts into practice. *Ecological Economics* **77**, 27–35.

Norgaard, R. B. (2010). Ecosystems services: from eye-opening metaphor to complexity blinder. *Ecological Economics* **69**, 1219–1127.

Norgaard, R. B. & Bode, C. (1998). Next, the value of God, and other reactions. *Ecological Economics*, **25**(1), 37–39.

Odum, E. P. & Odum, H. T. (1972). Natural areas as necessary components of man's total environment. Wildlife Management Institute, North American Wildlife and Natural Resources Conference, Washington, DC, Proceedings 37.

Ojea, E., Martin-Ortega, J., & Chiabai, A. (2012). Defining and classifying ecosystem services for economic valuation: the case of forest water services. *Environmental Science & Policy* **19–20**, 1–15.

Pavese, H., Ceotto, P., & Ribeiro, F. (2012). *TEEB for the Brazilian Business Sector: Preliminary Report.* Conservation International.

Pearce, D. W. & Turner, R. K. (1989). *Economics of Natural Resources and the Environment.* Johns Hopkins University Press, Baltimore, MD.

Peterson, M. J., Hall, D. M., Feldpausch-Parker, A. M., *et al.* (2010). Obscuring ecosystem function with application of the ecosystem services concept. *Conservation Biology* **24**(1), 113–119.

Porras, I., Aylward, B., & Dengel, J. (2013). *Monitoring Payments for Watershed Services Schemes in Developing Countries.* International Institute for Environment and Development. http://pubs.iied.org/pdfs/16525IIED.pdf (last accessed 21 October 2014).

Porras, I., Dengel, J., & Aylward, B. (2012). Monitoring and evaluation of Payment for Watershed Service Schemes in developing countries. In: *Proceedings of the 14th Annual BioEcon Conference on 'Resource Economics, Biodiversity Conservation and Development', 18–20 September 2012, Kings College*, Cambridge, United Kingdom.

Proops, J. L. (1989). Ecological economics: rationale and problem areas. *Ecological Economics* **1**(1), 59–76.

Raymond, C. M., Singh, G. G., Benessaiah, K., *et al.* (2013). Ecosystem services and beyond: using multiple metaphors to understand human–environment relationships. *American Institute of Biological Sciences* **63**(7), 536–546.

Schomers, S. & Matzdorf, B. (2013). Payments for ecosystem services: a review and comparison of developing and industrialized countries. *Ecosystem Services* **6**, 16–30.

Scott, A., Carter, C., Hölzinger, O., *et al.* (2014). Tools – applications, benefits and linkages for ecosystem science (TABLES). Report for the National Ecosystem Assessment follow-on.

Spangenbergh, J. H. & Settele, J. (2010). Precisely incorrect? Monetising the value of ecosystem services. *Ecological Complexity* **7**(3), 327–337.

Study of Critical Environmental Problems (1970). *Man's Impact on the Global Environment*. MIT Press, Cambridge, MA and London.

UK National Ecosystem Assessment (2011). *The UK National Ecosystem Assessment: Synthesis of the Key Findings*. UNEP-WCMC, Cambridge.

UK National Ecosystem Assessment (2014). *The UK National Ecosystem Assessment Follow-on: Synthesis of the Key Findings*. UNEP-WCMC, Cambridge.

Wallace, K. J. (2007). Classification of ecosystem services: problems and solutions. *Biological Conservation* **139**(3), 235–246.

Waylen, K. & Martin-Ortega, J. (2013). Report on Knowledge Exchange Workshops on an Ecosystem Services Approach. WATERS Project: Towards Equitable Resource Management Strategies, Malawi. Available at www.hutton.ac.uk/research/themes/managing-catchments-and-coasts/ecosystem-services/water-futures (last accessed 8 July 2013).

Waylen, K. A., Blackstock, K., & Holstead, K. L. (2013). Exploring experiences of the Ecosystem Approach. Research report, available at www.hutton.ac.uk/sites/default/files/files/Report on EcA review Final.pdf (last accessed 21 October 2014).

Waylen, K. A., Hastings, E., Banks, E., *et al.* (2014). The need to disentangle key concepts from Ecosystem Approach jargon. *Conservation Biology* (in press).

WBCSD (2014). Website available at www.wbcsd.org/home.aspx (last accessed 10 April 2014).

Westman, W. E. (1977). How much are nature's services worth? *Science* **197**(4307), 960.

Wunder, S., Engel, S., & Pagiola, S. (2008). Taking stock: a comparative analysis of payments for environmental services programs in developed and developing countries. *Ecological Economics* **65**(4), 834–852.

Part I
Addressing global challenges

3 Assessing climate change risks and prioritising adaptation options using a water ecosystem services-based approach

Samantha J. Capon and Stuart E. Bunn

3.1 INTRODUCTION

Climate change poses a significant threat to the capacity of the world's freshwater ecosystems to provide critical water ecosystem services upon which both human and non-human systems rely (Bates et al. 2008). High degrees of exposure and sensitivity to climate change effects, amplified by the position of freshwater ecosystems in the landscape and constraints on their adaptive capacity due to intensive human use and modification, imply a high level of vulnerability among freshwater ecosystems to climate change (Vörösmarty et al. 2010; Capon et al. 2013). Rising temperatures, sea-level rise, and changes to the hydrologic cycle are all expected to alter the distribution and extent of goods and services supplied by freshwater ecosystems (Palmer et al. 2009). At the same time, dramatic increases in human demands for water ecosystem services, particularly provisioning services (e.g. water supplies for irrigation), are widely anticipated (UNESCO 2012; Salman and Martinez, this book). The importance of many regulating and supporting ecosystem services (e.g. water purification) for both human and non-human needs are also likely to grow under a changing climate (Capon et al. 2013). Indeed, the role of some water ecosystems services can be anticipated to become increasingly critical in relation to climate change mitigation (e.g. climate regulation) and adaptation (e.g. flood control). Effective human responses to climate change must therefore be underpinned by an understanding of climate change risks to water ecosystem services and options for their protection, restoration, or enhancement.

In this chapter we propose an ecosystem services-based approach to climate change adaptation to enable an integrated assessment of climate change risks that accounts for both human and non-human systems, as well as their interactions. Adopting ecosystem services as targets for managed adaptation can guide the prioritisation of adaptation measures so that low-regret options with multiple benefits are highlighted and perverse outcomes avoided. An ecosystem services-based approach also offers a basis for communication and education to engender public and political engagement in climate change adaptation decision-making. Here, we discuss the risks posed by climate change to water ecosystem goods and services globally and present a synthesis of relevant adaptation measures within a water ecosystem services-based approach. We also explore the efficacy and limitations of taking such an approach to adaptation decision-making.

3.2 RISKS TO WATER ECOSYSTEM SERVICES UNDER A CHANGING CLIMATE

3.2.1 The supply of water ecosystem services

Water ecosystem services derive directly from the world's freshwater ecosystems, the suite of which includes rivers, lakes, floodplains, and an enormous range of wetlands as well as their adjoining riparian areas (Millennium Ecosystem Assessment 2005). While freshwater ecosystems globally clearly encompass considerable ecological and geographic diversity, their vulnerability to climate change effects shares a number of common elements with respect to their exposure and sensitivity to changing climatic stimuli as well as their capacity to adapt to such changes.

First, freshwater ecosystems typically have high levels of exposure to changes in climatic stimuli. In addition to CO_2 enrichment and warming, freshwater ecosystems, by their nature, are particularly exposed to changes in precipitation and runoff as well as the impacts of these on other elements of the hydrologic cycle (e.g. groundwater replenishment). Although the direction and degree of projected changes to rainfall vary considerably with location and entail significant uncertainties, for the most part the global trend is towards intensification, with wetter areas becoming wetter and drier areas becoming drier, and with more precipitation projected to fall in intense rainfall events in most areas (Intergovernmental Panel on Climate Change 2007; Bates et al. 2008). Because of their topographic position in the

17

landscape, freshwater ecosystems are also often highly exposed to extreme events (e.g. floods, droughts, and intense storms), the frequency and intensity of which are projected to increase in the future in many regions (Intergovernmental Panel on Climate Change 2007; Bates *et al* 2008; Capon *et al*. 2013). Freshwater ecosystems in coastal areas are further affected by sea-level rise, while those of alpine freshwater ecosystems are influenced by reduced snow fall (Vicuna & Dracup 2007).

Second, freshwater ecosystems are highly sensitive to changes in climatic stimuli, especially those which impact on hydrological regimes, as these are generally the major driver of freshwater ecosystem structure and function (Poff & Zimmerman 2010). Hydrologic regimes themselves are highly sensitive to changes in precipitation, as well as evapotranspiration, with small declines in precipitation usually resulting in much larger reductions in stream flow while, conversely, proportionately greater increases in mean stream flow and even larger rises in flood discharges can occur in response to relatively small gains in annual precipitation (Goudie 2006). Hydrologic regimes are also directly sensitive to temperature. For example, a significant proportion (i.e. 15%) of recent major declines in annual inflows to Australia's Murray-Darling Basin has been attributed solely to warming of 1 °C (Cai & Cowan 2008). Water quality, as well as quantity, is sensitive to changes in climatic stimuli, with many biogeochemical processes (e.g. decomposition) being directly influenced by temperature as well as CO_2 concentrations and hydrologic conditions. Drying, for instance, may result in the transformation of some wetland soils from sinks to sources of potentially harmful solutes (e.g. nitrate, sulfate, sodium etc.; Freeman *et al*. 1993). Physical geomorphic processes are similarly sensitive to climate change, especially changes in precipitation, with significant implications for sediment dynamics and the physical form of freshwater ecosystems, particularly in fine-grained alluvial ecosystems (Goudie 2006). Aquatic organisms, as well as the assemblages which they form, are affected by changes in climatic stimuli both directly (e.g. physiological effects of temperature) and indirectly as a result of effects on hydrology, geomorphology, and biogeochemistry. Changes in the condition and abundance of organisms, species distributions, and assemblage structures are widely anticipated (e.g. Steffen *et al*. 2009; Kernan *et al*. 2010).

Finally, the vulnerability of many freshwater ecosystems to climate change is likely to be aggravated by significant constraints on their adaptive capacity due to pressures wrought by non-climatic stressors and the high levels of modification and degradation to which they have been subjected as a result of human activities (Palmer *et al*. 2007, 2008; Capon *et al*. 2013). While many freshwater organisms may be capable of adjusting to changes in climatic stimuli either in situ, through behavioural changes, plasticity or genetic change, or by moving to new locations with favourable climatic conditions, there is concern that the extent and

pace of current climate change exceed the limits of ecological adaptive capacity (Visser 2008). Additionally, the susceptibility of freshwater ecosystems to non-climatic threats is expected to rise as a result of climate change effects, with increasing potential for many complex feedback loops to drive future ecosystem structure and function (Rood *et al*. 2008; MacNally *et al*. 2011).

The capacity of freshwater ecosystems to supply ecosystem services depends on their ecological functions which, in turn, depend on their ecological components (e.g. biota) and processes (e.g. sediment dynamics; Capon *et al*. 2013). Since these are highly vulnerable to climate change impacts, water ecosystem services are almost certainly susceptible to significant climate change effects (Table 3.1). Although there is potential for some positive effects to occur in some locations (e.g. increased floodplain pasture growth in response to more frequent inundation), there is widespread consensus that the overall impact of climate change on the delivery of water ecosystem services will be negative in most situations (Gleick 2003; Bates *et al*. 2008; Dragoni & Sukhiga 2008; Palmer *et al*. 2008; Vörösmarty *et al*. 2010; Capon *et al*. 2013). Future provision of fresh water for drinking and sanitation as well as agricultural and, to a lesser degree, industrial uses represents a major concern to water managers (Gleick 2003). Even in areas where rainfall and stream flow are projected to increase (e.g. eastern and south-eastern Asia), water scarcity is likely to rise as a result of seasonal shifts and increased variability in rainfall and runoff regimes as well as declines in water quality (Bates *et al*. 2008). Other provisioning services (e.g. supply of fish, timber, etc.) will also be influenced because these rely on freshwater ecosystem components (i.e. organisms) that are strongly dependent on climate. Strong effects on regulating and supporting services provided by freshwater ecosystems can similarly be anticipated, especially where temperature and hydrology are key drivers (e.g. water filtration, nutrient cycling etc.; Table 3.1). Furthermore, cultural connections between people and freshwater ecosystems will be influenced. For example, changes in the aesthetics of freshwater ecosystems resulting from shifts in their species composition or physical character (e.g. salinisation of coastal freshwater wetlands due to sea-level rise) may affect their value as recreational areas. Similarly, spritiual connections to freshwater ecosystems may be weakened if climate change leads to a deterioration of elements which are held sacred (e.g. persistency of surface water). On the other hand, the importance of freshwater ecosystems as sources of spiritual or creative inspiration may increase in landscapes increasingly denuded by the combined impacts of climate change and anthropgenic influences.

3.2.2 Demand for water ecosystem services

Climate change can be expected to affect the demand for, as well as the supply of, freshwater ecosystem goods and services

Table 3.1 *A selection of key freshwater ecosystem goods and services and examples of potential climate change impacts on their supply and demand. Ecosystem services shown (and discussed in the text) follow the Millennium Ecosystem Assessment typology. Other ecosystem services classifications are discussed in Martin-Ortega* et al.*, this book.*

Water ecosystem goods and services	Example	Examples of potential climate change impacts	
		Supply-side	Demand-side
Provisioning			
Food	Fish production	Abundance and distribution of fish likely to shift	May increase in significance if surrounding landscape becomes drier and less productive
Fresh water	Storage and retention of water	Changes in riparian and aquatic vegetation, soils, and biogeochemistry will affect quantity and quality of stream, flood, and groundwaters	Greater importance due to increased frequency of intense rainfall and runoff events
Fibre and fuel	Wood production	Changes in regulating and habitat functions and biota will affect production of raw materials	May increase in significance if surrounding landscape becomes drier and less productive
Biochemical	Extraction of materials from aquatic biota		
Genetic materials	Extraction of genes from aquatic biota	Diversity of genetic resources will change with changed riparian biota	May increase in significance if surrounding regions become genetically homogenised
Regulating			
Climate regulation	Provision of sink for greenhouse gases	May switch to sources with warming and drying	Increased importance due to climate change mitigation
Water regulation (hydrology)	Groundwater recharge	Changes in topography and vegetation will affect runoff and flood patterns and ground water dynamics	Greater importance due to increased frequency of intense rainfall and runoff events
Water purification and waste treatment	Water filtration (e.g. removal of pollutants)	Changes to aquatic and riparian vegetation, soils, and biogeochemistry may limit capacity to breakdown compounds and act as solute sinks	Greater importance due to increased frequency of intense rainfall and runoff events
Erosion regulation	Retention of sediments	Changes in hydrology and vegetation will alter capacity of soils to support pasture growth	Greater importance due to increased frequency of intense rainfall and runoff events, especially in drying landscapes
Natural hazard regulation	Flood mitigation	Changes in wetland extent and distribution, aquatic vegetation, and topography will influence patterns of flooding	Greater importance due to increased frequency and intensity of extreme flood events
Pollination and propagule dispersal	Provision of habitat for pollinators and animal dispersal vectors	Pollination and dispersal will be affected by changes in aquatic habitat and biota	Increasing importance as pathways for migration in response to shifting climate (especially riparian zones)
Supporting			
Soil formation	Accumulation of organic matter	Changes in hydrology and vegetation will alter capacity of soils to support pasture growth	Greater importance due to increased frequency of intense rainfall and runoff events, especially in drying landscapes
Nutrient cycling	Storage of nutrients	Changes to aquatic and riparian soils and biota will affect nutrient cycling	
Cultural			
Spiritual and inspirational	Source of inspiration		May increase in significance if surrounding landscape is significantly altered

Table 3.1 (*cont.*)

Water ecosystem goods and services	Example	Examples of potential climate change impacts	
		Supply-side	Demand-side
Recreational	Provision of opportunities for recreation	Changes in climate, topography, soil, water, and biota will affect recreational value and capacity for use	May increase in significance if surrounding landscape is significantly altered
Aesthetic	Appreciated for aesthetic beauty	Changes in regulating and habitat functions will affect scenery	May increase in significance if surrounding landscape is significantly altered
Educational	Provision of opportunities for education and research	Opportunities for research and education will vary with other changes	Increased significance for adaptive learning and management

Adapted from Millennium Ecosystem Assessment (2005) and Capon *et al.* (2013).

(Table 3.1). Dramatic increases in demand for water resources are particularly anticipated in the agricultural sector, where rising temperatures and increased variability of rainfall are likely to lead to greater demands for higher security water supplies (UNESCO 2012). In comparison, relatively small increases in demand for water resources are projected for human settlements and industrial uses as a direct result of climate change, although considerable growth in water resources demand is forecast in relation to population growth and economic development (Coates *et al.* 2012). A greater demand for water to cool buildings and infrastructure, for example, may occur in response to warming. Climate change mitigation activities, such as hydropower generation and carbon sequestration via the growth of plantations, may also result in greater demands for the provision of water resources (Pittock 2011).

The demand and importance of other provisioning services supplied by freshwater ecosystems are also likely to shift as a result of local as well as broader, regional climate change effects. Climate change-induced declines in fish health or the quality of riparian timbers, for instance, may reduce demands for harvesting. Alternatively, reductions in the production of food and other raw materials (e.g. fuel and fibre) elsewhere in drying landscapes, for example, could foreseeably increase demands on freshwater ecosystems to provide such resources. The significance of some regulating and supporting ecosystem services might similarly be expected to grow under many climate change scenarios (Table 3.1). The role of freshwater ecosystems in flood mitigation and storm protection, for instance, is likely to become increasingly important to human communities in areas subject to more frequent and intense storms and floods (Capon *et al.* 2013). Habitat functions of some freshwater ecosystems, for both aquatic and terrestrial

organisms, will also grow in significance under climate change (e.g. provision of drought refuge or corridors for migration in response to shifting climatic conditions; Seavy *et al.* 2009; Capon *et al.* 2013).

3.3 ADAPTATION OPTIONS ADDRESSING CLIMATE CHANGE RISKS TO WATER ECOSYSTEM SERVICES

A wide range of approaches to climate change adaptation have been proposed and are currently being implemented by various sectors globally (Hansen & Hoffman 2011). Numerous frameworks have been put forward to categorise climate change adaptation options in relation to, for example, their level of planning (i.e. autonomous vs. planned), timing in relation to impacts (i.e. reactive vs. proactive), or use of physical engineering structures (i.e. hard vs. soft; Smith *et al.* 2000; Gleick 2003; Hallegatte 2009; Füssell 2007; Capon *et al.* 2013). Here, we propose that many adaptation options might also be considered in terms of their aims with respect to ecosystem services, since adaptation is often triggered by perceived risks to a particular ecosystem service or suite of services (e.g. the provision of potable water). Adaptation options of relevance to water ecosystem services might therefore be categorised initially as either supply-side options or demand-side measures (Table 3.2).

Relevant supply-side adaptation options will essentially aim to either protect, restore, enhance, or replace water ecosystem services with respect to the risks posed by climate change to their future provision (Table 3.2). Minimising other threats to freshwater ecosystems by, for example, increasing the area

Table 3.2 *Examples of adaptation options categorised in relation to their aims with respect to the supply of water ecosystem services.*

Adaptation aim	Examples of adaptation options
Protection of water ecosystem services	Management of existing, non-climatic stressors
	Expansion of protected area network
	Restriction of development
Restoration of water ecosystem services	Riparian vegetation restoration
	Flow regime restoration
	Dam removal
	Removal of river bank armouring
	Restore connectivity, e.g. fish ladders
Enhancement of water ecosystem services	Planting of fast-growing, high-shade riparian trees
	Storage and delivery of dilution flows to address water quality problems
	Species translocations, e.g. fish stocking
Replacement of water ecosystem services	Construction of new water storage infrastructure, e.g. dams
	Construction of artificial ecosystems, e.g. wetlands
	Construction of sea walls

protected by reserve networks, represents a major approach to climate change adaptation in conservation and natural resources management sectors that may protect many vulnerable water ecosystem services (Hansen & Hoffman 2011). Where climate change risks exceed the capacity of protected area networks to protect water ecosystem services, novel approaches to conservation might include locating new reserves in areas projected to have high future value (Fuller *et al.* 2010). Temporal, as well as spatial, flexibility can also be introduced into threat management strategies (e.g. protection of stream flows from water extraction following long periods of drought). At their most extreme, supply-side adaptation options aimed at protecting water ecosystem services may involve complete or partial retreat via the removal of infrastructure or activity (e.g. dam removal or floodplain restoration; Stanley & Doyle 2003; Pittock 2009).

Some adaptation strategies might seek to restore the supply of vulnerable water ecosystem services in the face of projected climate change risks. A need for greater flood mitigation by freshwater ecosystems upstream of human settlements, for example, may require restoration of wetland and riparian habitats (e.g. vegetation regeneration). Similarly, climate change risks to aquatic biota strengthen the need to restore connectivity between

habitats, through flow restoration or retrofitting physical infrastructure (e.g. fish ladders in weirs), to enable the movement of organisms (and genes) and the persistence of viable populations. Riparian restoration is likely to be particularly important for climate change adaptation since it often represents a low-regret strategy with multiple, far-reaching benefits (Seavy *et al.* 2009; Capon *et al.* 2013).

In situations where climate change risks to highly valued water ecosystem services are great, adaptation strategies might aim to enhance some water ecosystem services. Over-restoring riparian vegetation (e.g. with fast-growing, densely canopied trees), for example, could be conducted to increase the capacity of riparian vegetation to reduce water temperatures and minimise risks to aquatic biota associated with warming (Davies 2010). Such 'ecological engineering' options may attempt to enhance some ecosystem services to address anticipated growth in their demand (e.g. fish stocking or planting of different riparian or aquatic plants that may minimise bank erosion or enable greater retention of pollutants). The construction or re-operation of anthropogenic wetlands for flood mitigation or waste management, for instance, represent the more extreme adaption options in this category.

Finally, supply-side adaptation options may seek to replace water ecosystem services altogether with hard-engineered alternatives (e.g. dams, barrages, artificial ecosystems, etc.). While these are often associated with a high probability of failure and maladaptation and large opportunity costs (Barnett & O'Neil 2010; Nelson 2010; Capon *et al.* 2013), such approaches may be the only option in areas of extremely high risk or where supporting services underpinning many highly valued water ecosystem services are threatened. Hard approaches may also be useful as temporary 'band-aid' measures that enable other adaptation pathways to be implemented. Safety margins that account for climatic extremes and regular reviews of works are recommended (Pittock & Harmann 2011).

Demand-side adaptation options include 'soft' approaches that employ communication, educational, or institutional change to reduce or shift the demand for water ecosystem goods and services (e.g. conversion of permanent abstraction licences to temporary licences). Some 'hard' engineering strategies that aim to increase water efficiency (e.g. conversion to more efficient irrigation equipment) might also serve to reduce demand, although such effects are debatable (e.g. Crase & O'Keefe 2009). The effectiveness of many supply-side options is likely to be enhanced by the implementation of accompanying demand-side options (e.g. changes to property rights to enable planned retreat). A shift in orientation of water resources management from supply-side to demand-side may promote the implementation of robust adaptation options that are better suited to the uncertainties associated with climate change and the need for flexibility (Wilby & Dessai 2010).

3.4 AN ECOSYSTEM SERVICES-BASED APPROACH TO MAKING ADAPTATION DECISIONS

Adaptation decision-making is highly complex and occurs in the context of many ecological, socioeconomic, cultural, and institutional factors (Füssell 2007). In practice, the development of adaptation strategies involves a myriad of drivers, aims, and stakeholders – all operating across multiple scales and the latter with varying roles and responsibilities. Consequently, climate change adaptation itself can be associated with high levels of risk (i.e. potential for maladaptation or perverse outcomes). Maladaptation is particularly likely where the goals of climate change adaptation seek only to address socio-economic needs without consideration of ecological consequences, or where a narrow focus on environmental outcomes is adopted without taking into account human aspects (Hulme 2005; Hadwen et al. 2012; Capon et al. 2013). An ecosystem services-based approach (see Chapter 2) to climate change adaptation decision-making, however, could ensure that both ecosystem functioning and human wellbeing are considered as well as the complex interactions between these. Undertaking adaptation decision-making in an adaptive management framework with a focus on flexibility, learning, and integration and collaboration among decision-makers and sectors is also widely espoused as critical for effective adaptation to climate change in socio-ecological systems (Pahl-Wostl 2007; Kingsford et al. 2011). Here, we outline the potential benefits of an ecosystem services approach in key stages of adaptation decision-making: i.e. framing the problem; identifying options; prioritising options; implementation; and monitoring and evaluation (Randall et al. 2012).

3.4.1 Framing the problem

Taking an ecosystem services-based approach enables an integrated assessment of the triggers and goals underpinning the development of adaptation strategies and ensures consideration of risks to both human and non-human systems, as well as their interaction. Many adaptation actions, especially in the water sector, focus predominantly on risks to humans (e.g. sanitation, drinking water, etc.). Without consideration of the role of ecological systems, some supply-side interventions with very narrow objectives (e.g. increasing potable water supplies through dam construction) may result in perverse ecological outcomes (e.g. declining downstream water quality) that eventually lead to maladaptation in the very area that was the original goal (e.g. undrinkable water downstream of dams). A water ecosystem services-based approach facilitates a broader formulation of the goals of adaptation. Similarly, a more comprehensive range of relevant decision-makers and stakeholders are likely to be identified as important (i.e. not solely water managers or consumers; Gleick 2003).

3.4.2 Identifying adaptation options

Generating options for adaptation to climate change is a major element of adaptation planning and one that may be daunting to many managers faced with seemingly insurmountable risks (e.g. sea-level rise threats to coastal wetlands). By framing the problem through the lens of ecosystem services, decision-makers can identify which water ecosystem services are most at risk from projected climate change in their sector or region and determine whether or not these risks are due to projected changes in ecosystem supply or demand. As a first step, this might influence an emphasis on identifying supply-side or demand-side measures. Where supply-side measures are deemed significant, decision-makers could then determine whether the current supply of water ecosystem services can be better protected from non-climatic stressors or restored. Management of existing threats and restoration of freshwater ecosystems are widely recommended as effective adaptation strategies since they typically entail low risks of maladaptation, have a high potential for multiple benefits, involve limited opportunity costs and are often reversible (Capon et al. 2013). In contrast, if very high-value water ecosystem services are considered to be extremely vulnerable to climate change impacts, temporary measures that aim to enhance or replace some ecosystem services might be required. In areas supporting dense human settlements where dramatic increases in extreme flooding and storm surge are projected, for example, the capacity of freshwater ecosystems to adequately mitigate hazardous flooding is likely to be outstripped by climate change. Where retreat is not possible in the short term, hard-engineering options such as sea walls, armouring, artificial wetlands, levee banks, and barrages may present the only options to address such significant risks.

3.4.3 Prioritising adaptation options

The prioritisation and selection of suitable adaptation options can benefit from a water ecosystem services-based approach in several ways. First, adaptation options associated with both highly valuable and highly vulnerable ecosystem services can be identified and prioritised for attention. Similarly, adaptation options that distribute benefits and minimise risks of maladaptation can also be determined, many of which may be low-regret options that might be implemented regardless of the uncertainties involved. An ecosystem services-based approach can also facilitate a broader consideration of the costs associated with an adaptation action, especially with respect to opportunity costs, i.e. the costs of sacrificing alternative adaptation options.

3.4.4 Implementation

An ecosystem services-based approach might benefit the implementation phase of adaptation actions, including their longer-term maintenance, by promoting public and political support for selected actions. Clear communication to a wide diversity of stakeholders of the choice and desired benefits of adaptation actions will be aided by demonstrating these in relation to an ecosystem services framework. Similarly, cross-sectoral collaborations through the implementation phase might be engendered by an ecosystem services-based approach since the goals and approaches to adaptation will necessarily entail more than those relevant to a single sector.

3.4.5 Monitoring and evaluation

Finally, an ecosystem services-based approach can provide a suitable framework for evaluating the effectiveness of adaptation actions with respect to their desired outcomes and their effects, both positive and deleterious, on other ecosystem services.

3.4.6 Potential limitations of an ecosystem services-based approach to adaptation

Ecosystem services represents but one approach to integrating human and ecological values and needs with respect to natural resources management, and choosing this approach over others (e.g. triple bottom-line approach) will necessarily entail some risks. Some such limitations might include the potential for impacts of climate change or adaptation actions on ecological components or processes to be overlooked where their links to ecosystem goods and services are not clearly understood or recognised (e.g. values of rare, cryptic, or poorly known species). Similarly, some human interests (e.g. those of marginalised groups) might also be underappreciated if socio-ecological links are not well known. Ecosystem services-based approaches are also criticised for diminishing the importance of non-market values (e.g. spiritual values) while promoting market values (e.g. harvestable resources), especially when they are used to calculate dollar values for goods and services provided. Such fiscal valuation of nature has resulted in an aversion to the ecosystem services concept among some environmentalists that may stymy collaborative adaptation planning efforts. An awareness of the potential limitations of ecosystem services-based approaches ecosystem services approach to adaptation decision-making, however, may enable such risks to be minimised or at least managed.

3.5 CONCLUSIONS

Water ecosystem services depend on the ecological functions, components, and processes of the world's freshwater ecosystems. Due to high levels of exposure and sensitivity to changes in climatic stimuli as well as constraints on their adaptive capacity as a result of their human modification and degradation, freshwater ecosystems are highly vulnerable to climate change impacts. Both the supply of and demand for provisioning, regulating, supporting, and cultural services provided by freshwater ecosystems can therefore be expected to change in response to projected climate change, with the majority of impacts anticipated to be negative. Consequently, there is a need for urgent planned adaptation to climate change with respect to water ecosystem services. Adaptation actions themselves, however, can entail a high degree of risk and have the potential to result in a wide range of perverse outcomes. Such maladaptation is particularly likely where adaptation strategies are limited to socio-economic objectives and methods without sufficient consideration of ecological impacts or approaches (or vice versa). Taking a water ecosystem services-based approach to adaptation decision-making can minimse risks of maladaptation by contributing to the development of holistic, integrated, and collaborative adaptation strategies that take into account the wellbeing of people as well as the freshwater ecosystems on which we all rely. As a result, a water ecosystem services-based approach to adaptation decision-making can increase the opportunity for identifying low-regret adaptation options that have

---

Box 3.1 A water ecosystem services-based approach to adaptation decision-making

- A water ecosystem services-based approach can help define climate adaptation problems in an inclusive and integrated framework.
- Such an approach can aid in identifying appropriate adaptation options by focusing attention on highly valued and/or vulnerable ecosystem services and whether risks associated with climate change are due to their supply and/or demand.
- Adaptation options can then be prioritised in relation to their potential to distribute benefits and minimise risks of perverse outcomes in relation to non-target ecosystem goods and services.
- An ecosystem services-based approach to adaptation can engender public and political support during planning and implementation by communicating cross-sectoral linkages and dependencies.
- Monitoring and evaluation of adaptation actions (e.g. indicator selection) can be guided by an ecosystem services-based approach to ensure results are assessed and reported in relation to integrated goals.

Box 3.2 Key messages

- Freshwater ecosystems are highly vulnerable to climate change impacts due to their high levels of exposure to changes in climatic stimuli, high degree of sensitivity to these, and constrained adaptive capacity due to human modification and degradation.

- Both the supply of and demand for water ecosystem services are likely to be significantly affected by climate change over the coming century, with most impacts likely to be negative.

- A wide range of adaptation options are available to manage climate change impacts on water ecosystem services including supply-side and demand-side options.

- Supply-side adaptation options generally seek to either protect, restore, enhance, or replace water ecosystem services.

- Demand-side adaptation mainly comprise 'soft' measures associated with institutional change, e.g. improved communication, education, and stakeholder engagement.

- Adaptation actions risk perverse consequences and maladaptation, especially where their orientation is narrowly focused on either ecological or socio-economic outcomes and approaches.

- Adaptation decision-making can be aided by taking a water ecosystem services-based approach that explicitly recognises connections between ecosystems and human wellbeing and vice versa.

a high potential to achieve multiple benefits. Future research in this domain therefore depends on improvements in the integration of knowledge, values, and methods across the natural and social sciences as well as the humanities.

References

Barnett, J. A. & O'Neill, S. (2010). Maladaptation. *Global Environmental Change* **20**, 211–213.

Bates, B. C., Kundzewicz, Z. W., Wu, S., & Palutikof, J. P. (eds) (2008). *Climate Change and Water*. Technical Paper of the Intergovernmental Panel on Climate Change, IPCC Secretariat, Geneva.

Cai, W. & Cowan, T. (2008). Evidence of impacts from rising temperature on inflows to the Murray-Darling Basin. *Geophysical Research Letters* **35**(7), L07701.

Capon, S. J., Chambers, L. E., Mac Nally, R., *et al.* (2013). Riparian ecosystems in the 21st century: hotspots for climate change adaptation? *Ecosystems* **16**, 359–381.

Coates, D., Connor, R., Lecler, L., Rast, W., Schumann, K., & Webber, M. (2012). Water demand: what drives consumption? In *Managing Water Under Uncertainty and Risk; the United Nations World Water Development Report 4*, vol. **1**. UNESCO, Paris.

Crase, L. & O'Keefe, S. (2009). The paradox of national water savings: a critique of 'Water for the Future'. *Agenda* **16**(1), 45–60.

Davies, P. M. (2010). Climate change implications for river restoration in global biodiversity hotspots. *Restoration Ecology* **18**, 261–268.

Dragoni, W. & Sukhiga, B. S. (2008). Climate change and groundwater: a short review. *Geological Society, London, Special Publications* **288**, 1–12.

Freeman, C., Lock, M. A., & Reynolds, B. (1993). Climatic change and the release of immobilized nutrients from Welsh riparian wetland soils. *Ecological Engineering* **2**, 367–373.

Fuller, R. A., McDonald-Madden, E., Wilson, K. A., *et al.* (2010). Replacing underperforming protected areas achieves better conservation outcomes. *Nature* **466**, 365–367.

Füssell, H. (2007). Adaptation planning for climate change: concepts, assessment, approaches, and key lessons. *Sustainability Science* **2**, 265–275.

Gleick, P. H. (2003). Global freshwater resources: soft-path solutions for the 21st century. *Science* **302**, 1524–1528.

Goudie, A. S. (2006). Global warming and fluvial geomorphology. *Geomorphology* **79**(3), 384–394.

Hadwen, W. L., Capon, S., Kobashi, D., *et al.* (2012). *Coastal Ecosystems Responses to Climate Change: A Synthesis Report*. National Climate Change Adaptation Research Facility, Gold Coast, Australia.

Hallegatte, S. (2009). Strategies to adapt to an uncertain climate change. *Global Environmental Change* **19**, 240–247.

Hansen, L. J. & Hoffman, J. R. (2011). *Climate Savvy: Adapting Conservation and Resource Management to a Changing World*. Island Press, Washington, DC.

Hulme, P. E. (2005). Adapting to climate change: is there scope for ecological management in the face of a global threat? *Journal of Applied Ecology* **42**, 784–794.

Intergovernmental Panel on Climate Change. (2007). *Climate Change 2007: Synthesis Report*. Cambridge University Press, Cambridge.

Kernan, M. R., Battarbee, R. W., & Moss, B. (eds). (2010). *Climate Change Impacts on Freshwater Ecosystems*, vol. **314**. Wiley-Blackwell, Oxford.

Kingsford, R., Biggs, H., & Pollard, S. (2011). Strategic adaptive management in freshwater: 16 protected areas and their rivers. *Biological Conservation* **144**, 1194–1203.

MacNally, R., Cunningham, S. C., Baker, P. J., Horner, G. J., & Thomson, J. R. (2011). Dynamics of Murray-Darling floodplain forests under multiple stressors: the past, present, and future of an Australian icon. *Water Resources Research* **47**, W00G05.

Millennium Ecosystem Assessment. (2005). *Ecosystems and Human Well-being: Wetlands and Water Synthesis*. World Resources Institute, Washington, DC.

Nelson, D. R. (2010). Adaptation and resilience: responding to a changing climate. *Wiley Interdisciplinary Reviews: Climate Change* **2**(1), 113–120.

Pahl-Wostl, C. (2007). Transitions towards adaptive management of water facing climate and global change. *Water Resources Management* **21**, 49–62.

Palmer, M. A., Allan, J. D., Meyer, J., & Bernhardt, E. S. (2007). River restoration in the twenty-first century: data and experiential knowledge to inform future efforts. *Restoration Ecology* **15**, 472–481

Palmer, M. A., Reidy Liermann, C. A., Nilsson, C., *et al.* (2008). Climate change and the world's river basins: anticipating management options. *Frontiers in Ecology and the Environment* **6**, 81–89.

Palmer, M. A., Lettenmaier, D. P., Poff, N. L., Postel, S. L., Richter, B., & Warner, R. (2009). Climate change and river ecosystems: protection and adaptation options. *Environmental Management* **44**(6), 1053–1068.

Pittock, J. (2009). Lessons for climate change adaptation from better management of rivers. *Climate and Development* **1**, 194–211.

Pittock, J. (2011). National climate change policies and sustainable water management: conflicts and synergies. *Ecology & Society* **6**, 1–20.

Pittock, J. & Hartmann, J. (2011). Taking a second look: climate change, periodic re-licensing and better management of old dams. *Marine and Freshwater Research* **62**, 312–320.

Poff, N. L. & Zimmerman, J. K. (2010). Ecological responses to altered flow regimes: a literature review to inform the science and management of environmental flows. *Freshwater Biology* **55**(1), 194–205.

Randall, A., Capon, T., Sanderson, T., Merrett, D., & Hertzler, G. (2012). *Making Decisions Under the Risks and Uncertainties of Future Climates*. National Climate Change Adaptation Research Facility, Gold Coast, Australia.

Rood, S. B., Pan, J., Gill, K. M., Franks, C. G., Samuelson, G. M., & Shepherd, A. (2008). Declining summer flows of Rocky Mountain rivers:

changing seasonal hydrology and probable impacts on floodplain forests. *Journal of Hydrology* **349**, 397–410.

Seavy, N. E., Gardali, T., Golet, G. H., *et al.* (2009). Why climate change makes riparian restoration more important than ever: recommendations for practice and research. *Ecological Restoration* **27**, 330–338.

Smith, B., Burton, I., Klein, R. J., & Wandel, J. (2000). An anatomy of adaptation to climate change and variability. *Climatic Change* **45**(1), 223–251.

Stanley, E. H. & Doyle, M. W. (2003). Trading off: the ecological effects of dam removal. *Frontiers in Ecology and the Environment* **1**(1), 15–22.

Steffen, W., Burbidge, A. A., Hughes, L., *et al.* (2009). *Australia's Biodiversity and Climate Change*. CSIRO Publishing, Collingwood.

UNESCO. (2012). *Managing Water under Uncertainty and Risk. The United National World Water Development Report 4*, vol **1**. UNESCO, Paris.

Vicuna, S. & Dracup, J. A. (2007). The evolution of climate change impact studies on hydrology and water resources in California. *Climatic Change* **82**(3–4), 327–350.

Visser, M. E. (2008). Keeping up with a warming world: assessing the rate of adaptation to climate change. *Proceedings of the Royal Society B: Biological Sciences* **275**(1635), 649–659.

Vörösmarty, C. J., McIntyre, P. B., Gessner, M. O., *et al.* (2010). Global threats to human water security and river biodiversity. *Nature* **467**, 555–561.

Wilby, R. L. & Dessai, S. (2010). Robust adaptation to climate change. *Weather* **65**(7), 180–185.

4 Operationalizing an ecosystem services-based approach for managing river biodiversity

Catherine M. Febria, Benjamin J. Koch, and Margaret A. Palmer

4.1 INTRODUCTION

Covering less than 1% of the Earth's surface, freshwater – streams, rivers, ponds, wetlands, and lakes – supports as much as 10% of all animal species, including one-third of all vertebrates (Strayer & Dudgeon 2010; Figure 4.1). While being among the most biologically diverse, freshwater ecosystems are also among the most imperiled on the planet. In developed regions such as Europe and USA, more than 30% of freshwater species are now thought to be threatened or extinct. As with the majority of the world's ecosystems, accelerated rates of human population growth, industrialization, and agricultural intensification are driving these dramatic species losses, with invasive species introductions, over-harvesting, and loss of habitats being the primary causes (Dudgeon 2013).

Reducing freshwater biodiversity loss is at odds with managing many other ecosystem services. Much of the conflict lies in the fact that freshwaters are 'hot spots' for both human needs (i.e. drinking, irrigation, transportation) as well as biodiversity (i.e. the number or suite of species native to a given area; Strayer & Dudgeon 2010; Leisher, this book). Moreover, the majority of the world's population is facing, and will continue to face, increasing water scarcity (Vörösmarty et al. 2010). Humans rely heavily on freshwater, and in particular from large rivers for hydropower, transportation, and fisheries. Water abstraction further strains the integrity of a river's biophysical components, including biodiversity. While some aspects of freshwater ecosystem services are renewable (e.g. water supply), biodiversity is not.

This chapter seeks to clarify how ecosystem service-based approaches can be put into practice ('operationalized') for the purpose of prioritizing freshwater biodiversity alongside other river services. We propose a framework that is meant to serve as a way to explore trade-offs in ecosystem services delivery. We base this operationalization on three typical perspectives associated with freshwater biodiversity: inherent value as a final service, biodiversity–ecosystem function (an intermediate or provisioning service), and some combination of the two. We have chosen to emphasize the biophysical aspects of biodiversity loss from rivers (deepening the discussion of core element 2 in

Chapter 2), where a great deal of scholarly work is urgently needed (Palmer & Filoso 2009), leaving other freshwater ecosystems and the valuation aspects of biodiversity for those better qualified in economics and social policy.

To help inform policy and research needs, we provide an overview of the factors responsible for freshwater biodiversity loss, the impacts on biodiversity of using river services, and the management actions required to mitigate biodiversity loss. We recognize that the provision of ecosystem services may be influenced by aspects of biodiversity that extend beyond the level of species (i.e. genetic, functional, and habitat/ecosystem diversity) but much of the scientific understanding needed to link natural ecosystems with humans for the purposes of decision-making is still incomplete. Therefore, we focus on species-level biodiversity to illustrate the trade-offs that can arise when targeted management actions affect river biodiversity. For example, building dams for hydropower and extracting water for drinking and irrigation has degraded many downstream aquatic ecosystems and the biota those ecosystems support (Vörösmarty et al. 2010). By applying a framework based as an ecosystem services-based approach that prioritizes freshwater biodiversity alongside other services, it may be possible to manage for river biodiversity and minimize loss.

4.2 THE ROLE OF FRESHWATER BIODIVERSITY IN AN ECOSYSTEM SERVICES-BASED APPROACH

The ecosystem services concept and related approaches introduced an effective set of communication tools for scientists and conservationists to raise awareness of freshwater biodiversity loss (see Chapter 2 of this book). The Millennium Ecosystem Assessment generated a list of freshwater ecosystem services (Aylward et al. 2005) that included: clean water for drinking, sufficient water for irrigation or hydropower generation, flood protection, food and related products (algae, rice, fish, invertebrates), recreation (fishing, swimming, water sports), aesthetics, and biodiversity (existence of species and ecosystems). By focusing on the benefits provided to humans, the concept provides a platform for

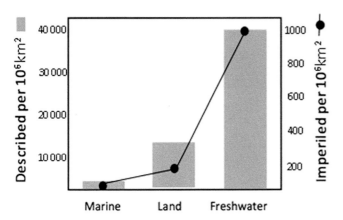

Figure 4.1 The number of described and imperiled species of eukaryotes is extremely high in freshwater ecosystems compared to the area of the globe occupied by freshwater (<1%), and high compared to other major ecosystem types. Modified from Strayer and Dudgeon (2010).

Table 4.1 *Key motivations for ecosystem service-based approaches used to manage river biodiversity*

(1) As a final ecosystem good or service. Biodiversity is an ecosystem good or service when there is the provision of food, recreation, and/or aesthetics. Management must focus intensively on the needs of the species of interest or on protecting/restoring the historic ecological conditions that support the full suite of native species.

(2) Supports biodiversity–ecosystem functioning. In some ecosystems, biodiversity is known to play a role in facilitating or driving ecological processes that are critical to the delivery of goods and services. If this is true in river ecosystems then biodiversity itself (or individual species known to be important to the desired services) must be fully integrated into management plans.

(3) Co-managing and valuing both (1) and (2). Ideally managers will attempt to balance management for delivery of ecosystem services with protection and/or recovery of biodiversity. This can be challenging when desired services and species needs are at odds.

explaining the roles of natural ecosystems and their species in enhancing daily lives and overall wellbeing.

Individually, freshwater services may be tightly coupled (e.g. clean water is required for both drinking and recreation) while others represent trade-offs (e.g. flow regulation using dams reduces native biodiversity above and below the dam). As with many ecological attributes, full scientific understanding of the complex interactions between humans and freshwater biodiversity remains incomplete, but is steadily improving (e.g. www.living-rivers.org). As ecosystem service-based approaches have moved beyond communication into the realm of natural resource management, policy making, and economics, managing for river biodiversity has become more challenging to put into practice. Some of the challenges have been due to a lack of consistency around defining ecosystem goods and services (see Chapter 2 and Nahlik *et al.* 2012) as well as due to a mismatch between social values and concrete, measurable attributes in a given ecosystem (Reyers *et al.* 2013; Ringold *et al.* 2013). Furthermore, from an ecosystem service perspective biodiversity can be treated in multiple ways. Biodiversity can be viewed as a 'final' ecosystem service if it is consumed or its mere existence is valued for human benefit (Boyd & Banzhaf 2007); however, it is also an 'intermediate' ecosystem service if biodiversity is essential to understanding, predicting, and managing other final goods or services (Ringold *et al.* 2013). Often, it is the final goods and services that impact human wellbeing most rapidly and dramatically that are valued more than the less tangible, intermediate services. It is critical to understand a stakeholder's motivation for managing biodiversity in order to distinguish it as either an intermediate or final service.

In this light, biodiversity can motivate ecosystem service-based management in one of three ways: (1) as a final ecosystem good or service (e.g., provision of commercially important species or as a cultural service related to bequest and existence

values); (2) in supporting biodiversity–ecosystem functioning as an intermediate service; or (3) some combination of the two (Table 4.1).

Distinguishing among those roles is a social decision (described in Figure 4.2) and may be difficult in practice. Nonetheless, drawing such a distinction is a critical starting point and a heuristic tool for analyzing the complex array of factors associated with ecosystem service-based approaches and for setting conservation priorities for a given river ecosystem. Therefore, a framework for prioritizing biodiversity alongside other services may be a fruitful management strategy for all stakeholders.

4.3 AN ECOSYSTEM SERVICES FRAMEWORK FOR MANAGING RIVER BIODIVERSITY

In this chapter we propose a framework that is meant to serve as a structured way to explore the trade-offs between ecosystem services delivery and river biodiversity loss and should help with formulating and prioritizing management goals for a particular river ecosystem (Figure 4.2).

To begin, a representative group of scientists, managers, and stakeholders must identify and prioritize one or multiple ecosystem services (Step 1). Geography and other site-specific constraints assist stakeholders in placing biodiversity alongside other ecosystem services of interest. A shared research effort between natural and social scientists is required to assess the biophysical and social contexts within which a certain ecosystem good or

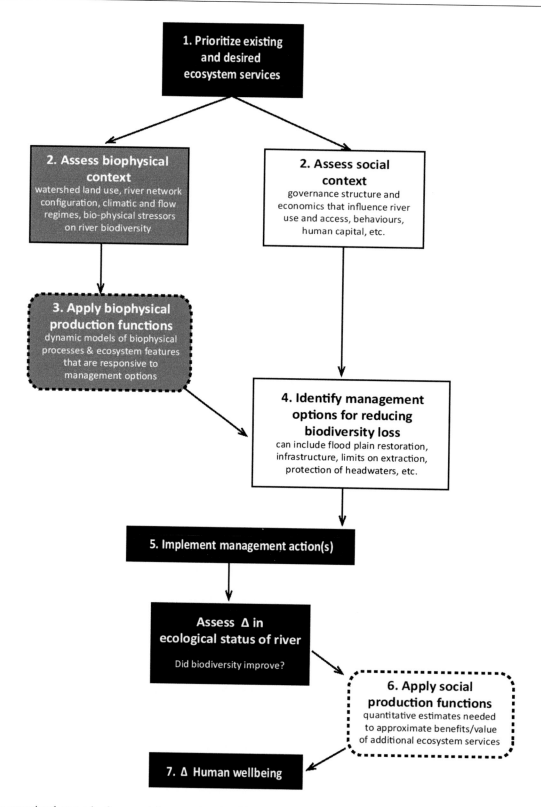

Figure 4.2 An operational, step-wise framework for managing river biodiversity based on an ecosystem services-based approach. The framework can be viewed as using a series of steps shared between biophysical data needs (left) and social data needs (right). General operational steps (arrows) identify specific actions to be undertaken by diverse groups of scientists, managers, and stakeholders.

service can be generated (Step 2 in Figure 4.2). Key components of the framework can be linked quantitatively as production functions (i.e. ecological, social) through a series of relationships that independently warrant substantial technical and/or academic expertise (Steps 3 and 6, respectively, in Figure 4.2). These scientists must develop quantitative relationships that describe how biophysical processes, ecosystem attributes, and other factors important to the delivery of an ecosystem service (i.e. 'intermediate' ecosystem services that are inputs into ecological 'production functions') inform and constrain a suite of potential management options (Step 4). The implementation of a management action (Step 5) and resultant changes in ecosystem functions or status are factored into social (or socio-ecological) production functions (Step 6). Any changes in the status of human wellbeing (Step 7) in turn might introduce additional feedbacks, requiring stakeholders to re-visit and re-assess their prioritization in Step 1.

Steps of the described framework (i.e. Steps 3 and 6) that link ecological or social information with the management actions often encompass advanced technical or academic data, analyses, or knowledge in order to generate robust quantitative relationships. These relationships are called ecological (or social) production functions and they are paramount in this framework. Production functions quantitatively link ecological and social data with management actions and depend greatly on the data available, and can be technically complex to develop (Kareiva *et al.* 2011; Seppelt *et al.* 2011; Ringold *et al.* 2013; see Figure 4.3 for an example). The production function component can also be useful in demonstrating trade-offs between management options and the costs to implement a management option for managing river biodiversity (e.g. Kondolf *et al.* 2008).

To date, ecosystem service based-approaches generally extract results from basic research on ecosystem dynamics to produce the necessary production functions used in predicting the delivery of an ecosystem service (Nelson *et al.* 2009; Kareiva *et al.* 2011). The form of the production function is most often a process-based model or statistical relationship based on field and literature data, which may be applied to economic or other social assessments (Vigerstol & Aukema 2011; Keeler *et al.* 2012). For river ecosystems, many of the available tools for characterizing production functions are based on hydrologic models that route water, sediment, and pollutants as a function of local climate, topography, soils, and land use (Vigerstol & Aukema 2011). Such models can include feedbacks to surface water dynamics mediated by vegetative cover, evapotranspiration, and groundwater exchange, but they are largely limited to predicting water availability and quality with respect to a small set of pollutants (e.g. nitrogen, phosphorus, sediments). River biodiversity per se is rarely modelled. In fact, there are no formal ecosystem service tools that

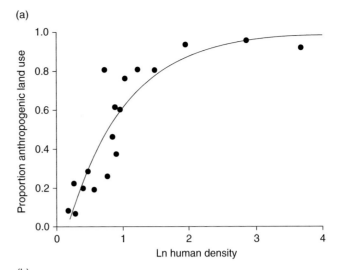

(a)

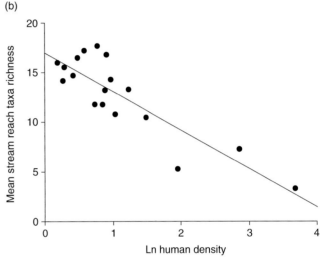

(b)

Figure 4.3 Example of data collected to quantitatively estimate stream invertebrate taxa richness across land use intensification and human density across watersheds in Connecticut, USA. Here the authors describe patterns in (a) anthropogenic land use across a gradient of human density (based on number of households) which itself is correlated to (b) biodiversity (measured as mean taxa richness) within a watershed. Figures are shown only to indicate that highly quantitative data are required to develop ecological production functions used to predict levels of biodiversity (and ecosystem goods and services more generally). Taxa richness was calculated as the mean number of taxa identified from two replicate samples in each of three different stream habitat types (pools, runs, riffles). Modified from Urban *et al.* 2006.

explicitly include components of river biodiversity – either as drivers or as an intermediate service (i.e. as the product of ecological attributes and biophysical processes). Compared to other ecological features in rivers, even less is known about how to incorporate biodiversity or species functional traits into generalizable relationships in socio-ecological systems (Dole-dec & Bonada 2013; Leisher, this book).

4.4 MANAGING RIVER BIODIVERSITY: TRADE-OFFS AND EXTERNALITIES

On the ground, implementation of ecosystem service-based approaches to help reverse the decline of biodiversity loss is context-specific, and rates of success vary when the consequences of local decisions extend across space and time. Inevitably, spatial and temporal externalities arise and can influence availability of ecosystem services in distant regions and ultimately influence services at the local level via complex feedback loops (Seppelt *et al.* 2011). For example, when the water needs for a village are met by groundwater pumping, this activity can reduce water availability in other villages that rely on the same aquifer, and can potentially impact freshwater biodiversity in groundwater-fed streams and rivers. If over-extraction continues, future water needs may not be met at either place (Brozovic *et al.* 2010) and regional freshwater biodiversity will most certainly suffer. Fully understanding place-based externalities requires a cross-scale analysis to determine how future availability of water may depend on larger watershed or aquifer processes and changes in availability and use in the future. Below, we re-visit the role of biodiversity in an ecosystem services-based approach, drawing on real-world examples to illustrate the trade-offs and challenges that arise when operationalizing ecosystem services-based approaches for the reduction of freshwater biodiversity decline specifically.

4.4.1 Biodiversity has inherent value as a final good or service

If the most valued ecosystem service in a region is a single aquatic species or suite of species, then operationalizing an ecosystem services-based framework is reduced to managing only the biophysical factors most critical to that or those species. Narrowing the focus to one or a few related species does not suggest that management is easy, but it does focus the scope of management objectives and tools. If data are available on the focal species' environmental requirements and if maintenance of the species is a top priority, ecosystem services for that ecosystem could be managed to benefit that species alone (Kondolf *et al.* 2008). Such prioritization may come at the detriment of other species or services. For example, given the dominant threats to river biodiversity (Figure 4.4), a likely single-species management strategy may prohibit dam building and impose limits on fishing, land conversion, and water abstraction.

As stated earlier, managing river biodiversity alone can be detrimental to human needs because decisions often incur trade-offs with other highly valued goods and services (Figure 4.5). One common example is associated with

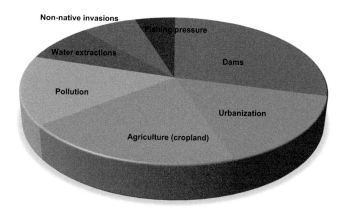

Figure 4.4 Most important threats to river biodiversity based on global-scale data from Vörösmarty *et al.* (2010). At local scales, impacts from various sources will vary as a function of land use, population, status of development, and lifestyles, and will influence stakeholder prioritization of freshwater goods and services. A black and white version of this figure will appear in some formats. For the colour version, please refer to the plate section.

hydropower. Maintaining free-flowing water or some close proximity to the historic flow regime is necessary for the persistence of many in-stream and riparian organisms (Poff & Zimmerman 2010). Accordingly, dams must either not be built (e.g., forego the hydropower benefit) or must be built to ensure native biota can respond to the cues of a natural hydrograph and can move freely upstream at the appropriate times. The latter is very difficult to accomplish, although there are examples of successes through the use of fish ladders as a management tool. More generally, managing for biodiversity typically involves accepting trade-offs in a fundamental conflict; namely, stressors that have been shown to be associated with high losses of native biodiversity must be reduced and many of those stressors are associated with meeting human needs (Vörösmarty *et al.* 2010; Dodds *et al.* 2013).

In contrast, managing rivers for a single taxon such as salmon that requires clean, free-flowing water can result in unanticipated benefits for several other species; thus overall benefits to river biodiversity can be great. For example, salmonids are very sensitive to a variety of stressors and thus managing rivers to support wild runs of anadromous salmon in the Pacific Northwest of the USA involves preserving and restoring many components of riparian biodiversity that help maintain cool water temperatures, buffer the stream from pollutants, and provide woody debris for in-stream habitat and complexity. The salmon life cycle requires gravel beds for spawning adults, backwater habitats for newly hatched salmon fry, a diverse assemblage of macroinvertebrates and fish to serve as a food base, and a specific set of varying flow conditions that help maintain particular geomorphic, chemical,

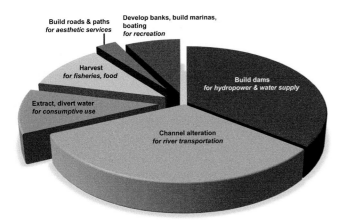

Figure 4.5 Impacts of human actions to river ecosystem services. Accessing river ecosystem services other than biodiversity can have unintended negative impacts on freshwater biodiversity. Here the impacts of individual human actions (or stressors) are reported by individual stressor (shown here as a percentage of all stressors; data from Vörösmarty *et al.* 2010. In order to benefit from or gain access to certain river ecosystem services, humans have altered fundamental biophysical processes and ecosystem attributes. For example, the use of rivers for transportation by large ships or barges has involved extensive alteration of the channel to ensure it is wide and deep enough for passage. To ensure sufficient water for consumptive use, humans have extracted large amounts from rivers, diverted water flows to agricultural fields, and built dams to store water. Even cultural ecosystem services such as those associated with aesthetics, spiritual values, and recreation may require the building of roads near waterways or the construction of marinas; however, those impacts on biodiversity are modest relative to the other categories. All of these actions have negative consequences for biodiversity. At least one ecosystem service provided by rivers – flood protection – requires no action unless the region is developing. In that case, actions that support or enhance freshwater biodiversity (i.e. preservation of floodplains and riparian corridors) may be necessary. A black and white version of this figure will appear in some formats. For the colour version, please refer to the plate section.

and temperature conditions (Poff & Zimmerman 2010). Wild salmon also preserve genetic diversity that is greatly reduced in hatchery-reared populations (Phelps *et al.* 1994,; Eldridge & Naish 2007; Neville *et al.* 2007). The existence of healthy salmon runs indirectly supports many additional components of freshwater biodiversity, including in-stream microbes, riparian predators, and plants that rely on the marine-derived energy and nutrients contained in spawning adults (Willson & Halupka 1995; Willson *et al.* 1998; Gende *et al.* 2002; Naiman *et al.* 2002; Holtgrieve & Schindler 2011; Levi *et al.* 2013). A diverse assemblage of riparian animals – including bears, raptors, river otters, and mink – depend on annual salmon runs for food (Willson & Halupka 1995; Ben-David *et al.* 1998a, 1998b; Hilderbrand *et al.* 1999). Furthermore, predators transport salmon and the nutrients to the terrestrial

landscape, thereby creating hotspots of fertilization that favor certain plant species (Helfield & Naiman 2001; Hocking & Reynolds 2011).

4.4.2 Managing river biodiversity to support ecosystem functioning

Broadly, ecological production functions that incorporate river biodiversity are lacking and represent a growing area of ecological research. This deficiency represents perhaps the weakest part of any argument for ecosystem service-based management – that not enough is known to support evidence-based management decisions – as discussed also by Leisher (Chapter 8). Numerous studies have explored relationships between biodiversity and ecosystem functions in freshwater (Covich *et al.* 2004; Cardinale *et al.* 2012), revealing varying spatio-temporal patterns across gradients to environmental change or stressors, but rarely in response to management actions and the feedbacks that may arise. The existing studies have not established robust patterns or dependencies. Relying on observations and experiments examining the link between biodiversity and ecosystem processes in terrestrial plant communities, investigators have postulated that increased freshwater biodiversity (species richness) will enhance productivity (biomass). There is limited support from a handful of studies on algal biodiversity and biomass (Cardinale *et al.* 2012) and some evidence from laboratory flume studies that greater nitrogen uptake occurs at higher levels of algal diversity (Cardinale 2011). Based on limited studies, it is tempting to assume that one can manage for higher food yields from rivers if biodiversity is maintained at high levels and that algal diversity can enhance water purification in eutrophic rivers. To date, however, there is no evidence supporting either of those conclusions. For example, as Baulch *et al.* (2011) correctly point out, results from a manipulative flume experiment are difficult to extrapolate to larger field settings where the link between a service such as water purification and biodiversity is much more complex.

Ecosystem services-based approaches for river management are largely driven by physical data and processes (e.g. infiltration of rain and the flux of that water from soils to groundwater to rivers is primarily a function of climate, soils, topography, and underlying hydrogeology). While living organisms and different functional groups certainly have a large impact on infiltration – because vegetation, microbes, and soil organisms influence evapotranspiration, soil porosity, and other factors (Belnap *et al.* 2005) – it has not yet been established that such effects depend broadly on specific species. Further, there is no evidence that infiltration rates sufficient to maintain river water levels at desired stages depend on the number of plant or soil-dwelling species nor that river flows sufficient to generate hydropower or transport timber or vessels downriver depend on biodiversity. Simply put, scientific understanding of the relationships between freshwater species diversity and specific services is incomplete. We know more about the role of functional groups (e.g. denitrifying bacteria) in supporting ecosystem services such as water purification and nutrient cycling, but whether or not more functional groups or some specific combination of them enhance purification services is unknown (Sims *et al.* 2013). Therefore, understanding of the biophysical underpinnings related to supporting ecosystem functioning and ecosystem service provisioning is a salient area of research.

4.4.3 Co-managing for biodiversity as a final good or service and to support ecosystem functioning

Ideally, management solutions that maintain the full suite of native species and ecosystem goods and services that river ecosystems provide would be the priority. This is an extremely appealing perspective but may be especially difficult to achieve because river ecosystems are hot spots of both human reliance and biodiversity (Strayer & Dudgeon 2010). As discussed above, current societal needs often conflict with biodiversity, and, freshwater services often involve actions known to eliminate or threaten species. To understand what is involved in maximizing the provision of freshwater ecosystem services while minimizing biodiversity loss, it is useful to compare the management actions necessary to ensure that rivers are able to provide desired ecosystem services with the management actions needed to maintain biodiversity (Table 4.2).

Generally, managing for ecosystem services related to provision of water (water quantity) poses severe threats to river biodiversity due to the structural features that alter the flow regime and limit passage of organisms. In contrast, maintaining and managing for clean, drinkable water, or preserving floodplains and riparian areas for flood protection present fewer direct threats to biodiversity. Ecosystem service-based management also necessitates actions outside the river channel and in the watershed where recharge areas, riparian forests, wetlands, and other physical features promote pollutant attenuation and water infiltration. As a result, the social context and management options (Steps 2 and 4 on our proposed framework, Figure 4.2) will change. Cultural and aesthetic services that benefit from river biodiversity exert minimal to modest threats to biodiversity, while recreational activities can be, but are not necessarily, detrimental.

4.5 CONCLUSIONS

Rapid declines in freshwater biodiversity are thought to be an inevitable consequence of development (Vörösmarty *et al.* 2010), and potential ecosystem services that include freshwater biodiversity are already threatened globally (Dodds *et al.* 2013).

Box 4.1 Key messages

- Managing freshwater often presents conflicts between the reliance on freshwater for direct ecosystem goods and services such as power, irrigation, and drinking supply over services linked with biodiversity.
- Biodiversity can motivate ecosystem services-based management in one of three ways: (1) as a final good or service; (2) in supporting biodiversity–ecosystem functioning as an intermediate service; or (3) some combination of the two.
- An ecosystem services-based approach allows stakeholders and natural and social scientists to discern the motivation behind river biodiversity management. It is a critical starting point for analysing the complexity of factors associated with decision-making.
- Ecological production functions, the quantitative relationships describing biophysical processes and ecological attributes that contribute to or are a product of biodiversity, are essential for the framework to achieve successful outcomes.
- Ecosystem service-based approaches for managing biodiversity are not a panacea, but may enhance conservation efforts under certain circumstances.

As the ecosystem services concept moves from a communication tool to a suite of approaches that involve a certain degree of operationalization, trade-offs must be weighed at each step of the process. Ecosystem services-based approaches can serve as a complement – not a replacement – for conservation-based ones. Moreover, research on biophysical processes that underpin freshwater biodiversity, ecosystem function, and ecosystem services are urgently needed. Until then, for highly valued ecosystem services that are only loosely linked to a few components of biodiversity, we expect, similarly to Leisher (this book), ecosystem services approaches may not adequately address freshwater biodiversity loss. When highly valued services are tied to multiple levels of biodiversity, an ecosystem services-based approach may succeed where traditional site- and species-based conservation measures might otherwise fail. Finally, supporting better management that takes into account both human needs and biodiversity will require much more than ecological research. It will require a great deal of collaboration across disciplines, with natural and social scientists, resource managers, and policy makers at the table to identify governance structures and modes of public engagement that can facilitate the challenging decisions that communities must face when balancing nature with human needs and values.

Table 4.2 *Comparisons of the management actions needed to maintain biophysical processes and ecosystem features sufficient to support desired final ecosystem services, the actions humans often use to access those services, and the management actions needed to minimize biodiversity losses given the need for the ecosystem service.*

Ecosystem service	Management actions to maintain the ecosystem service	Human actions associated with use of the service	Impacts on River biodiversity (+/−)	Management actions to maintain biodiversity
Ecosystem services related to provision of water (water quantity)				
Hydropower production	Maintain vegetated land in watershed sufficient for recharge	Dam construction	Severe (−)	Avoid building large dams or manage dam for environmental flows
Consumptive	Maintain vegetated land in watershed sufficient for recharge	Dams, water extractions and diversions	Modest to severe (−)	Keep total water extractions below recharge rate to ensure maintenance of healthy water table
Water flows for transportation	Maintain vegetated land in watershed sufficient for recharge.	Channelization, regulate water levels	Severe (−)	Limit size of ships and timing of passage when water is adequate
Drinking water (water quality)	Determine primary recharge areas and flow paths to rivers and maintain health vegetation and soils in them. Preserve riparian forests, wetlands; limit any upstream pollution	Water extractions, diversions, redistribution	Modest to intermediate (−)	
Fisheries and food products – subsistence or markets	Depending on species or food, water quality and quantity will need to be maintained at the appropriate level using those actions listed above	Harvesting products	Modest to intermediate (−)	Limit total harvest and minimize pollution from boats; prohibit destructive fishing practices such as dredging
Flood control – protect infrastructure on land	Preserve floodplains and riparian wetlands; ensure significant setback from river is protected from development	None	Protective (+)	None
Aesthetics, spiritual benefits – enjoy natural ecosystems	For maximum benefits preserve entire river ecosystem in its natural state including headwater regions	Provide roads and paths for access	Minimal	Minimize road density and road width; maintain buffer between roads and river
Recreation – boating and swimming	While less stringent actions needed compared to water quality for drinking, many are the same	Development of river banks, marinas, boat use	Modest to severe (−)	Limit use of marinas – require boats on trailers when possible; prohibit fuel or waste disposal from boats; limit recreation catch

ACKNOWLEDGEMENTS

Funding for this work was provided by the US Environmental Protection Agency (Grant # GS-10F-0502N).

References

Aylward, B., Bandyopadhyay, J., & Belausteguigotia, J-C. (2005). Freshwater ecosystem services. In R. Hassan, R. Scholes, and N. Ash (eds), *Ecosystems and Human Well-being: Current State and Trends*. Vol. 1. Island Press, Washington, DC.

Baulch, H. M., Stanley, E. H., & Bernhardt, E. S. (2011). Can algal uptake stop NO_3^- pollution? *Nature* **477**, E3.

Belnap, J., Welter, J. A., Grimm, N. B., Barger, N., & Ludwig, J. A. (2005). Linkages between microbial and hydrologic processes in arid and semiarid watersheds. *Ecology* **86**(2), 298–307.

Ben-David, M., Bowyer, R. T., Duffy, L. K., Roby, D. D., & Schell, D. M. (1998a). Social behavior and ecosystem processes: river otter latrines and nutrient dynamics of terrestrial vegetation. *Ecology* **79**, 2567–2571.

Ben-David, M., Hanley, T. A., & Schell, D. M. (1998b). Fertilization of terrestrial vegetation by spawning Pacific salmon: the role of flooding and predator activity. *Oikos* **83**, 47–55.

Boyd, J., & Banzhaf, S. (2007). What are ecosystem services? The need for standardized environmental accounting units. *Ecological Economics of Coastal Disasters* **63**, 616–626.

Brozovic, N., Sunding, D. L., & Zilberman, D. (2010). On the spatial nature of the groundwater pumping externality. *Resource and Energy Economics* **32**(2), 154–164.

Cardinale, B. J. (2011). Biodiversity improves water quality through niche partitioning. *Nature* **472**, 86–113.

Cardinale, B. J., Duffy, J. E., Gonzalez, A., *et al.* (2012). Biodiversity loss and its impact on humanity. *Nature* **486**, 59–67.

Covich, A. P. M., Austen, C., Bärlocher, F., *et al.* (2004). The role of biodiversity in the functioning of freshwater and marine benthic ecosystems. *BioScience* **54**, 767–775.

Dodds, W. K., Perkin, J. S., & Gerken, J. E. (2013). Human impact on freshwater ecosystem services: a global perspective. *Environmental Science & Technology* **47**, 9061–9068.

Doledec, S. & Bonada, N. (2013). Implications of loss of biodiversity for ecosystem functioning. In: S. Sabater and A. Elosegi (eds), *River Challenges and Opportunities*. Fundacion BBVA.

Dudgeon, D. (2013). Anthropocene extinctions: global threats to river biodiversity and the tragedy of freshwater commons. Chapter 6 in: *River Challenges and Opportunities*. Edt. S. Sabater and A. Elosegi. Fundacion BBVA, Bilbao.

Eldridge, W. H. & Naish, K. A. (2007). Long-term effects of translocation and release numbers on fine-scale population structure among coho salmon (*Oncorhynchus kisutch*). *Molecular Ecology* **16**, 2407–2421.

Gende, S. M., Edwards, R. T., Willson, M. F., & Wipfli, M. S. (2002). Pacific salmon in aquatic and terrestrial ecosystems. *BioScience* **52**, 917–928.

Helfield, J. M. & Naiman, R. J. (2001). Effects of salmon-derived nitrogen on riparian forest growth and implications for stream productivity. *Ecology* **82**, 2403–2409.

Hilderbrand, G. V., Hanley, T. A., Robbins, C. T., & Schwartz, C. C. (1999). Role of brown bears (*Ursus arctos*) in the flow of marine nitrogen into a terrestrial ecosystem. *Oecologia* **121**, 546–550.

Hocking, M. D. & Reynolds, J. D. (2011). Impacts of salmon on riparian plant diversity. *Science* **331**:1609–1612.

Holtgrieve, G. W. & Schindler, D. E. (2011). Marine-derived nutrients, bioturbation, and ecosystem metabolism: reconsidering the role of salmon in streams. *Ecology* **92**, 373–385.

Kareiva, P., Tallis, H., Ricketts, T. H., Daily, G., & Polasky, S. (eds) (2011). *Natural Capital: Theory and Practice of Mapping Ecosystem Services*. Oxford University Press, Oxford.

Keeler, B. L., Polasky, S., Brauman, K. A., *et al.* (2012). Linking water quality and well-being for improved assessment and valuation of ecosystem services. *Proceedings of the National Academy of Sciences* **109**(45), 18619–18624.

Kondolf, M., Angermeier, P. L., Cummins, K., *et al.* (2008). Projecting cumulative benefits of multiple river restoration projects: an example from the Sacramento–San Joaquin River system in California. *Environmental Management* **42**, 933–945.

Levi, P. S., Tank, J. L., Tiegs, S. D., Chaloner, D. T., & Lamberti, G. A. (2013). Biogeochemical transformation of a nutrient subsidy: salmon, streams, and nitrification. *Biogeochemistry* **113**, 643–655.

Nahlik, A. M., Kentula, M. E., Fennessy, M. S., & Landers, D. H. (2012). Where is the consensus? A proposed foundation for moving ecosystem service concepts into practice. *Ecological Economics* **77**, 27–35.

Naiman R. J., Bilby, R. E., Schindler, D. E., & Helfield, J. M. (2002). Pacific salmon, nutrients, and the dynamics of freshwater and riparian ecosystems. *Ecosystems* **5**, 399–417.

Nelson, E., Mendoza, G., & Regetz, J., *et al.* (2009). Modeling multiple ecosystem services, biodiversity conservation, commodity production, and tradeoffs at landscape scales. *Frontiers in Ecology & the Environment* **7**, 4–11.

Neville, H., Isaak, D., Thurow, R., Dunham, J., & Rieman, B. (2007). Microsatellite variation reveals weak genetic structure and retention of genetic variability in threatened Chinook salmon (*Oncorynchus tshawytscha*) within a Snake River watershed. *Conservation Genetics* **8**, 133–147.

Palmer, M. A. & Filoso, S. (2009). Restoration of ecosystem services. *Science* **325**, 575–576.

Phelphs, S. R., Leclair, L. L., Young, S., & Blankenship, H. L. (1994). Genetic diversity patterns of chum salmon in the Pacific-Northwest. *Canadian Journal of Fisheries and Aquatic Sciences* **51**, 65–83.

Poff, N. L. & Zimmerman, J. K. H. (2010). Ecological responses to altered flow regimes: a literature review to inform the science and management of environmental flows. *Freshwater Biology* **55**, 194–205.

Ringold, P. L., Boyd, J., Landers, D., & Weber, M. (2013). What data should we collect? A framework for identifying indicators of ecosystem contributions to human well-being. *Frontiers in Ecology and the Environment*, **11**, 98–105.

Reyers, B., Biggs, R., Cumming, G. S., Elmqvist, T., Heinowicz, A. P., & Polascky, S. (2013). Getting the measure of ecosystem services: a social–ecological approach. *Frontiers in Ecology and the Environment* **11**, 268–273.

Seppelt, R., Dormann, C. F., Eppink, F. V., Lautenbach, S., & Schmidt, S. (2011). A quantitative review of ecosystem service studies: approaches, shortcomings and the road ahead. *Journal of Applied Ecology* **48**, 630–636.

Sims, A., Zhang, Y., Gajarai, S., Brown, P. B., Hu, Z. (2013). Toward the development of microbial indicators for wetland assessment. *Water Research* **47**, 1711–1725.

Strayer, D. L. & Dudgeon, D. (2010). Freshwater biodiversity conservation: recent progress and future challenges. *Journal of the North American Benthological Society* **29**, 344–358.

Urban, M. C., Skelly, D. K., Burchsted, D., Price, W., & Lowry, S. (2006). Stream communities across a rural–urban landscape gradient. *Diversity and Distributions* **12**, 337–350.

Vigerstol, K. L. & Aukema, J. E. (2011). A comparison of tools for modeling freshwater ecosystem services. *Journal of Environmental Management* **92**, 2403–2409.

Vörösmarty, C. J., McIntyre, P. B., Gessner, M. O., *et al.* (2010). Global threats to human water security and river biodiversity. *Nature* **467**, 555–561.

Willson, M. F. & Halupka, K. C. (1995). Anadromous fish as keystone species in vertebrate communities. *Conservation Biology* **9**, 489–497.

Willson, M. F., Gende, S. M., & Marston, B. H. (1998). Fishes and the forest. *BioScience* **48**, 455–462.

5 Water for agriculture and energy

The African quest under the lenses of an ecosystem services-based approach

Maher Salman and Alba Martinez

5.1 INTRODUCTION

According to the Millennium Ecosystem Assessment (2005), most changes in ecosystems have been made to meet a dramatic growth in the demand for food, water, timber, fiber, and energy (e.g. fuelwood and hydropower). As the global population grows, demand for these provisioning ecosystem services will continue to increase, especially the demand for food and energy. In terms of water, the global challenge is to develop these resources to their full potential to sustain provisioning services such as food and energy, while minimizing the impact on other ecosystem services.

This chapter looks at this longstanding and ever-increasing emergency under new lenses, that of ecosystem services-based approaches (as defined in this book), and reflects on the opportunities and limitations that are arising from this issue. In the first section of the chapter, we reflect on how the provisioning of services for food and energy have been developed to the detriment of other ecosystem services, such as erosion control, water regulation, and purification, and we explore how the impacts derived from this could be minimized. The second section grounds the previous discussion by placing the focus on the use of water for the production of food and energy in the African continent.

5.2 FINDING THE RIGHT BALANCE ACROSS ECOSYSTEMS SERVICES

Water plays a critical role in sustaining provisioning services, notably in the form of food and energy (see Box 5.1). The Millennium Ecosystem Assessment (2005) identified that growth in agriculture has been responsible for much of the ecosystem degradation witnessed. Increased agricultural production has been achieved with both the intensification (Figure 5.1) and expansion (Figure 5.2) of agriculture, both contributing to the loss of other provisioning, regulating, cultural, and supporting services. In this section we reflect on the need to find the right balance across ecosystem services, looking at the trade-offs between food, energy production, water, and other ecosystem services, and on some of the proposals that have been made to address these trade-offs.

5.2.1 Agriculture for food and biofuel production versus other ecosystem services

The intensification of agriculture is characterized by the introduction of improved high-yielding varieties, together with the use of high levels of inputs such as irrigation, fertilizers, and pesticides. The most visible impact of intensification with regards to the use of high levels of water for irrigation is salinization. This is especially important in arid and semi-arid areas where surface irrigation is practiced. This is generally a very inefficient practice by which large amounts of water are used. The filtering capacity of soils is overwhelmed by the irrigation rate, therefore, water stagnates on the ground until it evaporates, accumulating salt behind. In addition, irrigation can raise the water table considerably, providing capillary access to dissolved salts accumulated in the groundwater from mineral weathering and rainfall. It has been observed that irrigation in general, and salinity in particular, can damage soil structure (Murray & Grat 2007). The degradation of soil structure can lead to a reduction of soil fertility. When used inappropriately, fertilizers and pesticides can have a negative impact not only on soil properties and its supporting services, but also on the provision of good freshwater quality in an ecosystem. Maintaining good soil properties is essential for the supporting services of an ecosystem, like primary production, nutrient cycling, and soil formation. Soil properties also have a role in regulating services such as erosion control, water regulation and purification, and waste treatment.

The expansion of agriculture, often at the expense of forested areas (Figure 5.2), has led to land use changes that have also had a negative impact on ecosystem services, mainly on those related to the regulation of ecosystem processes. For example, forest ecosystems can play an important role in climate regulation. They can affect both temperature and precipitation and they

Box 5.1 The role of water in sustaining provisioning services: food and energy

There are considerable variations between countries, but in general terms it is safe to state that agriculture is a major consumer of water all over the world and that water is the major limiting factor for crop production (Comprehensive Assessment of Water Management in Agriculture 2007). To produce enough food to satisfy a person's daily dietary needs takes about 3000 liters of water. The general rule of thumb is that to produce one calorie, about one liter of water is required (Comprehensive Assessment of Water Management in Agriculture 2007). Irrigated agriculture accounts for around 20% of the world's arable land, but for more than 40% of total agricultural production. This is because yields from irrigated agriculture can be three times higher than those from rainfed agriculture (FAO 2011b).

The last 50 years have seen remarkable progress in the development of water resources for agriculture. The development of water storage infrastructure and irrigation schemes, coupled with the use of improved varieties, fertilizers, and other agricultural inputs, boosted agricultural production. As a matter of fact, in 2007 it was calculated that world food production outstripped population growth (Comprehensive Assessment of Water Management in Agriculture 2007), leading to a decline of food prices.

Agriculture development not only contributes to food security but also to energy security. Ethanol and biodiesel can be derived from crops such as corn, sugarcane, wheat, or soybeans, and can be used as fuel. According to the International Energy Agency (2014a), global production of biofuels has increased over six-fold in the last ten years, going from 16 billion liters in 2000 to more than 100 billion liters in 2011.

When burnt, biofuels release fewer greenhouse gases and pollutants than fossil fuels and they are a renewable source of energy. Therefore, they can play a key role in climate change mitigation by reducing CO_2 emissions and in increasing self-sufficiency in energy production. However, the use of agricultural land for biofuel production can have a detrimental impact on food production.

Water can also be a source of energy as long as it flows. Hydropower is globally the largest source of renewable energy and already accounts for about 20% of the world's electricity supply (World Bank 2013). As demand for cleaner and renewable energy grows, demand for hydropower will increase. Global hydropower production has increased almost three-fold in the past 40 years, rising from 1294 TWh in 1971 to 3566 TWh in 2011. However, its contribution in the total energy mix has been constant at around 10% (International Energy Agency 2013).

contribute to the capture of greenhouse gases. When forests are replaced by cropland, the sequestration capacity can be significantly reduced.

The effect on water regulation can also be negative with land use changes. The timing and magnitude of runoff, flooding, and aquifer recharge can be strongly influenced, with the replacement of forests by croplands. It could also be the case for other regulating services such as erosion control, water purification and waste treatment, biological and disease control, and pollination. Although some agriculture ecosystems have a cultural value, the expansion of crop land can also represent a threat for some recreational or spiritual areas.

Since there is less room for expansion of arable land, the Food and Agricultural Organization (FAO 2003) estimates that between 2015 and 2030, about 80% of the required food production increases will have to come from intensification in the form of yield increases and higher cropping intensities. However, intensification of crop production has to be done in a sustainable manner.

Sustainable crop production intensification is proposed as the way forward for agriculture to meet food and energy security challenges (FAO 2011a). Sustainable crop production intensification is based on conservation agriculture, uses good seeds of high-yielding adapted varieties, integrated pest management, plant nutrition based on healthy soils, efficient water management, and the integration of crops, pastures, trees, and livestock. All of these measures have to be supported by policies and governance arrangements that support their implementation. Major investment is needed to build research and technology transfer capacity on the above-mentioned measures in developing countries.

5.2.2 Food production versus biofuel production

Important food crops like corn and soybean are generally used for the production of biofuels. This has raised concerns that biofuel production may jeopardize the production of food and thus endanger food security all over the world. Other crops such as sugarcane and sugar beet, cassava, and rapeseed are also used for the production of biofuels (i.e. ethanol, biodiesel, and biomethane).

The High Level Panel of Experts on Food Security and Nutrition (2013) estimates that today's biofuel production accounts for around 2–3% of arable lands globally. Although their share in arable land is still not very significant, biofuels are still a competitor in terms of land and water resources and

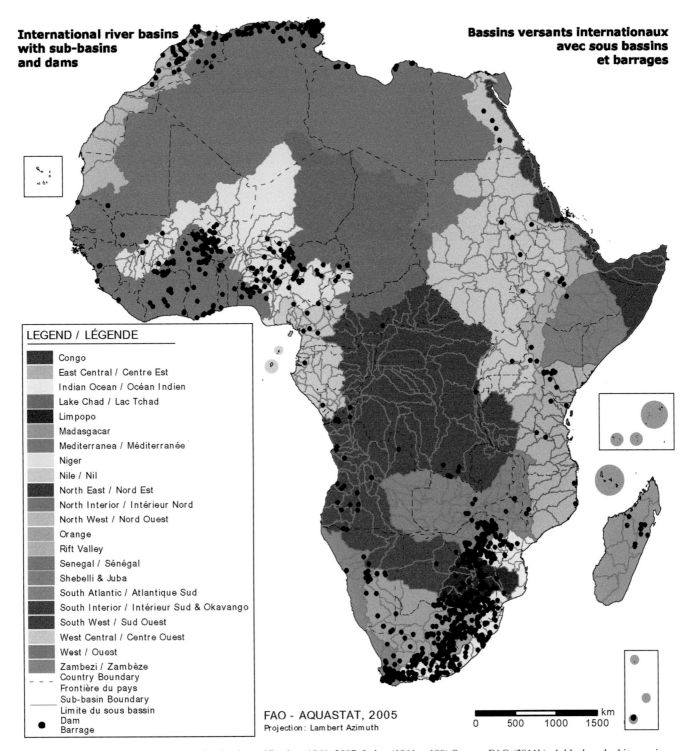

International river basins with sub-basins and dams

Bassins versants internationaux avec sous bassins et barrages

LEGEND / LÉGENDE

- Congo
- East Central / Centre Est
- Indian Ocean / Océan Indien
- Lake Chad / Lac Tchad
- Limpopo
- Madasgacar
- Mediterranea / Méditerranée
- Niger
- Nile / Nil
- North East / Nord Est
- North Interior / Intérieur Nord
- North West / Nord Ouest
- Orange
- Rift Valley
- Senegal / Sénégal
- Shebelli & Juba
- South Atlantic / Atlantique Sud
- South Interior / Intérieur Sud & Okavango
- South West / Sud Ouest
- West Central / Centre Ouest
- West / Ouest
- Zambezi / Zambèze
- – – – Country Boundary / Frontière du pays
- —— Sub-basin Boundary / Limite du sous bassin
- • Dam / Barrage

FAO - AQUASTAT, 2005
Projection: Lambert Azimuth

km
0   500   1000   1500

Figure 5.1 Indicators of global crop production intensification, 1961–2007. Index (1961 = 100) Source: FAO (2011b). A black and white version of this figure will appear in some formats. For the color version, please refer to the plate section.

other complementary inputs. Wherever there is enough land to grow food in surplus and biofuels crops in significant quantities, there should not be a conflict between both productive uses of an ecosystem. However, this is not the case in many countries of the world.

What experts envision as a promising option for the development of biofuels, without compromising food security, is the use of waste or byproducts of food crops and other crops to produce biofuels. Technologies for this purpose are available but still rather costly (Bill and Melinda Gates Foundation 2013).

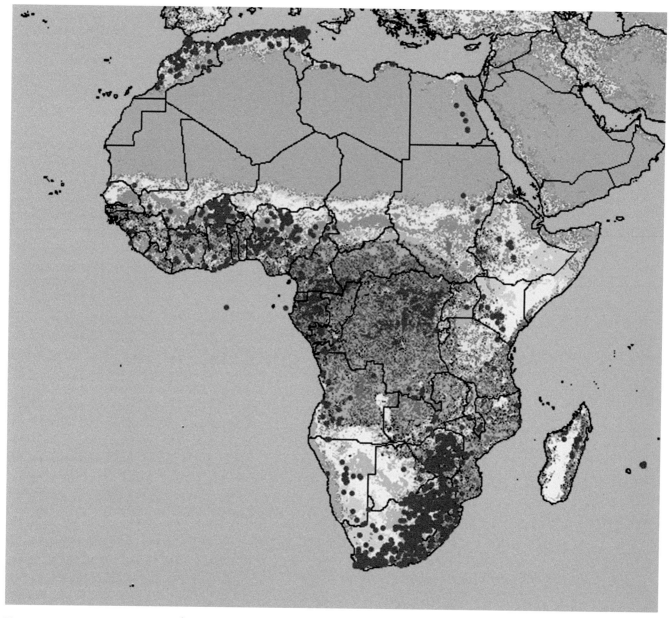

Figure 5.2 Increase of agricultural area,[1] 1961–2011 Source: FAO (2011c). A black and white version of this figure will appear in some formats. For the color version, please refer to the plate section.

The use of *Jatropha* could also help reduce the negative impacts of biofuel production on food security. This non-edible crop has numerous advantages for the production of biodiesel. It has low water and nutrient requirements and can grow on poor-quality land. It also has high oil yields per plant: it is estimated that one hectare of *Jatropha* could yield about 1892 liters of oil (Bill and Melinda Gates Foundation 2013). However, it is important to consider that *Jatropha* is a wild plant that has not been fully domesticated; therefore, there are still aspects of its growth and development that are unknown, such as its optimal growing environment or its overall growth potential.

5.2.3 Hydropower production versus other ecosystem services

The construction of dams to produce energy from the flow of water may lead to an increase in provisioning ecosystem services, e.g. hydropower and freshwater supply for agricultural, municipal, and industrial purposes. The regulating ecosystem

---

[1] Agricultural area is the sum of areas under (1) arable land, (2) permanent crops, and (3) permanent meadows and pastures.

services, such as flood control and drought protection, may also be enhanced through the construction of dams. However, all these improvements would be achieved at the detriment of other important ecosystem services.

Dams disturb natural river flow that can lower water quality and affect the morphology of the river bed. For instance, changes in temperature, the chemical composition of the water, levels of dissolved oxygen, etc., might occur, which in turn can have an impact on the flora and fauna of the river and other ecosystem services (International Rivers 2014b). In addition, dams alter the sediment transport downstream, causing significant damage. Usually, sediments coming from upstream catchment areas deposit in the reservoir bed, which stops the natural replenishment of sediments downstream. This leads to erosion on the downstream river bed to compensate for the sediment loss. This can have important environmental and economic implications when agricultural areas are located downstream. Sediments can be a natural source of fertilization and when these are held upstream, it forces farmers to use chemical fertilizers to compensate for the loss, if yields are to be maintained.

When wetlands are located downstream of a dam, the environmental impacts can be fatal. Dams have often caused the desiccation of wetlands, with its consequent loss of biodiversity and other ecosystem services. A dam can also block fish migrations and even separate spawning habitats from rearing habitats for certain species. This has led, in some cases, to the extinction of fish and other aquatic species. It may also lead to the outbreak of diseases and the invasion of algae, which is against the regulating services of an ecosystem (International Rivers 2014b).

On the other hand, small hydropower (generally defined as with a capacity of not more than 10–15 MW) is gaining momentum due to its lower environmental impacts, its short gestation period and low investment requirements. However, small hydropower can also have considerable impact on river ecosystems (World Wildlife Fund 2003). It is, therefore, recommended to not make general assumptions on the size of hydropower systems but to judge in relation to the site-specific situation.

5.2.4   Water for agriculture versus water for hydropower production

Rainfed agriculture produces about 60% of the food available in the world but accounts for around 80% of the cultivated area (FAO 2011d). The reason behind this is that productivity in rainfed systems is highly variable as it relies on rainfall fluctuations as a result of drought or floods. Therefore, water storage, even in relatively small volumes, is seen as the main means to secure water supply for agriculture as well as for hydropower production (and other uses).

Agriculture and energy are two highly interconnected sectors but at present these linkages are not fully taken into account in policy making. Agriculture is a key sector where the spillovers of energy production and consumption are pronounced. On the one hand, any further growth in the rural space will necessarily increase demand for energy and will be competing with the industrial and urban centers. On the other hand, accelerating access to electricity for the rural poor offers new opportunities for agriculture, including access to a source of energy for water pumping. If access can be improved, and energy needs for agriculture anticipated and met, then a potential roadblock to agricultural growth can be avoided (FAO 2008).

Whenever the irrigation potential is located upstream of the hydropower potential, the consumptive use of water can compromise existing generation capacity (FAO 2008). This would also call for a thorough socio-economic analysis under an ecosystem services-based approach (core elements 3 and 4).

To better tap the potential of a dam to produce hydropower and ensure freshwater supply for agriculture, operating rules should be consistent with the seasonal demands of irrigation. This is mainly the case whenever irrigation potential is situated downstream of the hydropower potential. Whenever countries face this situation they should also balance water use in order to optimize economic gains, but should also consider food security concerns.

5.3   THE AFRICAN QUEST

Around 850 million people in the world are undernourished (FAO et al. 2012) and 1317 million lack access to electricity (International Energy Agency 2014b). Of those, 239 million and 585 million, respectively, can be found in sub-Saharan Africa (FAO et al. 2012; International Energy Agency 2014b).

In Africa, increased water demands for drinking, industrial, and irrigation purposes coupled with the likelihood of reduced water availability as a consequence of climate change pose a great challenge for water resources management. Good water management in the case of agriculture means producing more crops per unit of water. As the population grows, it is not only the demand for food that increases, but also the demand for energy. The development of renewable energy is seen as a way to ensure the sustained supply of energy and to mitigate climate change. The challenge is even greater considering that the increase in food and energy production has to be carried out without impacting on other ecosystem services.

Projections from the Food and Agriculture Organization show that, in order to satisfy Africa's food demand, irrigated land in the continent will have to increase by 18.3% by 2030 (FAO 2008). Similarly, energy demand in Africa as a whole is expected to double by 2030 (FAO 2008).

Table 5.1 *Major land and water systems at risk in Africa*

Location	Risks
Rainfed cropping, dry tropics	
Smallholder farming in sub-Saharan African savannahs Agro-pastoral systems in the Sahel and the Horn of Africa	Overexploitation of natural resources, depletion of soil organic matter and fertility, soil acidification, poor soil moisture-holding capacity, wind and water erosion, biomass and biodiversity decline
Rainfed cropping, highlands	
Rift Valley, Ethiopian Plateau, Southern Africa	Erosion, land degradation, reduced productivity of soil and water
Rainfed cropping, subtropical	
Mediterranean basin	Overexploitation of land and water, erosion, loss of soil fertility, biodiversity decline
Intensive rainfed cropping, temperate	
Southern Africa	Soil health degradation (compaction, organic matter decline, sealing), pollution of soils and water, loss of biodiversity
Rangelands: subtropics	
Pastoral and grazing lands in Western Africa (Sahel) and North Africa	Desertification, land abandonment
Irrigated, other crops	
North Africa	Overexploitation of land and water, soil and water salinization, loss of biodiversity, desertification, loss of buffer role of aquifers
Forest	
Tropical forest–cropland interface in Central Africa	Cropland encroachment, introduction of invasive species, pests and diseases, loss of biodiversity, erosion, sedimentation, and land degradation

Source: adapted from FAO (2011d).

5.3.1 Ecosystems at risk in Africa

The total land area in Africa is around three billion hectares. Of this, 9% (0.26 billion ha) is currently in use for cultivation of agricultural crops, 22% (0.67 billion ha) is under forest, and 30% (0.9 billion ha) comprises permanent meadows and pastures (FAO 2011d). The rest is covered by sparsely vegetated and barren land, settlements, and other built-up areas. In Africa, agricultural area, which comprises land devoted to the cultivation of crops and permanent meadows and pastures, has increased by 11% in the past 50 years, whereas in the Asian continent, this increase has been up to 54% (FAO 2011d). Around 6% of the area under cultivation of crops is equipped for irrigation in Africa, with an even lower 3% in sub-Saharan Africa, compared to 41% in Asia (FAO 2011d). This means that rainfed systems are predominant, making Africa the continent with the highest yield gaps.[2] Taking the example of rainfed maize, yields in sub-Saharan Africa have remained at around 1 t ha^{-1} in the past 50 years, while in Latin America and the Caribbean yield tripled from 1 t ha^{-1} to 3 t ha^{-1} (FAO 2011d).

Notwithstanding the low input farming practices in Africa, especially in sub-Saharan Africa, some agricultural practices in

the continent have put ecosystems at risk of sustaining their services. The Food and Agriculture Organization publication *The State of the World's Land and Water Resources for Food and Agriculture* has identified major categories of systems at risk in the continent for which especial attention is needed (Table 5.1).

The Lake Chad case gives a concrete example of how human action coupled with climate change has put its basin's entire ecosystem at risk.

In the Lake Chad Basin (Box 5.2), agriculture is the main economic activity, with major crops being cotton, groundnuts, sorghum, cassava, millet, rice, and onions (Odada *et al.* 2006). Although rainfed farming in the basin is dominant, water use for irrigated agriculture has experienced a four-fold increase since the early 1980s. It is estimated that around 115 000 ha are irrigated, representing less than 10% of the irrigation potential of the basin (Odada *et al.* 2006). Agriculture has also contributed to the degradation of water quality in the lake. The commercial production of rice and cotton, known to use large quantities of agro-chemicals, has led to high water pollution levels (Salman & Casarotto 2010).

The impact of lower water quantity and quality is felt at all levels when considering ecosystem services. A collapse of some fisheries and recessional rice cultivation was observed, as well as biodiversity loss and the decreased viability of biological resources (FAO-WATER 2009). In addition, unsustainable farming practices led to the sedimentation of rivers and water courses that have produced

[2] Yield gap is defined by comparing current productivity with what is potentially achievable, assuming that inputs and management are optimized in relation to local soil and water conditions (FAO 2011d).

Table 5.2 *The impact of agriculture in ecosystem services of the Lake Chad.*

Provisioning services		Regulating services		Cultural services		Supporting services	
Food	↑	Climate regulation	–	Spiritual /religious	–	Soil formation	↓
Freshwater	↓	Disease regulation	↓	Recreation / tourism	↓	Nutrient cycling	↓
Energy (biofuels)	↑	Water regulation	↓	Aesthetic	↓	Primary production	↓
Fiber	↑	Water purification	↓	Inspirational	–		
Biochemicals	↓	Pollination	↓	Educational	–		
Genetic resources	↓			Sense of place	↓		
				Cultural heritage	↓		

↑ Increase of ecosystem service

↓ Loss of ecosystem service

Box 5.2 The Lake Chad Basin

The Lake Chad Basin stretches over an area of 2 397 423 km². It is distributed between Chad, Niger, the Central African Republic, Nigeria, Algeria, Sudan, Cameroon, and Libya. More than 30 million inhabitants live in the basin, where they sustain their livelihoods through activities like fishing, agriculture, and animal husbandry (FAO 2008). Lake Chad is the fourth largest lake in Africa. The lake itself, and the associated wetlands, harbor biodiversity that is of global importance; the area is the habitat of 176 species of fish and over 500 species of birds.

The arid climate of the basin coupled with the low depth levels of the lake result in high water evaporation losses. Despite this, the lake is not salty (Carmouze *et al.* 1983).

Annual and seasonal variability aside, the water surface of the lake has been continuously shrinking over the past 50 years. In 1964 the water surface covered 25 000 km². In 2008 it stretched over an area of less than 1000 km² during the annual lowest water levels of the region (FAO 2008).

changes in channel flow patterns, a reduction in the inflows to the lake and the proliferation of invasive species (Bdliya & Bloxom 2007). Increased irrigated cultivation tends to concentrate pests. Borers, caterpillars, locusts, crickets, quelea birds, and golden sparrows are already endemic in the area, forcing farmers to use pesticides (Bdliya & Bloxom 2007).

The impact of agriculture in ecosystem services of the Lake Chad Basin is illustrated in Table 5.2.

The institutional responses to challenges in the basin have not been adequate. As reported in the Lake Chad Basin Commission

Vision 2025 (2003), the management of the basin's natural resources has been characterized by unsustainable decision-making, lack of adequate water, environmental management policies, and political will on the part of member states, poor coordination mechanisms, low levels of stakeholder participation, weak institutions, poverty, and the fragile economic situation of the region.

Agriculture is the cause of many of the problems that affect Lake Chad, but it can also be part of the solution. The Lake Chad Basin Sustainable Development Program, which was designed in 2007 as a contribution to the implementation of the Lake Chad Strategic Action Plan and Vision 2025, highlights certain agricultural activities that can minimize impacts on other ecosystem services, including: soil conservation and soil moisture conservation, plan of optimal management of reservoirs and water supply points of the basin, and improvement of stakeholder skills.

5.3.2 Impacts and mitigation measures of hydropower production in Africa

According to the Food and Agriculture Organization-Aquastat (2013), there are 1562 dams in Africa, with a total storage capacity of 1043 billion m³. Of those, 401 dams are dedicated to the generation of hydropower, of which 55 are also used for irrigation purposes. Around 60% of the total dam capacity in Africa is located in the Nile and the Zambezi river basins. However, water storage per capita in the continent is the lowest in the world (White 2005) and it is one of the reasons why the continent is lagging behind in terms of agriculture and energy production.

The total installed hydropower capacity in the continent is 21 000 MW, 90% of which is concentrated in eight countries, namely, the Democratic Republic of the Congo, Egypt, Gabon, Ethiopia, Nigeria, Zambia, Madagascar, and Mozambique (FAO

Box 5.3 Impacts of large dams: Akosombo dam in the Volta basin

The Akosombo dam was put into operation in 1966 with the main purpose of generating hydropower. It produces 912 MW of electricity at its maximum operating capacity (Gyau-Boakye 2001). Other uses of the dam include fishing, farming, transportation, and tourism. Despite economic benefits, some negative impacts came with the construction and operation of the dam. Gyau-Boakye (2001) compiles some of these negative impacts and classifies them into three categories:

(1) **Physical**

Seismicity: The Akosombo dam resulted in the creation of the Volta lake. The weight of water creates significant weight in the underlying rocks. For instance, between 1964 and 1971 and when impoundment was progressing, several earthquakes occurred. However, the epicenters of these earthquakes were further from the dam and although it is suspected that the events were related to the construction of the dam, a direct attribution cannot be established.

Sediment and load changes: There has been a significant reduction in sediment concentration downstream of the dam. Taking the Volta River at Ajena as a reference, concentration decreased by almost 60% from 1956 to 2000 (Gyau-Boakye 2001).

Morphological changes: The construction of the dam has provoked an accelerated shift of the Volta River where it enters the sea. In addition, the dam has led to a reduced flood-flushing capacity of the river. Moreover, these changes have probably been the cause of the accelerated coastal erosion at Ada.

(2) **Biological**

Weeds: A proliferation of different plant species in the reservoir has been observed. These weeds are impeding transportation in Volta Lake; this is especially important in fishing villages, where it is difficult for boats to leave/approach the villages, even with outboard motors.

Environment-related health: After the construction of Akosombo dam, an increase in the incidence of urinary schistosomiasis, malaria, and onchocerciasis was noticed.

(3) **Social**

Resettlement: Among the negative social impacts was the relocation of populations that were settled in the reservoir area. About 80 000 people had to be relocated in 52 resettlement villages (Gyau-Boakye 2001). Resettlement is a traumatic experience as it goes hand in hand with the loss of land, the abandonment of shrines, graves, churches, and other buildings of cultural importance.

Lost lands/decline in primary economic activities: Volta Lake represents a loss of about 8500 km^2 (3% of the total surface of Ghana) that were once forests or cultivated land.

Breakdown of some cultural practices: Many sacred places were submerged with the impoundment of the dam. In addition to this, some cultural practices that were part of the identity of resettled communities drowned in Volta Lake.

2008). It has been estimated that Africa is currently producing only 20% of its total hydropower potential (FAO 2008), which is estimated at around 100 000 MW, the bulk of which can be found in the Inga and the Congo basin.

The focus over the years in many African countries has been on large-scale dams and hydropower schemes. Notwithstanding the economic benefits of large-scale dams, the social and environmental impacts associated with the construction and operation of large dams has been significant in many cases. Box 5.3 shows an African example of the impacts of large dams in ecosystem services.

Recent studies have shown that electricity generation through small hydropower[3] can be a viable alternative to large-scale hydropower schemes. Small-scale hydropower is gaining momentum due to its short gestation period, low investment, and lower environmental impacts. Also, economically viable small-scale hydropower technologies have been commercially developed and are available for generating both electrical and mechanical power for rural industrialization and development (FAO 2008). A number of national governments in Africa and international donors have acknowledged the great potential of small hydropower in securing energy access in remote rural areas in the continent. Despite its vast potential, there have been a relatively low number of small hydropower projects implemented in Africa. As said previously, it is recommended that small dams be judged in relation to site-specific situations.

5.3.3 Water for agriculture and energy in Africa

Sustainable social and economic growth in Africa has to be driven by the development of its agricultural sector, which is

---

[3] Small hydropower refers to those stations with 10–15 MW or less of installed capacity.

Box 5.4 Water storage for agriculture and hydropower in the Zambezi River Basin

With a flow of 2750 km, the Zambezi River is the fourth largest river in Africa. Its basin is shared among Angola, Botswana, Malawi, Mozambique, Namibia, Tanzania, Zambia, and Zimbabwe. It is a source of important ecosystem services to the region and is essential for food and energy security. Currently, it is estimated that the basin has almost 5000 MW of installed hydropower generation capacity and around 183 000 hectares equipped for irrigation (World Bank 2010). In addition, its development projects have vastly improved the economy of its region, characterized by extreme climatic variability and experiencing cycles of floods and droughts.

With its 78 dams (FAO-Aquastat 2013), the Zambezi is one of the most dammed rivers in Africa. However, the construction and operation of these dams have had negative impacts on other ecosystem services. Dams have disturbed the natural river flows, affecting wildlife and downstream populations. The Zambezi Delta has been severely damaged by all of these development projects (International Rivers 2014a). Floodplain recession agriculture, fish production, wildlife resources, cattle grazing, eco-tourism, biodiversity, natural products, and medicine are the main products and services of the wetlands that have been affected by dam construction upstream (Seyam *et al.* 2001).

It is very likely that conditions will continue to deteriorate unless key indicators of hydrological change can be improved (Beilfuss *et al.* 2005). Nevertheless, countries still aim to increase the storage capacity in the basin. According to the Plans for the Southern African Power Pool, if the full hydropower potential of the basin is to be developed it would require the implementation of 53 projects, including the rehabilitation of existing hydropower facilities and the construction of new ones, doubling the average energy production from 30 000 to around 60 000 GWh year^{-1}.

The extension of area under irrigation is also included in countries' plans. The World Bank (2010) identified around 100 irrigation projects in the pipeline. The implementation of these projects would add approximately 336 000 ha equipped under irrigation, tripling the area equipped for irrigation.

In light of these challenges, the World Bank conducted a study in 2010 on growth-oriented investments in the Zambezi River Basin, considering related sectors like the environment and the wetlands. The study concluded that restoration of natural flooding could be granted by modifying reservoir operating rules at Cahora Bassa Dam. Depending on the natural flooding scenario selected, hydropower production could be reduced between 3 and 33% for the Cahora Bassa Dam and between 4 and 34% for the planned Mphanda Nkuwa Dam. However, the study also points out that a more detailed analysis should be conducted. It was also concluded that improved coordination of hydropower and irrigation projects, such as developing irrigation infrastructure downstream of certain dams, could increase energy generation without compromising the area to be develop under irrigation.

the mainstay for 70% of its population and 80% of its poor. Under such conditions, significant efforts are needed to make African agriculture more productive and more efficient, but also more resilient to climate change and more environmental friendly. That calls for better control of production factors, especially water, which is the key to food security (FAO 2008). The development of the energy sector is also the key to ensuring socio-economic growth. Renewable energies have to be promoted if Africa is to sustain that growth over time.

The development of the agriculture and energy sector calls for increased water storage, considering that precipitation in the continent is seasonal and erratic. The likely trend is that dams will be multipurpose, with hydropower and agriculture being the major water users. However, this poses different challenges. Whenever the irrigation potential is located upstream of the hydropower potential, which is the predominant case in the African context, the consumptive use of water can compromise existing generation capacity (FAO 2008). On the other hand, wherever irrigation potential lies downstream of the hydropower potential, operating rules should be consistent with irrigation

demands. In both cases, countries may balance water use in order to optimize economic gains, but should also take into account food security concerns.

Box 5.4 illustrates how the use of a dam for hydropower and agriculture production purposes was analyzed in terms of optimizing economic returns and promoting food and energy security, in the Zambezi River basin.

5.4 CONCLUSIONS

This chapter has looked into a number of issues concerning the role of water ecosystem services regarding increasing demands for food and agriculture. Special attention has been given to the African challenge of meeting the need to produce more food and energy while preserving other ecosystem services

It can be drawn from the chapter that the agricultural sector plays a crucial role in Africa, and has a high potential for growth. The role of water is crucial to increasing yields. However, the continent's productivity levels are far from reaching its full potential since yields remain low. Furthermore, agriculture plays a role in the degradation of ecosystems. In Africa, some agricultural practices are threatening ecosystems. All this makes evident the need to establish decision-making processes that allow for sustaining agricultural growth while minimizing adverse environmental impacts on ecosystems. An ecosystem services-based approach as defined in this book represents a promising way of looking at the African quest. This approach would look at all the changes within the ecosystem through cost-effective and integrated management of human activity. This type of systems-thinking accepts the human activity in the ecosystem and tries to find sustainable ways to minimize its adverse effects on the functioning of the ecosystem while maintaining the activity. This type of approach would allow planning, development, and management of water and terrestrial systems in a more holistic way that can help address the various needs of humans while reducing the threat to the availability of resources in the future.

As shown in the case studies, the alteration of nature affects the entire system in different ways and it impacts the local communities. We suggest that in order to endure sustainable agricultural growth in Africa, an ecosystem services-based approach could prove useful because of the intrinsically interconnected nature of Earth's systems and their relation to human wellbeing (core element 1). In addition, the chapter proposes a number of viable solutions to produce both hydropower and ensure freshwater supply. The chapter also suggests that the case studies specifically could benefit from an ecosystem services-based approach. If the experience of the Zambezi River Basin is considered, for example, it can be observed that there were several negative environmental impacts throughout the system because of the obstruction on the natural river flow due to the dams. An ecosystem services-based approach could help restore the basin's ecosystem and build resilience to the communities affected by it because of its multi-dimensional nature which takes into account not only a single service, but the system as a whole.

It should be pointed out that ecosystem services-based approaches have their limitations. Given the rapid nature of human activities' demands, this type of approach is not always realistic given that often there is a lack of sufficient information and knowledge on a certain system. This type of approach requires a longer time to develop, which makes it more useful for long-term solutions. Often short-term solutions are required to which ecosystem services-based approaches might not be able to respond rapidly. In addition, ecosystem services-based approaches require constant monitoring and research on the ecosystem affected (to fulfill core element 2). Long-term monitoring of the ecosystem-based approach is necessary in order to see the response of the ecosystem over time, and this can be challenging in certain cases. Finally, ecosystem services-based approaches require the long-term commitment of stakeholders to be able to observe positive effects (core element 3). If not, they could be hindered over time.

Box 5.5 Key messages

- Ecosystems are enduring high levels of stress to meet the demands of food, water, timber, fiber, and energy (fuelwood and hydropower). Given the growth in population, this demand is only going to increase.
- Water plays a critical role in the sustainable long-term provision of food and energy.
- In Africa, the intensification and expansion of agriculture has led to land use changes that have a negative impact on ecosystem services, with high environmental costs. However, agriculture contributes to food and energy security and therefore solutions must be identified.
- Maintaining a healthy ecosystem is essential to sustaining the increasing demand in agricultural production.
- Ecosystem services-based approaches can help sustain agricultural growth while preserving the natural resources of the ecosystem.
- Existing interdisciplinary programs that analyze this type of approach should be supported and made available.
- Capacitating local communities on this type of approach would also be needed.

Building on these conclusions, concerted efforts should be made to overcome the aforementioned limitations. Funding in Africa should be made available for further research on ecosystem services. Furthermore, existing interdisciplinary programs that analyze this type of approach should be supported and made available. Capacitating local communities on this type of approach would also be needed. Given Africa's enormous agricultural productivity potential, an early intervention using ecosystem services-based approaches could contribute to ensuring long-term, sustainable growth.

References

Bdliya, H. H. & Bloxom, M. (2007). Transboundary diagnostic analysis of the Lake Chad Basin. Prepared for the LCBC–GEF project on the reversal of land and water resources degradation. Available at: http://lakechad.iwlearn.org/publications/reports/lake-cha-basin-tda-report-english (last accessed 21 October 2014).

Beilfuss, R. (2005). Natural and dam-induced patterns of hydrological change in the Zambezi river basin. Available at: http://bscw-app1.let.ethz.ch/pub/nj_bscw.cgi/d11577040/Beilfuss_2005_Natural_and.pdf (last accessed 21 October 2014).

Bill and Melinda Gates Foundation (2013). Biofuels and the poor: a project funded by Bill and Melinda Gates Foundation. Available at: http://biofuelsandthepoor.com/case-study-india (last accessed 21 October 2014).

Carmouze, J.-P., Durand, J.-R., & Lévêque, C. (1983). *Lake Chad: Ecology and Productivity of a Shallow Tropical Ecosystem*. Dr W. Junk Publishers, The Hague.

Comprehensive Assessment of Water Management in Agriculture (2007). *Water for Food, Water for Life: A Comprehensive Assessment of Water Management in Agriculture*. Earthscan and International Water Management Institute, London and Colombo. Available at: www.fao.org/nr/water/docs/Summary_SynthesisBook.pdf (last accessed 21 October 2014).

FAO (2003). *World Agriculture: Towards 2015/2030*. Earthscan and FAO, London and Rome.

FAO (2008). *Water for Agriculture and Energy in Africa – The Challenges of Climate Change*. Report of the ministerial conference 15–17 December 2008, Sirte, Libyan Arab Jamahiriya. Available at: www.fao.org/docrep/014/i2345e/i2345e01.pdf (last accessed 21 October 2014).

FAO (2011a). Save and grow: a policymaker's guide to the sustainable intensification of smallholder crop production. Available at: www.fao.org/docrep/014/i2215e/i2215e00.htm (last accessed 21 October 2014).

FAO (2011b). Water for agriculture and energy in Africa: the Challenges of climate change. Report of the Ministerial conference, 15–17 December 2008. Libya. Available at: www.fao.org/docrep/014/i2345e/i2345e.pdf (last accessed 21 October 2014).

FAO (2011c). FAOSTAT statistical database. Available at: http://faostat.fao.org/site/291/default.aspx (last accessed 21 October 2014).

FAO (2011d). *The State of the World's Land and Water Resources for Food and Agriculture*. FAO, Rome.

FAO, IFAD, & WFP (2012). The state of food insecurity in the world. Available at: www.fao.org/docrep/016/i3027e/i3027e.pdf (last accessed 21 October 2014).

FAO-Aquastat (2013). Geo-referenced dams database. Available at: www.fao.org/nr/water/aquastat/dams/index.stm (last accessed 21 October 2014).

FAOWATER (2009). Adaptive water management in the Lake Chad Basin. Seminar Proceedings, World Water Week, Stockholm, 16–22 August 2009.

Gyau-Boakye, P. (2001). Environmental impacts of the Akosombo dam and effects of climate change on the lake levels. *Environment, Development and Sustainability* **3**, 17–29.

High Level Panel of Experts on Food Security and Nutrition (2013). Biofuels and food security. A report by the High Level Panel of Experts on Food Security and Nutrition of the Committee on World Food Security, Rome 2013. Available at: www.fao.org/fileadmin/user_upload/hlpe/hlpe_documents/HLPE_Reports/HLPE-Report-5_Biofuels_and_food_security.pdf (last accessed 21 October 2014).

International Energy Agency (2013). Key world energy statistics. Available at: www.iea.org/publications/freepublications/publication/KeyWorld2013_FINAL_WEB.pdf (last accessed 21 October 2014).

International Energy Agency (2014a). Topics: biofuel. Available at: www.iea.org/topics/biofuels (last accessed 21 October 2014).

International Energy Agency (2014b). Topics: access to electricity. Available at: www.iea.org/publications/worldenergyoutlook/resources/energydevelopment/accesstoelectricity (last accessed 21 October 2014).

International Rivers (2014a). Zambezi, river of life. Available at: www.internationalrivers.org/campaigns/zambezi-river-of-life (last accessed 21 October 2014).

International Rivers (2014b). Environmental impacts of dams. Available at: www.internationalrivers.org/environmental-impacts-of-dams (last accessed 21 October 2014).

Lake Chad Basin Commission (2003). Integrated River Basin Management – Challenges of the Lake Chad Basin, Vision 2025. www.cblt.org/sites/default/files/vision_2025_en.pdf (last accessed 21 October 2014).

MA (Millennium Ecosystem Assessment) (2005). *Ecosystems and Human Well-being.*: World Resources Institute, Washington, DC.

Murray, R. S. & Grat, C. D. (2007). The impact of irrigation on soil structure. School of Earth & Environmental Sciences, University of Adelaide. Available at: http://lwa.gov.au/files/products/national-program-sustainable-irrigation/pn20619/pn20619.pdf (last accessed 21 October 2014).

Odada, E. O., Oyebande, L., & Oguntola, J. A. (2006). Lake Chad: experience and lessons learned brief. Available at: www.worldlakes.org/uploads/06_Lake_Chad_27February2006.pdf (last accessed 21 October 2014).

Salman, M. & Casarotto, C. (2010). Sustainable management of the Lake Chad Basin. Saving Lake Chad Conference, N'Djamena, Chad, 29–31 October 2010.

Seyam, I. M., Hoekstra, A. Y., Ngabirano, G. S., & Savenije, H. H. G. (2001). The value of freshwater wetlands in the Zambezi basin. International Institute for Infrastructural, Hydraulic, and Environmental Engineering.

White, W. R. (2005). *World Water Storage in Man-Made Reservoirs: A Review of Current Knowledge*. Foundation for Water Research, Marlow.

World Bank (2010). The Zambezi River Basin: a multi-sector investment opportunities analysis. http://siteresources.worldbank.org/INTAFRICA/Resources/Zambezi_MSIOA_-_Vol_1_-_Summary_Report.pdf (last accessed 21 October 2014).

World Bank (2013). Hydropower overview. Available at: www.worldbank.org/en/topic/hydropower/overview (last accessed 21 October 2014).

World Wildlife Fund (2003). Hydropower in a changing world: WWF's initiative – Living Waters, conserving the source of life. Available at: http://awsassets.panda.org/downloads/hydropowerfacts.pdf (last accessed 21 October 2014).

Part II
Applying frameworks for water management and biodiversity conservation under an ecosystem services-based approach

6   Using ecosystem services-based approaches in Integrated Water Resources Management

Perspectives from the developing world

Madiodio Niasse and Jan Cherlet

6.1   INTRODUCTION

Integrated Water Resources Management is a normative water management paradigm, widely applied at river basin and country level, which 'promotes the co-ordinated development and management of water, land and related resources, in order to maximise the resultant economic and social welfare in an equitable manner without compromising the sustainability of vital ecosystems' (Global Water Partnership 2000, p.22). In other words, the paradigm recognises, and tries to reconcile, the many competing uses of freshwater: water for human/domestic use, agriculture and industrial and energy production, and natural ecosystems.[1] Although the water needs of natural ecosystems are amply recognised in Integrated Water Resources Management *theory* (Global Water Partnership 2000), in *practice* they are sacrificed in favour of productive water uses – especially in developing countries.

Every year more and more water is withdrawn from the hydrological cycle for productive purposes.[2] As a result, the resource is increasingly commoditised and disputed. The competition for water is also more global than ever, which has led to frequent instances of land and water grabbing (Woodhouse 2012). In this context of international competition, the sidelining of natural ecosystems in water resources management is likely to be exacerbated. This will amplify the water crisis even further, given the central role of natural ecosystems in the provision, regulation, and recycling of water resources; according to the Millennium Ecosystem Assessment, forest and mountain ecosystems are the sources of 85% of the world's total freshwater runoff (Millennium Ecosystem Assessment 2005, p.167).

We argue that ecosystem services-based approaches can encourage Integrated Water Resources Management practice to pay adequate attention to the water needs of natural ecosystems, i.e. their water requirements for provisioning, regulating, supporting, and cultural services as well as the maintenance of the overall ecosystem health. Because natural ecosystems themselves are the major users of the world's freshwater resources, effective implementation of Integrated Water Resources Management is needed to sustain and enhance the services derived from these ecosystems.

The following section discusses the possible synergies between Integrated Water Resources Management and ecosystem services in theory, similarly to Blackstock *et al.* (this book) in the context of the European Water Framework Directive. The third section gives some examples from the developing world, where the concept of ecosystem services is implicitly or explicitly used to improve water resources management. We also discuss how emerging approaches and instruments such as dam reoperation and implementation of environmental flows can help strengthen the operational linkages between Integrated Water Resources Management and ecosystem services. The chapter ends with lessons learnt and policy recommendations regarding the use of ecosystem services-based approaches in Integrated Water Resources Management.

6.2   POTENTIAL SYNERGIES BETWEEN INTEGRATED WATER RESOURCES MANAGEMENT AND ECOSYSTEM SERVICES-BASED APPROACHES

6.2.1   Two paradigms, one goal

Integrated Water Resources Management and ecosystem services-based approaches share striking common features in terms of their genesis, underpinning assumptions, objectives, and approaches.[3] First, both approaches emerged in a context of increased awareness of the dramatic pressures on the world's

---

[1] The concept of natural ecosystems (as opposed to 'human-dominated ecosystems' such as agricultural, urban, and/or industrial systems) is used here to refer to 'a set of organisms living in an area, their physical environment and the interactions between them' (Daily *et al.* 1997).

[2] According to the Comprehensive Assessment of Water in Agriculture Programme, withdrawals of water for agriculture, industry, and domestic use, which were estimated at about 1% of total renewable resources in the early 1900s, increased to 5% in the 1960s and to 8.8% in 2000 (Molden 2007).

[3] Similar to the Ecosystem Approach of the Convention of Biological Diversity and ecosystem services-based approaches (see Box 2.1 in Martin-Ortega *et al.*, this book).

Box 6.1 The rise of the Integrated Water Resources Management paradigm

- During the 1980s – declared by the United Nations as the International Drinking Water Supply and Sanitation Decade – water managers realised that an integrated, rather than a sectoral, approach is needed for the conservation and development of water resources.
- In 1992, at the International Conference on Water and Environment in Dublin, in preparation of the Earth Summit in Rio, 28 UN agencies and 58 external organisations agreed on the concept of Integrated Water Resources Management and its four underpinning principles (see Box 6.2).
- In 1992, at the Earth Summit in Rio, the Integrated Water Resources Management concept was included in Agenda 21 and all states in the world were invited to develop a national Integrated Water Resources Management plan by 2005.
- In 1996 the Global Water Partnership and the World Water Council were created to further promote Integrated Water Resources Management.
- In 2002 the call for Integrated Water Resources Management plans was repeated at the Rio+10 conference in Johannesburg.
- In 2012 over 80% of all countries in the world had Integrated Water Resources Management principles in their water laws and two-thirds had developed a national Integrated Water Resources Management plan.

Box 6.2 The Dublin Principles

(1) Fresh water is a finite and vulnerable resource, essential to sustaining life, development, and the environment.
(2) Water development and management should be based on a participatory approach, involving users, planners, and policy makers at all levels.
(3) Women play a central part in the provision, management, and safe-guarding of water.
(4) Water has an economic value in all its competing uses and should be recognised as an economic good.

natural resources and of the risks associated with highly resource-intensive and unsustainable modes of production and consumption on human wellbeing.

The roots of Integrated Water Resources Management lie in the 1970s and 1980s, when the world was increasingly faced with severe water crises in the form of droughts, declining water quality, and competition for an ever scarcer resource (see Box 6.1). It is estimated, for example, that global water withdrawals (mainly for irrigated agriculture) more than doubled in the second half of the twentieth century (Millennium Ecosystem Assessment 2005), while the average discharge of river systems dramatically declined, especially in the tropical regions (Bates *et al.* 2008). During the same period, environmental perils multiplied while the alteration and degradation of forests, wetlands, soils, marine ecosystems, and associated biological diversity reached unprecedented levels. These environmental challenges led to the convening of the World Summit on Sustainable Development in Rio de Janeiro in 1992, where Integrated Water Resources Management was

formally endorsed by 172 nations. Around the same time, landmark conventions for the protection of the environment were adopted, such as those on climate change (United Nations Framework Convention on Climate Change), desertification (United Nations Convention to Combat Desertification), and biodiversity (Convention on Biological Diversity). With the aim of linking the need for safeguarding nature's ecological functions and the wellbeing of humans, the Millennium Ecosystem Assessment report popularised the ecosystem services concept and heralded the wide acceptance of ecosystem services-based approaches (Martin-Ortega *et al.*, this book).

Second, both approaches are grounded on the premise that the degradation of natural resources occurs because these resources and the services they provide to humans are undervalued, if not ignored. This premise is well summarised by Myers and Reichert (1997, p.xix) who, in support of an ecosystem services paradigm, observed: 'We don't protect what we don't value' (core element 4). This thesis is shared by the proponents of Integrated Water Resources Management, who believe that if water resources are being improperly managed, wasted, and degraded, it is because the services they provide to humans are undervalued (Jones-Walters & Mulder 2009). The fourth principle of Integrated Water Resources Management states that 'Water is a public good and has a social and economic value in all its competing uses' (core element 4). Similarly, the essential aim of using an ecosystem services-based approach is to accord a value (monetary and/or non-monetary) to different functions of the ecosystem so as to help 'prevent the erosion of natural capital' (Foresight Project 2011, p.4).

Third, ecosystem services-based approaches and Integrated Water Resources Management are both intended to be means for societal deliberations on trade-offs associated with resource management options and for negotiating fair arrangements for natural resource use and allocation (again, core element 4). While ecosystem services-based approaches adopt a systems perspective to resources management, recognising the interactions and interdependence between services provided by specific biomes (e.g. forests, freshwater, marine ecosystems), Integrated Water Resources Management uses an integrated, holistic perspective to water management, recognising also the interactions between humans, water, land, and natural ecosystems. These similarities should, in principle, facilitate the bridging of efforts to enhance and sustain ecosystem services and improve governance of water resources. As shown below, however, in practice the challenges are many.

6.2.2 How Integrated Water Resources Management can benefit from ecosystem services-based approaches

There is a disconnect between theory and practice with regard to considerations of environmental issues in Integrated Water Resources Management. In *theory*, Integrated Water Resources Management recognises the crucial importance of natural ecosystems in maintaining and providing resources and services to humanity. Two sources form the foundational pillars of Integrated Water Resources Management theory: the Dublin Principles, agreed at UN level in the run-up to the World Summit on Sustainable Development in 1992 (see Box 6.2), and the General Framework for Integrated Water Resources Management, proposed by the Global Water Partnership (2000). The first Dublin Principle on Integrated Water Resources Management claims that 'fresh water is a finite and vulnerable resource, essential to sustain life, development and the environment'. The Global Water Partnerships's theoretical framework evaluates the sustainability of water management against the three criteria of social equity, economic efficiency, and environmental sustainability – treating these as equally important (Global Water Partnership 2000, p.31).

These two sources of Integrated Water Resources Management theory display an ambiguous stance towards the relationship between humans and the ecosystem. The first Dublin Principle and some aspects of the Global Water Partnership's framework put the water needs of natural ecosystems at the same level as the water needs of people and economic development. Humans and the ecosystem are considered to constitute a single whole.

In other instances, however, natural ecosystems are relegated to the role of service provider: nature is an asset that needs to be maintained and protected 'to ensure that the desired services it provides are sustained' (Global Water Partnership 2000, p.4). In particular, the natural environment is of 'critical importance for resource availability and quality' (Global Water Partnership 2000, p.23). Water management should not compromise the use of the services of natural ecosystem by future generations (Global Water Partnership 2000, p.30). In other words, this face of Integrated Water Resources Management theory suggests that the water needs of the natural environment are to be taken into account because the ecosystem provides services to humans: the ecosystem regulates the quantity and quality of the water

available for productive use and consumption by humans. This idea also comes clearly to the fore in the fourth Dublin Principle.

These two views on the role and position of the ecosystem within Integrated Water Resources Management fundamentally conflict at a conceptual level: one is holistic and recognises the water needs of the ecosystem *an sich*; the other is anthropocentric and views the water needs of the natural environment through the eyes of humans. This is also very much related to the discussion – still open in the Integrated Water Resources Management community – about the meaning of 'integration' and how it is achieved (Biswas 2008). These theoretical callisthenics would not be much of an issue if it were not for the disregard of the ecosystem reflected in Integrated Water Resources Management *practice*.

This neglect has to do with the continued limited understanding of the water needs of ecosystems for the provision of services to humans (core element 2) and for the maintenance of ecosystem health, although significant progress has been made on assessing environmental flow requirements in river systems (Dyson *et al.* 2003; Richter *et al.* 2003; Tharme 2003; Arthington *et al.* 2010). The disregard also results from the fact that the ecosystem is absent or weakly represented in water allocation and management decision-making processes (core element 4). Integrated Water Resources Management, being a framework for negotiated decision-making on water allocation and management, is challenged because its outcomes are highly influenced by power imbalances. For this reason, nature, which is voiceless, tends to remain unheard in Integrated Water Resources Management-based decision processes. Similarly, in spite of its stated aim of achieving equity in water sharing, Integrated Water Resources Management practice does not always help improve access to water for women or for common property users such as indigenous peoples or pastoralists, who typically play an invisible and marginal role in water allocation and management decisions.

Applying an ecosystem services-based approach to water management can help give visibility to, and raise awareness of, the multiple values of services provided by ecosystems, and hence help society realise that water allocated or left to the ecosystem is not wasted (UNESCO 2009). By providing an economic rationale for ecosystem maintenance, it helps communicate the value of ecosystem services in a commonly accessible language (Kumar *et al.* 2013). It also helps broaden the constituency in support of conservation and enhancement of ecosystem functions (Ingram *et al.* 2012). Ecosystem services-based approaches can be used to establish a consultative and decision-making framework that brings to the fore, and gives voice to, the poor as custodians of these ecosystems (core elements 3 and 4). In this regard Integrated Water Resources Management, by using an ecosystem services-based approach, can be more suited to developing country contexts where the poor (especially the rural

poor) depend heavily upon ecosystem services for their livelihoods (Ferraro 2009). The debates surrounding the annual flood
regime of the Senegal River in the late 1980s and 1990s (see
Section 6.3.1) is a good illustration of how an ecosystem
services-based approach can steer water management decisions
towards greater consideration of the diversity of ecosystem services and, therefore, the livelihoods of people who directly
depend on them.

Adopting an ecosystem services-based approach can also
enable Integrated Water Resources Management to fulfil its
ambition of using a truly systemic approach in which interactions
between ecosystems are recognised (core elements 2 and 3) and
factored into decision processes (core element 4). Indeed, water
scarcity – which manifests itself in the form of reduced relative
and absolute availability and of water resource quality degradation – cannot be solved within the confines of the water resource
sector. It requires solutions from functions performed and services provided by ecosystems such as soils and forests. For
example, green water (moisture in soil), which is estimated at
$120\,000\,\text{km}^3$ of water per year (or two-thirds of total freshwater)
is supplied to humans from rainfall through soils, hence the
importance of maintaining the organic content of soils, i.e. their
ability to conserve water (Daily *et al.* 1997). The same functions
are played by forest biomes for blue water (natural runoff
through rivers and groundwater). The functions of provision,
maintenance, purification, and recycling of water, which are
played by ecosystems of different kinds, cannot be ignored in
Integrated Water Resources Management decision processes.

6.2.3 How ecosystem services-based approaches can benefit from an Integrated Water Resources Management perspective

Water plays a central role in the functioning of natural ecosystems and their ability to deliver services (Global Water Partnership 2000). Natural ecosystems are hence the main users of water
(Falkenmark & Rockström 2004). It is estimated that the maintenance of ecosystem health allowing it to sustainably provide
regulating and supporting services requires 75% of total freshwater use (blue- and greenwater) while direct withdrawal of
water for human use (provision of water, food, energy, etc.)
represents only 25% (Falkenmark & Rockström 2004; UNEP
2012). As demand for water for multiple human uses increases,
the competition between nature and humans intensifies (UNEP
2012). Also, part of the water requirements of natural ecosystems
is for the restoration of damage already caused by human
demand. A combined ecosystem services and Integrated Water
Resources Management approach could help change this competition into complementarity, in the sense that the way water is
managed affects the health of ecosystems, i.e. their ability to
deliver services to humans (UNESCO 2009).

6.3 EXPERIENCES WITH IMPLICIT ECOSYSTEM SERVICES-BASED APPROACHES IN WATER MANAGEMENT IN THE DEVELOPING WORLD

Despite the possible synergies that can derive from associating
Integrated Water Resources Management and ecosystem
services-based approaches, in practice the two have evolved
largely in separate ways and continue to operate in parallel.
There are, however, a number of water management experiences
at country, basin, or sub-basin level where the basic elements of
ecosystem services-based approaches have been implicitly
embraced. And although an ecosystem services-based approach
has been imperfectly implemented, these cases provide important
insights on the possible synergies with Integrated Water
Resources Management – probably more valuable than insights
offered by more recent projects where ecosystem services-based
approaches were explicitly implemented.

6.3.1 Senegal River

Under the auspices of the Senegal River Basin Development
Authority, the natural flood regime of the Senegal River came
under scrutiny when an ambitious plan was devised in the 1970s–
1980s to tame its waters, with an upstream multipurpose hydropower dam (Manantali) and a downstream anti-salt intrusion
dam (Diama).

With a reservoir of 11 billion cubic metres, the Manantali Dam
was conceived as the pillar of the programme, with three key
objectives: (1) expansion of irrigated areas (with a target of
375000 hectares of irrigated land); (2) hydropower generation
(800 GWh per year); and (3) year-round navigability of the lowest
800 km of the river. The hydrological and economic feasibility
studies for the project concluded that achieving these objectives
required the termination of the river's annual flooding system.
Under natural (pre-dam) conditions, the river's annual flow
reached its peak in August–October each year, inundating between
100 000 and 500 000 hectares of land. As the floodwaters receded,
riparian farmers grew sorghum and many other crops. This farming
system, known as *waalo* agriculture, was the backbone of agrarian
economies in the Senegal valley. The key argument for terminating
the annual flood, i.e. not allowing the generation of artificial floods
from the Manantali Dam, was that it would reduce electricity
production by 20%, resulting in an annual loss of US$12 million.
According to the promoters of the programme, this lost electricity
production could not be compensated for by the revenues generated by agriculture in the flooded area – estimated at US$5.6
million per annum (Senegal River Basin Development Authority
1987; Salem-Murdock *et al.* 1994; Ficatier & Niasse 2008).

The terms of the debate changed in the 1990s, when new
studies were launched using principles and methods that today

are associated with an ecosystem services-based approach. These studies (Salem-Murdock *et al.* 1994) showed that the annual flood of the river contributed more than flood recession agriculture alone: for each hectare inundated, it also contributed about 70 kg of fish and 0.35 Tropical Livestock Units.[4] According to these studies, when these benefits are added to the outputs of flood recession agriculture, the generation of an artificial annual flood becomes economically justified.

The Senegal River Basin Development Authority Water Charter adopted by riparian states of the Senegal River in 2002 illustrates the triumph of a wider perspective on water management, recognising the multiple services attached to the natural flow regime of the river ecosystem. Referring to the management of the Manantali Dam, the Charter states that 'in allocating available water, the necessary hydraulic conditions should be created for the flooding of the river valley and for supporting traditional recession agriculture' (Senegal River Basin Development Authority 2002, art. 5). In addition to reviving the floodplain economy in its diversity, the annual flood – which is today generated, except in years of severe hydrological deficits – has hence allowed the river ecosystem to continue to play a cleansing role (i.e. preventing the proliferation of invasive species and related disease vectors), while contributing to the recharge of groundwater as well as to the maintenance of forests.

6.3.2 Komadugu River

The Komadugu River on the border between Niger and Nigeria forms part of the Lake Chad Basin. In its mid-valley is the Hadejia Nguru wetland, a floodplain rich in biodiversity. As a result of the large upstream Tiga and Challawa Dams, commissioned in the mid 1970s and early 1990s, respectively, and the effects of climate change and variability, the river discharge had dramatically declined and its flow no longer reached Lake Chad. A series of studies in the late 1980s analysed the value of the services associated with the floodplain (agriculture, grazing, fishing, forest products, groundwater recharge, etc.), and the opportunity cost of allocating water for upstream irrigation at the expense of the floodplain (Barbier *et al.* 1991; Hollis *et al.* 1993). These studies were updated in the mid 2000s in the form of a water audit, taking into account reduced rainfall and river discharge, increased demand for water for irrigation and domestic use, and the floodplain production system in a general context of intensified competition between upstream and downstream states in Nigeria, and between Nigeria and riparian countries of the Lake Chad Basin. This water audit supported a series of multi-stakeholder consultations in all riparian states, and also served as the basis for the formulation of a framework for managing the basin waters (a Water Charter and a Catchment Management Plan). It subsequently led to the establishment of a Trust Fund jointly financed by the Nigerian Federal Government and riparian states of the Komadugu River (Barchiesi *et al.* 2011). The combined Integrated Water Resource Management and ecosystem services-based approach used in the Komadugu Basin facilitated a consensus on water allocation modalities among the six Nigerian riparian states, upstream users (the city of Kano, irrigation schemes), and downstream communities dependent on floodplain services, while working towards the restoration of the wetland ecosystem.

6.3.3 Environmental flows at river basin level

The debate about the essential water needs of the natural environment of impounded river systems has been strongly influenced by trends in the understanding and valuing of services provided by natural ecosystems to humans, and the need to negotiate trade-offs when dealing with competing demands for ecosystem services (core element 4). Initially the focus was on guaranteeing a 'minimum flow', i.e. ensuring that a minimum amount of water is left in the river system all year round, in order to avoid discontinuity in its flow that would affect, for example, the upstream–downstream movements of fish and other aquatic species. Hence, this initial notion of 'minimum flow' did not consider the plethora of services a river system pays to humans.

This notion of minimum flow became unsatisfactory as we improved our understanding of the multiplicity and importance of other functions played by the natural flow regime of a river (core element 2) – functions that are affected by changes in the quantities of water flowing, but also by its seasonal variations, water temperature, sediment load, etc. The notion of 'environmental flow' has therefore been adopted in recent years to better accommodate the complexity and dynamic nature of water needs of river ecosystems, especially the water requirements that are essential for ecosystems to be able to sustainably provide regulating and supporting services such as purification of polluted waters, flood control, groundwater recharge, nutrient recycling, and maintenance of soil fertility, etc. (King *et al.* 1999; Tharme 2003; Niasse & Lamizana 2004; Forslund 2009). According to the Brisbane Declaration of 2007: 'Environmental flows describe the quantity, quality and timing of water flows required to sustain freshwater and estuarine ecosystems and the human livelihoods and well-being that depend on these ecosystems.' The need to sustain the ability of nature to provide ecosystem services to humans is therefore central to the concept of environmental flow.[5] Today, the need to preserve

---

[4] The Food and Agriculture Organization of the United Nations defines one standard Tropical Livestock Unit as one head of cattle with a body weight of 250 kg.

[5] Declaration made at the 10th International River Symposium and International Environmental Flows Conference, held in Brisbane, Australia, 3–6 September 2007. See: www.eflownet.org/download_documents/brisbane-declaration-english.pdf (last accessed 12 December 2014).

and/or generate an environmental flow is considered a norm in large water infrastructure projects – as exemplified by the European Water Framework Directive (Acreman *et al.* 2009) – and has been more or less comprehensively taken into consideration in recent river basin development and management plans, as in the cases of the Senegal River, the Orange-Senqu River (in South Africa/Lesotho), the lower Mekong, etc.[6] Where large water control projects (dams, inter-basin transfer schemes, etc.) exist or are planned, efforts are made to carry out an environmental flow assessment[7] to serve as a basis for determining the environmental flow regime required to preserve the essential services of existing ecosystems as well as newly created ones (e.g. reservoirs) and to devise measures needed to ensure that water infrastructure projects are operated accordingly. A well-defined and effectively implemented environmental flow regime is, therefore, clearly supportive of the Integrated River Basin Management approach (the application of the Integrated Water Resource Management approach in a river basin context).

6.3.4 Dam reoperation to recover and optimise river ecosystem services

When preparing its recommendations for improving the planning and operation of large dam projects, the World Commission on Dams was confronted with the fact that the world had already invested massively in the impoundment of river systems. In such a context, any meaningful suggestion for behavioural change had to consider the existing global stock of 45 000 large dams (those more than 15 metres high). Most of these were designed in contexts of much lower standards with regards to environmental protection, equitable sharing benefits, respect of rights of communities, etc. In order, for example, to restore the health and biodiversity of highly impounded and fragmented river systems, existing dams needed to be operated differently.

Therefore, the World Commission on Dams recommended the *reoperation* of existing dams, in order to recover economic, social, and environmental benefits and to restore ecosystems – dimensions that might have been ignored, overlooked, or suboptimally considered in the operation of existing dams (World Commission on Dams 2000). Building on the Commission's recommendations, a number of dam reoperation initiatives have been launched in recent years, including notably by the Natural Heritage Institute in Africa and China (Richter & Thomas 2007; Thomas & DiFrancesco 2009). Dam reoperation is a strategy

that works best if supported by sound environmental assessment. It is a way of operationalising Integrated Water Resources Management and an ecosystem services-based approach in basin contexts where past water infrastructure investments have had a narrow perspective on water resource allocation and its effect on the ecosystem and on people.

6.3.5 Environmental considerations in country-level Integrated Water Resources Management strategies and plans

Most of the examples in this chapter on actual or possible linkages between Integrated Water Resources Management and ecosystem services-based approaches relate to river and lake basins. As shown in the cases of the Senegal River and the Komadugu Yobe Basin, environmental considerations are typically given considerable attention in water management efforts at basin level.

This is not the case for country-level water policies and strategies, which are particularly weak in addressing environmental aspects of freshwater management. The review carried out in 2008 by UN-Water to assess progress made in the development and implementation of Integrated Water Resources Management and Water Efficiency Plans (covering 104 countries) found that for developing countries in Africa, Asia, and Latin America the formulated plans scored very low on their coverage of programmes and policies to address ecosystem degradation (UN-Water 2008). In line with this finding, a more recent review by the African Ministers' Council on Water noted that in Africa it is only in rare cases (e.g. Tunisia, Mauritius, Algeria) that national IWRM plans explicitly factor in water allocations for environmental requirements (African Ministers' Council on Water 2012). None of these plans identified aquatic ecosystems as 'natural water infrastructures' that can, if properly managed, contribute to achieving Integrated Water Resources Management goals such as the provision of quality drinking water or water for irrigation. In fact, this is one of the key areas where countries engaged in Integrated Water Resources Management planning processes need clearer and more operational guidance, which requires Integrated Water Resources Management discourse and guiding material (Global Water Partnership 2000; Cap-Net 2005) to be strengthened with inputs based on ecosystem services-based approaches.

6.4   CONCLUSIONS

Integrated Water Resources Management and ecosystem services-based approaches are responses to the alarming levels of degradation of the world's natural resources base. Both are grounded on the premise that unsustainable and unwise resource use practices occur because these resources and the services

---

[6] See case examples documented by International Union for Conservation of Nature's Water and Nature Initiatives: www.iucn.org/about/work/ programmes/water/resources/toolkits/flow (last accessed 12 December 2014)
[7] An environmental flow assessment can be defined as the 'assessment of how much of the original flow regime of a river should continue to flow down it and onto its floodplains in order to maintain specified, valued features of the ecosystem' (Tharme 2003).

they provide to humans are undervalued, if not simply ignored. Both are meant to be decision-support frameworks for societal deliberations on trade-offs associated with resource management options and for negotiating fair arrangements for natural resource use and allocation. Despite their similarities, however, Integrated Water Resources Management and ecosystem services-based approaches have evolved in parallel, with minimal interactions.

Although ecosystem services-based approaches are inevitably anthropocentric and rather short-sighted because placing too much emphasis on humans' current and often immediate needs, they help recognise, identify, analyse, and value the multiple services of ecosystems. In this chapter we show that Integrated Water Resources Management and ecosystem services-based approaches can mutually strengthen each other when combined at the operational level. We argue that ecosystem services-based approaches can help Integrated Water Resources Management to operationalise its environmental sustainability pillar. Applying ecosystem services-based approaches to water management can help raise awareness on and better communicate the importance of the multiple values of services provided by ecosystems. It can also help put in place inclusive consultative platforms involving the poor as citizens and as custodians of ecosystems. Integrated Water Resources Management using an ecosystem services-based approach can be more suited to developing country contexts where the poor tend to be heavily dependent upon ecosystem services for their

livelihoods. approaches can help Integrated Water Resources Management establish much-needed bridges between water and other ecosystems: solutions to current water scarcity challenges cannot ignore the roles of ecosystems such as mountains, soils, and forests.

Conversely, ecosystem services-based approaches as defined in Chapter 2 can benefit from an Integrated Water Resources Management perspective. Because maintaining ecosystem structure and functions uses over three-quarters of the world's green- and bluewater, improved water management positively affects the health of ecosystems and associated functions and services.

A combined Integrated Water Resources Management and ecosystem services-based approach can build on the significant advances made in recent years in applying methods and instruments such as dam reoperation and implementation of environmental flows.

We observe that environmental considerations are typically given considerable attention in water management efforts at basin level. In some cases, key principles of what is today known as the ecosystem services-based approach have been applied and have contributed to fairer and more environmentally sensitive water allocation arrangements, as illustrated in the examples of the Senegal and Komadugu Yobe Basins. This is not the case, however, for country-level water policies and strategies, which are particularly weak in addressing environmental aspects of freshwater use. A common characteristic of Integrated Water Resources Management plans that have been formulated in recent years is their disregard of the need to factor in the water

Box 6.3 Key messages

- As ecosystems are the principal users of freshwater, improved management of water positively affects the health of ecosystems and associated functions and services.
- Integrated Water Resources Management can benefit from an ecosystem services-based approach to better operationalise its environmental sustainability pillar, as it helps: to communicate the importance of the multiple values of services provided by ecosystems; set in place inclusive consultative platforms involving the poor as custodians of ecosystems; bridge water and ecosystem management.
- A combined Integrated Water Resources Management and ecosystem services-based approach can capitalise on progress made in recent years in applying methods and instruments such as dam reoperation and implementation of environmental flows.
- Notable progress has been made in recent years in addressing ecosystem issues in river basin contexts through the application of values and principles that are recognised today as key features of ecosystem services-based approaches.
- Including an ecosystem services-based component in national Integrated Water Resources Management strategies and planning can help countries to better understand and factor in their natural capital; in particular their natural water infrastructure and their water requirements for maintaining and enhancing associated ecosystem services.
- Ecosystem services-based approaches, contrary to Integrated Water Resources Management, risk being overly anthropocentric and short-sighted, as they quantify the water needs of the ecosystem only in terms of the services provided by ecosystems to humans in the short term.
- A joint Integrated Water Resources Management and ecosystem services-based approach helps strengthen the sustainability dimension of water and ecosystem management.

requirements of ecosystems and of the importance of these ecosystems as natural water infrastructures. Promoting an ecosystem services-based approach at country level, therefore, deserves greater attention in efforts to develop, update, and implement national Integrated Water Resources Management plans.

References

Acreman, M., Aldrick, J., Binnie, C. *et al.* (2009). Environmental flows from dams: the Water Framework Directive. *Engineering Sustainability* **162**, 13–22.

African Ministers' Council on Water (2012). *2012 Status Report on the Application of Integrated Approaches to Water Resources Management in Africa.* AMCOW, Addis Ababa.

Arthington, A. H., Naiman, R. J., McClain, M. E., & Nilsson, C. (2010). Preserving the biodiversity and ecological services of rivers: new challenges and research opportunities. *Freshwater Biology* **55**(1), 1–16.

Barbier, E. B., Williams, W. H., & Kimmage, K. (1991). *Economic Valuation of Wetlands Benefits: The Hadejia-Jama'are Floodplain, Nigeria.* IIED, London. Available at: http://pubs.iied.org/pdfs/8022IIED.pdf (last accessed 21 October 2014).

Barchiesi, S., Cartin, M., Wellling, R., & Yawson, D. (2011). *Komadugu Yobe Basin, Upstream of Lake Chad, Nigeria. Case Study.* IUCN/Water and Nature Initiative, Gland, Switzerland. Available at: https://portals.iucn.org/library/efiles/documents/2011-097.pdf (last accessed 21 October 2014).

Bates, B., Kundzewicz, Z., Wu, S., & Palutikof, J. (eds) (2008). *Climate Change and Water. IPCC Technical Paper VI.* Intergovernmental Panel on Climate Change (IPCC) Secretariat, Geneva. Available at: http://digital.library.unt.edu/ark:/67531/metadc11958 (last accessed 21 October 2014).

Biswas, A. K. (2008). Integrated water resources development: is it working? *Water Resources Development* **24**(1), 5–22.

Cap-Net (2005). *Integrated Water Resources Management Plans – Training Manual and Operational Guide.* Cap-Net/Global Water Partnership/UNDP. Available at: www.cap-net.org/sites/cap-net.org/files/Manual_english.pdf (last accessed 21 October 2014).

Daily, G. C., Matson, P. A., & Vitousek, P. M. (1997). Ecosystem services supplied by soil. In G. C. Daily (ed.), *Nature's Services: Societal Dependence on Natural Ecosystems.* Island Press, Washington, DC.

Dyson, M., Bergkamp, G., & Scanlon, L. (eds) (2003). *Flow: The Essentials of Environmental Flows.* IUCN/Water and Nature Initiative, Gland, Switzerland.

Falkenmark, M. & Rockström, J. (2004). *Balancing Water for Humans and Nature: The New Approach in Ecohydrology.* Earthscan, London.

Ferraro, P. J. (2009). Regional review of payments for watershed services: sub-Saharan Africa. *Journal of Sustainable Forestry* **28**, 525–550.

Ficatier, Y. & Niasse, M. (2008). *Evaluation rétrospective – Volet social et environnemental du barrage de Manantali.* Agence Française de Développement (AFD), Paris.

Foresight Project (2011). *Maintaining Biodiversity and Ecosystem Services while Feeding the World: Foresight Project on the Global Food and Farming Futures.* UK Government Office for Science, London.

Forslund, A. (2009). Securing Water for Ecosystems and Human Well-being: The Importance of Environmental Flows. Stockholm: Stockholm International Water Institute. Available at: www.unepdhi.org/~/media/Microsite_UNEPDHI/Publications/documents/unep_DHI/Environmental%20Flows%20Report%2024%20-low-res.ashx (last accessed 21 October 2014).

Global Water Partnership (2000). *Integrated Water Resources Management.* Stockholm: GWP. Available at: www.gwp.org/Global/GWP-CACENA_Files/en/pdf/tec04.pdf (last accessed 21 October 2014).

Hollis, G. E., Adams, W. M., & Aminu-Kano, M. (1993). *The Hadejia-Nguru Wetlands. Environment, Economy and Sustainable Development of a Sahelian Floodplain Wetland.* IUCN, Gland, Switzerland.

Ingram, J. C., Redford, K. H., & Watson, J. E. M. (2012). Applying ecosystem services approaches for biodiversity conservation: benefits and challenges. *Sapiens* **5**(1), n.p.

Jones-Walters, L. & Mulder, I. (2009). Valuing nature: the economics of biodiversity. *Journal for Nature Conservation* **17**, 245–247.

King, J. M., Tharme, R. E., & Brown, C. A. (1999). *Definition and Implementation of Instream Flows: Thematic Report for the World Commission on Dams.* WCD, Cape Town.

Kumar, P., Brondizio, E., Gatzweiler, F., *et al.* (2013). The economics of ecosystem services: from local analysis to national policies. *Current Opinion in Environmental Sustainability* **5**, 78–86.

Millennium Ecosystem Assessment (2005). *Ecosystems and Human Well-being: Synthesis.* Island Press, Washington, DC.

Molden, D. (ed.) (2007). *Water for Food, Water for Life. A Comprehensive Assessment of Water in Agriculture.* International Water Management Institute (IWMI) and Earthscan, London.

Myers, J. P. & Reichert, J. S. (1997). Perspectives on nature's services. In G. C. Daily (ed.), *Nature's Services: Societal Dependence on Natural Ecosystems.* Island Press, Washington, DC.

Niasse, M. & Lamizana, B. (2004). La prise en compte de l'environnement et du social dans les politiques de l'eau en Afrique de l'Ouest: fondements juridiques et leçons de l'expérience. In M. Niasse, A. Iza, A. Garane, *et al.* (eds), *La Gouvernance de l'Eau en Afrique de l'Ouest : Aspects Juridiques et Institutionnels.* IUCN, Gland, Switzerland.

Organisation pour la mise en valeur du fleuve Sénégal (1987). Etude de la gestion des ouvrages communs de l'OMVS. Rapport Phase 1. Vol 1B: Optimisation de la crue artificielle. Sir Alexander Gibb and Partners.

Organisation pour la mise en valeur du fleuve Sénégal (2002). *Charte des eaux du fleuve Sénégal.* Dakar: OMVS.

Richter, B. D. & Thomas, G. A. (2007). Restoring environmental flows by modifying dam operations. *Ecology and Society* **12**(1). Available at: www.ecologyandsociety.org/vol12/iss1/art12 (last accessed 21 October 2014).

Richter, B. D., Mathews, R., Harrison, D. L., & Wigington, R. (2003). Ecologically sustainable water management: managing river flows for ecological integrity. *Ecological Applications* **13**(1), 206–224.

Salem-Murdock, M., Niasse, M., Magistro, J., *et al.* (1994). *Les barrages de la controverse. Le cas de la vallée du fleuve Sénégal.* Paris: L'Harmattan.

Tharme, R. E. (2003). A global perspective on environmental flow assessment: emerging trends in the development and application of environmental flow methodologies for rivers. *River Research and Applications* **19**, 397–441.

Thomas, G. A. & DiFrancesco, K. (2009). *Rapid Evaluation of the Potential for Re-optimizing Hydropower Systems in Africa: Final Report.* Natural Heritage Institute, San Francisco, CA. Available at: www.n-h-i.org/uploads/tx_rtgfiles/REOPS_Final_Report_04-15-09.pdf (last accessed 21 October 2014).

UNEP (2012). *Environment for the Future We Want: Global Environment Outlook 5.* UNEP, Nairobi.

UNESCO (2009). *Water in a Changing World: World Water Development Report 3.* UNESCO/Earthscan, Paris.

UN-Water (2008). *Status Report on IWRM and Water Efficiency Plans. Prepared for the 16th Session of the Commission on Sustainable Development.* United Nations, New York.

World Commission on Dams (2000). *Dams and Development: A New Framework for Decision-making. The Report of the World Commission on Dams.* Earthscan, London.

Woodhouse, P. (2012). Foreign agricultural land acquisition and the visibility of water resource impacts in sub-Saharan Africa. *Water Alternatives* **5**(2), 208–222.

7   Implementation of the European Water Framework Directive

What does taking an ecosystem services-based approach add?

Kirsty L. Blackstock, Julia Martin-Ortega, and Chris J. Spray

7.1   INTRODUCTION

The European Water Framework Directive (2000) provided a step-change in the way European waters are characterised, monitored, and managed. The Directive provides a framework to integrate multiple water environments and to coordinate a range of water-relevant legislation, while advocating for public participation and economic efficiency in water management. However, in the decade since the publication of the Directive, implementation has been problematic. As the Directive moves into its second implementation cycle, the European Commission and the member states are beginning to consider how it could be delivered using the concept of ecosystem services (Martin-Ortega 2012).[1] Therefore, this chapter considers what an ecosystem services-based approach might add to the different stages required as part of the River Basin Management Planning process in Europe, and whether the approach might improve the implementation of the Water Framework Directive.

7.2   THE WATER FRAMEWORK DIRECTIVE

7.2.1   Purpose

The objectives of the Water Framework Directive are to stop deterioration, improve the state of aquatic ecosystems, and promote the sustainable use of water by achieving 'good ecological status' in defined river basins (Box 7.1 collects the seven facts upon which the European Commission justifies and underpins the implementation of the Water Framework Directive). Under this norm, good ecological status is a composite assessment that measures the current state against the 'reference condition' for that type of water body (essentially, the state of the ecosystem before the impact of human pressures). Where the water body is at less than good status, measures (actions) must be

taken.[2] To achieve the good ecological status, cost-effective Programmes of Measures need to be set up. Where this is technically unfeasible or economically disproportionate then the objective can be reduced to moderate status or deferred to a later cycle (2021, 2027). Thus the plans provide an overview of the state of the ecosystem, the pressures on the ecosystem, and the actions that will be taken to remove the pressures and mitigate their impacts in an economically efficient way.

7.2.2   Problems

The third report on Water Framework Directive implementation (European Commission 2012) found only a 10% predicted increase in surface water bodies likely to reach good ecological status by 2015 compared to 2009; leaving almost half the surface waters in Europe likely to be less than good status in 2015. There are two major difficulties: scientific understanding and practical implementation. Scientifically, Hering *et al.* (2010) drew attention to the problems in developing a Europe-scale understanding of the state of the aquatic ecosystems and their restoration to good ecological status, concluding that the timescale of the Water Framework Directive is over-ambitious. Hering *et al.* (2010) highlighted difficulties in assessing good ecological status and setting appropriate thresholds between categories. The complex assessments amplify the uncertainties associated with any of the individual indicators making up ecological status – indeed the Directive is plagued by many different types of uncertainty (Sigel *et al.* 2010). These problems have knock-on implications for monitoring processes (van Hoey *et al.* 2010). These challenges may explain why its implementation continues to favour fixing point-source pollution over other pressures, such as diffuse pollution from agriculture (European Commission 2012).

These scientific difficulties are compounded when translated into management plans as there are further uncertainties and complexities when trying to understand the relationship between

---

[1] See also http://operas-project.eu/ and http://www.openness-project.eu (last accessed 9 December 2014).

[2] Sometimes existing obligations like Natura 2000 mean that the objective might be High Ecological Status to protect specific habitat or species. Where a water body has been altered from its natural form and cannot be restored, it is designated as 'heavily modified' and a target of 'good ecological potential' is set.

Box 7.1 European Commission's seven facts underpinning the Water Framework Directive

Fact 1: Europe's water is under pressure.

Fact 2: European Union action is necessary because river basins and pollution cross borders. The river basin approach is the best way to manage water.

Fact 3: Waters must achieve good ecological and chemical status, to protect human health, water supply, natural ecosystems, and biodiversity.

Fact 4: It is crucial to get people involved.

Fact 5: There is some progress already, but more needs to be done (the Commission checks each step of the implementation of the Directive).

Fact 6: Water management is linked to many policies: integration is the only way forward for sustainable water.

Fact 7: A changing environment creates challenges for the future, including climate change, floods, and drought.

Source: European Commission: http://ec.europa.eu/environment/pubs/pdf/factsheets/water-framework-directive.pdf (last accessed 9 December 2014).

Figure 7.1 Implementing an ecosystem service-based approach for River Basin Management Plans. Based on Spray & Blackstock (2013)

ecosystem status and impact of measures, particularly when there is a time-lag between the implementation of a measure and the response of the ecosystem. Many member states have failed to set up appropriate governance processes such that water use is not well regulated or incentivised and the River Basin Management Plans are not integrated into wider spatial plans (European Commission 2012). In particular, most River Basin Management Plans have not adopted suitable cost–benefit methodologies or appropriate water pricing that takes account of the environmental externalities (Martin-Ortega 2012). Many studies note the difficulties in selecting measures that will achieve the objectives in a social and economically acceptable manner (e.g. Volk *et al.* 2009). Public participation has also been implemented in a rather patchy way (De Stefano & Schmidt 2012), with many authors identifying problems with reconciling participation with other objectives of the Water Framework Directive (e.g. Blackstock 2009).

7.3 IMPLEMENTING THE WATER FRAMEWORK DIRECTIVE USING AN ECOSYSTEM SERVICES-BASED APPROACH

The 'added value' of adopting an ecosystem services-based approach is the provision of a means for conceptualising the link between the environment and the many ways in which people value ecosystem services. As such, it can help identify the value of ecosystems at a given scale (such as a catchment or water body), and assist in analysing trade-offs between land and water

management. In doing so, an ecosystem services-based approach identifies those stakeholders who will be most impacted by such decisions and thus who should be included within the decision-making process. Based on the definition in Chapter 2, we consider how an ecosystem services-based approach fits with the implementation of the Water Framework Directive. Although ecosystem services do not appear in the legal text as such, good ecological status provides a link between healthy functioning ecosystems and River Basin Management Planning, and there are a number of articles within the Directive that refer to elements of the idea of valuing ecosystems. Thus, good ecological status could directly support element 2 outlined in Chapter 2 – biophysical underpinning for the delivery of the service. Furthermore, the Water Framework Directive links ecological status with societal wellbeing and societal choices, and commits to public participation, which connects to core elements 1 (wellbeing) and 3 (trans-disciplinarity), respectively. Core element 4 (decision-making) is extremely important in the choice of measures, particularly how effective and proportionate such measures might be and for whom. Therefore, we have identified six main steps where an ecosystem services-based approach could improve the implementation of River Basin Management Planning processes within the current directive.

7.3.1 Selecting and engaging with stakeholders

While selecting and engaging with stakeholders is not a separate step within River Basin Management Planning, it is fundamental to the outputs and outcomes of an ecosystem services-based approach, particularly core element 3. Thus, it strengthens the Directive's article 14 calling for active involvement of interested parties, something that has not been well-implemented to date (De Stefano & Schmidt 2012). In fact, the elements feed into each other, with stakeholder engagement underpinning the identification of ecosystem services in catchment characterisation, objective setting, selection of measures, assessing disproportionality, and, finally, in monitoring and evaluation. Taking this approach to river basin planning might make some stakeholders more visible by highlighting those affecting, or affected by, less tangible aspects of the water environment. This would provide a stronger rationale for the involvement of water-based recreational users who are often left out of existing water stakeholder platforms (Ravenscroft & Church 2011). These changes could alter the stakeholder focus from those who negatively impact on the water environment, towards those responsible for the positive benefits of restoring and protecting a pristine water environment (Keeler et al. 2012). It is also an opportunity to balance environmentally driven targets with a concern for social, not just economic, wellbeing.

7.3.2 Characterising catchments in terms of ecosystem services delivered

The notion of ecosystem services can be used to better understand the pressures and impacts on the water environment, with implications for the services that provide human wellbeing. It could add a new dimension to economic characterisation by including all the services arising, not just those that can be monetised or linked to a specific pressure. This would shift the focus away from problems defined purely by the Water Framework Directive parameters, to explain how healthy rivers, lakes, and coastlines contribute to human wellbeing (Volker & Kistemann 2011). However, while it is important to understand the mutual interdependence between function, service, and benefit, there are many difficulties in operationalising these concepts (Chan et al. 2012). The more holistic perspective for water body, catchment, and river basin management can indicate where multiple benefits can be achieved, or where conflicting priorities need addressing during a decision-making process. Understanding the range of perspectives on what the environment does, and should do, for society can help to manage controversies and prevent expensive and time-consuming legal challenges to decisions. An ecosystem services-based approach might stimulate a more integrated land and water management approach, as the focus on linking ecological function with human wellbeing links water bodies to the wider catchment setting in which they belong (Newsom 1982).

7.3.3 Identifying outcomes and setting objectives

An ecosystem services-based approach can encourage a more transparent and participatory understanding of the Directive's objectives beyond the environmental parameters of ecological status to include ecosystem services in decision-making. This balances the assumptions about returning the water environment to 'natural' conditions with a counter-focus on how the water environment provides services to people. Objective setting would require the identification of a common vision for the catchment in terms of ecosystem service delivery so that objectives reflect the range of outcomes desired by the stakeholders. This inclusion of non-statutory goals desired by stakeholders has many benefits. It helps build ownership and resilience into the river basin planning process by aligning the technical legislative targets with things that matter to local people (Koontz 2005), encouraging local support and resource allocation. It provides opportunities to align with complementary policy objectives (e.g. flood risk management) (Vlachopoulou et al. 2014), helping to implement other supplementary objectives of the Directive. It draws attention to where additional benefits could be realised (Gilvear et al. 2013).

7.3.4 Selecting a Programme of Measures most suitable to achieve ecological objectives

Cost-effectiveness analysis to select the right measures to achieve the objectives is a major challenge due to uncertainty over ecological response and interactions between measures at a catchment scale (Martin-Ortega 2012). There is debate about the 'right' institutional mix of legislative, economic, or advisory approaches to utilise when designing Programmes of Measures (Meyer & Thiel 2012). An ecosystem services-based approach would require the 'right mix' that takes account of all current and potential ecosystem services and their interactions with one another. A focus on ecosystem services highlights the potential for Payment for Ecosystem Service schemes as a way of implementing voluntary measures – particularly as catchments are suitable areas for these schemes (Wunder et al. 2008). Furthermore, core elements 1 and 3 would encourage option appraisal to also consider the acceptability of any proposed measure to land or water managers and those affected by its implementation (Buckley et al. 2012). An ecosystem services-based approach with broad stakeholder involvement provides access to local and experiential knowledge about what is effective in specific situations. Thus, it extends technical economic instruments into participatory processes that take account of multiple values and preferences.

7.3.5 Considering disproportionality

Although critical, disproportionality has been poorly defined and implemented. Martin-Ortega et al. (2014) suggest four steps – cost–benefit analysis; consideration of distributional effects; affordability; and wider benefits. Using transdisciplinary valuation of ecosystem services would make visible certain less tangible benefits arising from ecosystem services, potentially changing the results from the cost–benefit analysis, in turn potentially altering the decision outcome. Vlachopoulou *et al.* (2014) claim that quantifying and monetising ecosystem services can shift cost–benefit analyses from 'grey' to 'green' infrastructure solutions. The consideration of 'wider benefits'[3] can be used to judge how different measures can best deliver multiple ecosystem service benefits; or identify where conflicts might occur. These points illustrate how an ecosystem services-based approach could help identify the equity effects of how costs and benefits are distributed. Once the distribution of a wider range of effects is identified, the affordability of the measures for particular sectors or geographic locations can be better assessed.

7.3.6   Monitoring and evaluation

The cyclical nature of the Water Framework Directive, with yearly monitoring of the status of water bodies, provides the foundation for assessing progress over time and against other member states. Monitoring is currently aimed at the state of the environment rather than ecosystem service delivery – the challenge is to measure and predict changes in ecosystem services resulting from improving ecological status. An ecosystem services-based approach expands monitoring from the Directive's parameters to include indicators linking ecological function to ecosystem service and the benefits derived, including monetary values and how ecosystem services are perceived, used, and valued by society. There is no agreed methodology for measuring service flows (Haines-Young & Potschin 2009). Using existing data to characterise current ecosystem services provision may provide a baseline against which to monitor change, but for many services modelling may be required to predict likely changes, and bespoke monitoring required to fully describe the outcomes. Some services are very poorly represented, even with proxies, thus the development of new indicators is a priority (Chan *et al.* 2012). Monitoring the cost-effectiveness, acceptability, and uptake of measures, and whether stakeholder engagement processes have helped integrate multiple forms of knowledge, is also required.

7.4   OPPORTUNITIES AND CHALLENGES

Many countries allowed the implementation of the Water Framework Directive to focus on the technical and organisational challenges of assessing compliance with the Directive (Vlachopoulou

et al. 2014). An ecosystem services-based approach should strengthen the attempt to shift water management from a focus on chemical quality to a more systemic interest in ecology. This should intensify the scientific enquiry stimulated by the Directive itself and help to answer some of the pressing research questions regarding how to understand, monitor, and measure the links between ecosystem status, function, service provision, and tipping points (Nicholson *et al.* 2009). Highlighting linkages between the environment and human wellbeing realigns the Water Framework Directive to the vision of third-phase European Union policies focused on sustainable development, integration, and subsidiarity (Kallis & Butler 2001). An ecosystem services-based approach could reinvigorate the societal focus of the Directive and its reposition within European spatial planning.

An ecosystem services-based approach provides a coherent framework for illustrating and valuing the multiple benefits arising from protecting or restoring the ecological function of water bodies. First, this links the environmental, economic, and social benefits arising from protecting 'natural capital'. Second, frameworks like the Millennium Ecosystem Assessment (2005) and the Convention on Biological Diversity's Malawi Principles (2000) can help implement this approach within a globally agreed focus on human wellbeing underpinned by sustainable use of natural capital. Therefore, taking an ecosystem services-based approach to the Water Framework Directive prevents diverging technical and social approaches by illustrating the dependence of human wellbeing on a healthy aquatic ecosystem. However, there are also challenges.

7.4.1   Challenge: trans-disciplinary decision-making

A systemic approach highlights the need to use decision-support methodologies that take account of complex interactions of often incommensurable entities in decision-making. Furthermore, it is unclear to what extent member states can deviate from the legal stipulations in Article 4 covering derogations, possibly negating any ability to take a wider perspective on objective setting to maximise further benefits. Thus, an ecosystem services-based approach could make the Directive more true to its sustainability ethos, but this requires an acceptance of potential divergence away from strict ecological status targets to wider visions for sustainable catchment management. An ecosystem services-based approach adds complexity and compounds uncertainty, further complicating setting up Europe-wide standards for ecological function (see Hering *et al.* 2010) by adding the relationship between function, service, and benefit arising. However, complex problems also stimulate new dialogue between policy makers, scientists, and stakeholders. Thus an ecosystem services-based approach could generate a more inclusive governance culture that has been lost in the implementation of the Water Framework Directive (Steyeart & Oliver 2007). However, both

---

[3] By wider benefits we mean the effects of measures beyond the immediate water environment. For example, tree planting in buffer strips to mitigate water diffuse pollution may enhance biodiversity and landscapes, and sequester carbon (Borin *et al.* 2010).

the Water Framework Directive and ecosystem services-based approaches can be criticised for their technocratic language, which make engaging the public difficult, and can make decisions more complex and conflicts more visible. While exposing conflicts in a deliberative process may generate opportunities for resolution (Huitema *et al.* 2009), participation should not be seen as an easy or inexpensive process.

7.4.2 Challenge: valuation

One of the most contested, but ultimately useful, aspects is the valuation of ecosystem services. Valuation does not have to mean monetisation, but covers a range of techniques to describe, quantify, and/or rank services in terms of their importance to society to aid decision-making. Valuation clarifies synergies and conflicts in people's views of ecosystem services; shows how overall environmental change is perceived and the environmental costs of human choices and behaviours. There are many potential techniques for environmental, social, and economic valuation of ecosystem services, with different strengths and weaknesses (Gómez-Baggethun *et al.* 2010). There is still contention over how to separate and measure function–service–benefit linkages and the most appropriate methods for characterising these (Maskell *et al.* 2013). Most economic appraisals have no absolute values because they are sensitive to omitted/included data and explicit/implicit assumptions, so trade-offs are further complicated. Finally, some people have a moral aversion to placing monetary values on the ecosystem (Van Hecken & Bastiaensen 2010). As generally with the European Directive (Steyart & Oliver 2007), there is no way to guarantee that the approach will not be subverted, with individual services being picked out and prioritised and the more difficult to measure intangible services being ignored (Rouquette *et al.* 2011).

7.4.3 Challenge: time and space

An ecosystem services-based approach draws attention to time and space through its focus on biophysical function, but cannot escape the problems of temporal and spatial mis-fits between natural and social/institutional processes (Hein *et al.* 2006). It includes a focus on dynamic ecosystems (core element 2), but the timescales in River Basin Management Planning are not long enough for ecosystem recovery (Hering *et al.* 2010). Indeed, an ecosystem services-based approach must consider both current delivery of services and the feedback loops regarding how measures in the River Basin Management Plans might affect service delivery in future. Thus, it might push river basin management processes to use participatory scenarios to consider how different options (measures) might behave under different conditions (e.g. Haines-Young & Potschin 2009). This would be a big step forward given that many member states, e.g. Spain, set

objectives by extrapolating 'business as usual' trends – resulting in targets no longer applicable to current conditions (Gómez-Limón & Martin-Ortega 2013). Using scenarios of possible futures to consider trade-offs can be less confronting than describing current options, and allow more creative solutions to be identified (Frijns *et al.* 2013). Furthermore, scenarios that encourage adaptive management solutions should make River Basin Management Plans more resilient to future changes, particularly climate change.

The challenge is to plan and manage at the geographical scale at which ecosystem services operate. Many ecosystem services are delivered through biophysical processes operating within a catchment (such as sediment transfer); however, some services, such as carbon sequestration, are not represented by catchments. Even surface and groundwater catchments are not necessarily congruent. There are different approaches to assessments of ecosystem service provision (Medcalf *et al.* 2012). Thus, the choice of scale(s) must fit with ecosystem functions, water bodies, and units that are meaningful to stakeholders. Currently, it is possible to implement River Basin Management Planning by addressing the issues pressure by pressure, water body by water body, meaning the cumulative issues are therefore not addressed. Utilising core elements 2 and 4 ensures a more systemic view.

An ecosystem services-based approach could use geographical information systems layers to illustrate the distribution of both service provision and the benefits that flow from them, and to highlight where these are not co-located, in order to link people, places, and the benefits from restoring and protecting water (see, for example, Mulligan *et al.* in this book). However, these maps are only as good as the data available, often relying on proxies and tending to map the most tractable services, not the full range (Raymond *et al.* 2009). Furthermore, Smith *et al.* (2013) have critiqued the focus on mapping within river basin planning for failing to recognise problems with how uncertainty and with multiple (and possibly conflicting) perceptions are made invisible by a single map. Assessment of ecological status is currently gathered at the water body scale, but the governance and reporting is done at a river basin scale. Thus the approach would require scaling up water body information to the catchment level but devolving planning from the basin scale to a local level, while retaining a strategic overview at the level of the member state. Such arrangements add another layer to the existing complex multi-scale governance.

7.4.4 Challenge: resource requirements

The Water Framework Directive has already required complex institutional and attitudinal changes (Meyer & Thiel 2012), resulting in uneven implementation (Liefferink *et al.* 2011). An ecosystem services-based approach would make an already steep learning curve even more demanding and require increased resources, which could be problematic in the current economic climate. This needs resources for the procedural aspects, as locally devolved participatory system assessments need careful design and facilitation to ensure they can be integrated into a common pan-European reporting system. Implementation of core element 4 requires valuation, which is expensive; in addition, benefit transfer mechanisms, while cheaper, remain problematic (Johnston & Duke 2010). Furthermore, additional data sources (and monitoring processes) will be required and the measures identified and agreed still require financing. There is already a mis-fit between the Water Framework Directive and the main source of funding for rural land-based measures stemming from the European Common Agricultural Policy (Somma 2013); taking a wider ecosystem services-based approach focus may strain this relationship even further. However, a counter-argument could be that an ecosystem services-based approach to the Water Framework Directive would better allow money, skills, data, and management processes to be shared in a collaborative spatial planning process (Stead & Meijers 2009); and more decentralisation may stimulate voluntary collective action to protect common pool resources (Carmona-Torres *et al.* 2011). Making ecosystem services visible would help the implementation of 'biodiversity off-setting' within River Basin Management Plans, using developer funds to implement river restoration in priority areas. There is an additional problem of when and how to combine the approach with the Directive – although the objectives of the second cycle are not due until 2021, the foundations for the Second River Basin Management Plans are already being laid in many cases. Thus, it might already be too late to implement an ecosystem services-based approach to river basin planning in the second cycle but steps can be taken to pilot processes for the third cycle.

7.5 CONCLUDING DISCUSSION

This chapter has highlighted aspects of the Water Framework Directive where an ecosystem services-based approach could be best introduced for its third cycle and indicated that this could help to improve our understanding of the benefits for human wellbeing from restoring or protecting the water environment, involve a greater section of society in water management, and focus trans-disciplinary efforts at the catchment scale.

Many of the advantages outlined above were already laid out in the tenets of Integrated Water Resource Management (Cook & Spray 2012; Niasse and Cherlet, this book). Also there are issues that these approaches does not explicitly address, such as scale mis-fits and adaptive management. Thus, an ecosystem services-

---

Box 7.2 Key messages

Ecosystem services-based approaches can:

- help the Water Framework Directive deliver with wider policy imperatives of sustainability, integration and subsidiarity, and live up to its original ambition.
- help illustrate how human wellbeing is dependent on ecological health; widening the focus from good ecological status as an end in itself to showing how it supports societal goals.
- Address some of the problems with Water Framework Directive implementation through:
  - widening and deepening stakeholder engagement;
  - shifting attention from water quality to aquatic ecosystem functions and the benefits these provide to people;
  - enhancing the role of economic instruments (cost-effectiveness analysis and analysis of disproportionality) to align achieving good ecological status with human wellbeing.

Challenges include:

- supporting trans-disciplinary decision-making about a socio-ecological system rather than technical decisions about individual environmental issues;
- placing values (monetary and non-monetary) on the full range of ecosystem services, which is difficult and expensive;
- taking account of different timescales for human and ecological processes, and reconciling the different geographic scales for assessing ecology and reporting on progress;
- providing additional monitoring, methodologies, and mapping, which may be expensive.

It requires implementation of all four of the nested are elements and should not be used to focus on the most easily measured and monitored immediate benefits to society, ignoring the less visible or less immediately relevant factors within the system.

based approach cannot and will not address all the problems with scientific underpinnings of the Water Framework Directive or the practical challenges faced in its operationalisation. Water management will continue to require long-term institutional and societal change, regardless of what frameworks and approaches are employed. However, an ecosystem services-based approach confronts these socio-ecological complexities and stimulates trans-disciplinary dialogue, making these difficulties visible rather than reducing the environment to a technical problem.

Where ecosystem services-based approaches add most value is the nested configuration of the core elements proposed in this book. It draws attention to the primary aim – human wellbeing – and how it is dependent on biophysical underpinnings, such that the Water Framework Directive should set good ecological status within the context of societal preferences for the water environment, but society cannot ignore the need to conserve and protect aquatic natural capital. This adds to the scientific challenges and also helps with the practical aspects of implementation. The Directive's implementation is also helped by recognising how understanding the link between environment and services to people requires wider and deeper trans-disciplinary participation, providing a new impetus for Article 14 (on public participation). In turn, the final nested core element – assessing ecosystem services – helps implementation through improved economic analyses and more comprehensive monitoring processes. None of these changes to the Water Framework Directive are easily achieved, but all will make it live up to its initial promise.

Overall, there are grounds for optimism for an ecosystem services-based approach within river basin planning given that the catchment is particularly well suited to linking ecosystem function with human wellbeing. While there are many benefits, there are also risks and costs arising from adapting a more complex framework, not least when several member states struggled with the existing requirements during the first cycle. Thus, it should be implemented as part of a wider focus on collaborative spatial planning to achieve greater efficiency, efficacy, and equity from the process. Such an approach might guard against the potential for an ecosystem services-based approach to be subverted into a technocratic exercise and sustain the focus on the link between natural capital and societal wellbeing.

ACKNOWLEDGEMENTS

This work has been funded by Scottish Government Rural Affairs and the Environment Portfolio Strategic Research Programme 2011–2016 (Theme 1: Ecosystem Services and Biodiversity). The work was also enriched by the discussions with SEPA on implementing River Basin Management Planning using an ecosystems services approach – see www.crew.ac.uk/call-down/optimising-wfd-delivery-rbmp-using-ecosystem-services-approach and FP7 REFRESH PROJECT: Adaptive Strategies to Mitigate the Impacts of Climate Change on European Freshwater Ecosystems www.refresh.ucl.ac.uk

References

Blackstock, K. L. (2009). Between a rock and a hard place: incompatible objectives at the heart of river basin planning? *Water Science and Technology* **59**(3), 425–431.

Buckley, C., Hynes, S., & Mechan, S. (2012). Supply of an ecosystem service: farmers' willingness to adopt riparian buffer zones in agricultural catchments. *Environmental Science and Policy* **24**, 101–109.

Carmona-Torres, C., Parra-López, C., Groot, J. C. D., *et al*. (2011). Collective action for multi-scale environmental management: achieving landscape policy objectives through cooperation of local resource managers. *Landscape and Urban Planning* **103**(1), 24–33.

Chan, K. M. A., Satterfield, T., & Goldstein, J. (2012). Rethinking ecosystem services to better address and navigate cultural values, *Ecological Economics* **74**, 8–18.

Convention on Biological Diversity (2000). CBD Report of the Workshop on the Ecosystem Approach. UNEP/CBD/COP/4/Inf.9.

Cook, B. R. & Spray, C. J. (2012). Ecosystem services and integrated water resource management: different paths to the same end? *Journal of Environmental Management* **109**, 93–100.

De Stefano, L. & Schmidt, G. (2012). Public participation and water management in the European Union: experiences and lessons learned. In: B. Cosens (ed.), *The Columbia River Treaty Revisited: Transboundary River Governance in the Face of Uncertainty. A Project of the Universities Consortium on Columbia River Governance*. Oregon State University Press, Corvallis, OR.

European Commission (2012). *Report From the Commission to the European Parliament and the Council on the Implementation of the Water Framework Directive (2000/60/Ec) River Basin Management Plans*, European Commission, Brussels. http://ec.europa.eu/environment/water/water-framework/pdf/COM-2012–670_EN.pdf (last accessed 28 June 2013).

Frijns, J., Büscher, C., Segrave, A., *et al*. (2013). Dealing with future challenges: a social learning alliance in the Dutch water sector. *Water Policy* **15**(2), 212–222.

Gilvear, D. J, Spray, C. J., & Casas-Mulet, R. (2013). River rehabilitation for the delivery of multiple ecosystem services at the river network scale. *Journal of Environmental Management* **126**, 30–43.

Gómez-Baggethun, E., De Groot, R. S., Lomas, P. L., *et al*. (2010). The history of ecosystem services in economic theory and practice: from early notions to markets and payment schemes. *Ecological Economics* **69**(6), 1209–1218.

Gómez-Limón, J. A. & Martin-Ortega, J. (2013). The economic analysis in the implementation of the Water-Framework Directive in Spain. *International Journal of River Basin Management* **11**(3), 301–310.

Haines-Young, R. & Potschin, M. (2009). Methodologies for defining and assessing ecosystem services. Nottingham University. Available at: www.nottingham.ac.uk/cem/pdf/JNCC_Review_Final_051109.pdf (last accessed 16 August 2012).

Hein, L., van Koppen, K., de Groot., R. S., *et al*. (2006). Spatial scales, stakeholders and the valuation of ecosystem services. *Ecological Economics* **57**(2), 209–228.

Hering, D., Borja, A., Carstensen, J., *et al*. (2010). The European Water Framework Directive at the age of 10: a critical review of the achievements with recommendations for the future. *Science of the Total Environment* **408**(19), 4007–4019.

Huitema, D., Mostert, E., Egas, W., *et al*. (2009). Adaptive water governance: assessing the institutional prescriptions of adaptive (co-)management from a governance perspective and defining a research agenda. *Ecology and Society* **14**(1).

Johnston, R. J. & Duke, J. M. (2010). Socioeconomic adjustments and choice experiment benefit function transfer: evaluating the common wisdom. *Resource and Energy Economics* **32**, 421–438.

Kallis, G. & Butler, D. (2001). The EU water framework directive: measures and implications. *Water Policy* **3**, 125–142.

Keeler, B. L., Polasky, S., Brauman, K., *et al.* (2012). Linking water quality and well-being for improved assessment and valuation of ecosystem services. *Proceedings of the National Academy of Sciences of the United States of America* **109**(45), 18619–18624.

Koontz, T. M. (2005). We finished the plan, so now what? Impacts of collaborative stakeholder participation on land use policy. *Policy Studies Journal* **33**(3), 459–481.

Liefferink, D., Wiering, M., & Uitenboogaart, Y. (2011). The EU Water Framework Directive: a multi-dimensional analysis of implementation and domestic impact. *Land Use Policy* **28**(4), 712–722.

Martin-Ortega, J. (2012). Economic prescriptions and policy applications in the implementation of the European Water Framework Directive. *Environmental Science and Policy* **24**, 83–91.

Martin-Ortega, J., Skuras, D., Perni, A., Holen, S., & Psaltopoulos, D. (2014). The disproportionality principle in the WFD: how to actually apply it? In: T. Bournaris, J. Berbel, B. Manos, & D. Viaggi (eds), *Economics of Water Management in Agriculture*. Taylor and Francis, Hoboken, NJ.

Maskell, L. C., Crowe, A., Dunbar, M., *et al.* (2013). Exploring the ecological constraints to multiple ecosystem service delivery and biodiversity. *Journal of Applied Ecology* **50**, 561–571.

Medcalf, K., Small, N., Finch, C., *et al.* (2012). *Spatial Framework for Assessing Evidence Needs for Operational Ecosystem Approaches*. JNCC-DEFRA, Peterborough. Available at: http://jncc.defra.gov.uk/page-6241 (last accessed 12 May 2013).

Meyer, C. & Thiel, A. (2012). Institutional change in water management collaboration: implementing the European Water Framework Directive in the German Odra river basin. *Water Policy* **14**(4), 625–646.

Millennium Ecosystem Assessment (2005). *Ecosystems and Human Well-being, Synthesis*. Island Press, Washington, DC.

Newson, M. D. (1992). *Land, Water, and Development: River Basin Systems and their Sustainable Management*. Routledge, London.

Newson, M. D. (2008). *Land, Water and Development: Sustainable and Adaptive Management of Rivers*. Routledge, London.

Nicholson, E., Mace, G. M., Armsworth, P. R., *et al.* (2009). Priority research areas for ecosystem services in a changing world. *Journal of Applied Ecology* **46**(6), 1139–1144.

Ravenscroft, N. & Church, A. (2011). The attitudes of recreational user representatives to pollution reduction and the implementation of the European Water Framework Directive. *Land Use Policy* **28**(1), 167–174.

Raymond, C. M., Bryan, B. A., Macdonald, D. H., *et al.* (2009). Mapping community values for natural capital and ecosystem services. *Ecological Economics* **68**(5), 1301–1315.

Rouquette, J. R., Posthumus, H., Morris, J., *et al.* (2011). Synergies and trade-offs in the management of lowland rural floodplains: an ecosystem services approach. *Hydrological Sciences Journal* **56**(8), 1566–1581.

Sigel, K., Klauer, B., & Pahl-Wostl, C. (2010). Conceptualising uncertainty in environmental decision-making: the example of the EU water framework directive. *Ecological Economics* **69**(3), 502–510.

Smith, H. M., Wall, G., & Blackstock, K. L. (2013). The role of map-based environmental information in supporting integration between river basin planning and spatial planning. *Environmental Science and Policy* **30**, 81–89.

Somma, F. (2013). River Basin Network on Water Framework Directive and agriculture: practical experiences and knowledge exchange in support of the WFD implementation (2010–2012). Reference Report by the Joint Research Centre of the European Commission. Available at: http://publications.jrc.ec.europa.eu/repository/bitstream/111111111/28687/1/lb-na-25978-en-n.pdf (last accessed 21 September 2013).

Spray, C. J. & Blackstock, K. L. (2013). Optimising Water Framework Directive River Basin management Planning Using an Ecosystem Services Approach. Available at: www.crew.ac.uk/publications (last accessed 21 October 2014).

Stead, D. & Meijers, E. (2009). Spatial planning and policy integration: concepts, facilitators and inhibitors. *Planning Theory and Practice* **10**(3), 317–332.

Steyaert, P. & Olivier, G. (2007). The European Water Framework Directive: how ecological assumptions frame technical and social change. *Ecology and Society* **12**(1): 25.

Van Hecken, G. & Bastiaensen, J. (2010). Payments for ecosystem services: justified or not? A political view. *Environmental Science and Policy* **13**(8), 785–792.

van Hoey, G., Borja, A., Birchenough, S., *et al.* (2010). The use of benthic indicators in Europe: from the Water Framework Directive to the Marine Strategy Framework Directive. *Marine Pollution Bulletin* **60**(12), 2187–2196.

Vlachopoulou, M., Coughlin, D., Forrow, D., *et al.* (2014). The potential of using the ecosystem approach in the implementation of the EU Water Framework Directive. *Science of the Total Environment* **470–471**, 648–694.

Volk, M., Liersch, S., & Schmidt, G. (2009). Towards the implementation of the European Water Framework Directive? Lessons learned from water quality simulations in an agricultural watershed. *Land Use Policy* **26**(3), 580–588.

Volker, S. & Kistemann, T. (2011). The impact of blue space on human health and well-being: Salutogenetic health effects of inland surface waters: a review. *International Journal of Hygiene and Environmental Health* **214**(6), 449–460.

Wunder, S., Engel, S., & Pagiola, S. (2008). Taking stock: a comparative analysis of payments for environmental services programs in developed and developing countries. *Ecological Economics* **65**(4), 834–852.

8 How useful to biodiversity conservation are ecosystem services-based approaches?

Craig Leisher

8.1 INTRODUCTION

De Groot (1992), Daily (1997), and Costanza *et al.* (1997) were seminal publications in developing an approach based on ecosystem services that has come to dominate biodiversity conservation (see Chapter 2 for an overview of the term's evolution). The thinking these authors helped to pioneer has become a success by many measures. There are now more than 100 universities with programmes focused on ecosystem services research, and in 2013 in Europe alone, 28 universities offered Master's degrees in Ecosystem Services (www.mastersportal.eu). The Nature Conservancy, World Wildlife Fund, Conservation International, Wildlife Conservation Society, International Union for the Conservation of Nature, and Fauna and Flora International all have ecosystem services programmes and projects.

The conservation world has a history of embracing new approaches that later prove to have fewer benefits than anticipated. Initial exuberance for an approach is often followed by a reassessment of the approach. Integrated conservation and development projects (Sanjayan & Shen 1997 versus McShane & Wells 2004), biodiversity hotspots (Myers *et al.* 2000 versus Kareiva & Marvier 2003), and a landscape-level approach (Noss 1983 versus Sayer *et al.* 2013) are several examples. Of the 21 approaches in use by conservation organizations in 2002 (Redford *et al.* 2003), only half are still prominently mentioned as approaches on the websites of the same conservation organizations that used them a decade ago (Box 8.1). Conservation is a fundamentally optimistic endeavour that welcomes big, new ideas. Ecosystem services are perhaps the biggest of the current new ideas in conservation.

The attractiveness of ecosystem services-based approaches comes, *inter alia*, from their potential value to conservation; the approaches highlight the often-overlooked goods and services that nature provides and helps inform environmental decision-making (Fisher *et al.* 2009). Ecosystem services-based approaches make the 'invisible' benefits of nature more visible and make the link between the wellbeing of nature and the wellbeing of people explicit (core element 1 in Chapter 2). They have the added benefit of offering more egalitarian alternatives to the criticized 'fortress conservation' approaches (Brockington

2002). The underlying assumption is that identifying and valuing an ecosystem service will make it more likely to be protected (core element 4).

As in Chapter 2, ecosystem services are defined here as per the Millennium Ecosystem Assessment (Millennium Ecosystem Assessment 2005), and biodiversity as per Article 2 of the Convention on Biological Diversity (1992). Ecosystem services themselves can be presented as a framework (e.g. Tallis *et al.* 2008), a concept (e.g. Farber *et al.* 2002), or even a paradigm (e.g. Gaodi *et al.* 2006), but here it is presented as a series of approaches (Turner & Daily 2008; Redford & Adams 2009) in keeping with the terminology of this book. Using 'approaches' has the benefit of inclusiveness and helps avoid tangential definitional issues.

A decade of ecosystem services experience has led to a growing number of critiques. Echoing a theme from the early days of conservation and the Muir-Pinchot debates (Leisher & Sanjayan 2013), Gómez-Baggethun *et al.* 2010), note that ecosystem services-based approaches often focus on monetary values while ignoring the non-monetary values derived from nature and that market-based policy design may attract political support for conservation, but the reliance on market forces to address environmental issues is highly problematic. Kosoy and Corbera (2010) argue that assigning a single value to an ecosystem service ignores the many potential values that can be attributed to a particular ecosystem service and reflects an oversimplification of ecological processes. Opdam (2013) points out that ecosystem services research does not generally provide the science needed for community-based landscape planning and conservation. Redford and Adams (2009) note that introduced species may be detrimental to biodiversity yet provide a particular ecosystem service better than native species, such as the role of zebra mussels (*Dreissena polymorpha*) in water filtration, and that the maximization of a single ecosystem service, such as carbon storage, can result in a simplified and less-resilient ecosystem.

In short, the antithesis to the ecosystem services thesis is beginning to emerge, and this chapter contributes towards this by asking how strong the links are between widely recognized ecosystem services and biodiversity. How are current efforts to

Box 8.1 Status of 2002 conservation organizations' main approaches 12 years later

Organization	Main approaches in 2002	Status in 2014
African Wildlife Foundation (AWF)	Heartland selection and heartland conservation planning	Changed
BirdLife International	Endemic-bird areas	Same
Conservation International (CI)	Biodiversity hotspots, major tropical wilderness areas, and designing sustainable landscapes	Changed
Natural England	Natural areas	Same
European Commission (Environment Directorate General)	Natura 2000	Same
Convention on Biological Diversity (CBD)	Ecosystem approach	Same
The Nature Conservancy (TNC)	Ecoregional conservation planning and site conservation planning	Changed
Ramsar Convention	Wetlands of international importance	Same
Wildlife Conservation Society (WCS)	Last of the wild, range-wide priority setting, and landscape-species approach	Changed
International Union for the Conservation of Nature (IUCN) Forest Conservation Programme	Landscape approach	Changed
World Wildlife Fund (WWF)	Global 200 Ecoregions and ecoregion conservation	Changed
World Resources Institute (WRI)	Global Forest Watch	Same

Sources: 2002 approaches from Redford *et al.* (2003) and 2014 status from organizational websites.

conserve ecosystem services impacted by social factors, and how willing and able are governments to prioritize ecosystem service protection?

This chapter could be viewed as discounting ecosystem services' contribution to biodiversity conservation. It is not, however, a 'thumbs down' to ecosystem services, but rather a 'heads up'. Its aim is a more pragmatic view of ecosystem service-based approaches within biodiversity conservation.

8.2 ABILITY TO DELINK ECOSYSTEM SERVICES AND BIODIVERSITY

Among the Millennium Ecosystem Assessment's (2005) four ecosystem service categories, supporting and cultural services are critical to many people's lives, yet provisioning and regulating services are more commonly the focus of work within biodiversity conservation and hence the focus here. While this categorization of ecosystem services has been criticized (Ojea *et al.* 2012), much of the ecosystem services literature makes use of these categories, and thus it provides the categorical framework here.

In a review of 238 ecosystem services studies, Cimon-Morin *et al.* (2013) find that provisioning services such as food, fibre, fodder, and fuel are weakly correlated with biodiversity. Regarding core element 2 of ecosystem services-based approaches, many biophysical underpinnings of ecosystem

service delivery have been modified. This is not surprising; since the advent of agriculture more than 10 000 years ago, humans have been reducing local biodiversity to maximize provisioning services. Just three plant species – rice (*Oryza sativa*), maize (*Zea mays*) and wheat (*Triticum* spp.) – provide the majority of calories consumed by people (FAO 1995); in 2010, 81% of global aquaculture production by weight came from carp, shellfish, shrimp, catfish, and salmon (FAO 2012). Within agriculture and aquaculture, ecosystems have been simplified and biodiversity has been largely decoupled from the near-term ecosystem service of food provisioning.

The provisioning of clean water also appears to have minimal dependence on biodiversity. The links between biodiversity and ecosystem services in groundwater have been little studied and subterranean biodiversity may yet prove to have a role in providing clean groundwater (Boulton *et al.* 2008), but in a review of 35 wetlands constructed to filter effluents and absorb excess nutrients, the average number of plant species used for filtration was three (Brisson & Chazarenc 2009). A more recent review of approximately 1700 ecosystem-services studies concluded that while there was evidence of links between several ecosystem services and biodiversity, there was minimal evidence linking freshwater purification and biodiversity (Cardinale *et al.* 2012). This is also in line with what was found by Febria *et al.* in this book.

Thus for food and water, the link between preserving the full range of biodiversity in a given location and the provisioning services on which most people depend is dubious. In other words, for most of the seven billion people on the planet, there is no direct link between biodiversity conservation and the food and water near-term provisioning services that sustain their lives (core element 1). Ecosystem services-based approaches to biodiversity conservation could be seen as largely irrelevant.

Within regulating services, three of the more commonly targeted ecosystem services are: water flow, carbon sequestration, and soil retention. Here, the picture becomes more nuanced. Deforestation can increase total water flow in a watershed, but the variations in water flow become greater when a forest's water-regulating function is lost (Andréassian 2004). A more diverse forest may intercept more rainfall and regulate water flow better than a simplified forest such as a plantation (Andréassian 2004), but grasslands and shrublands will yield greater total annual water flow – all else being equal – than a forest in the same location (Farley et al. 2005). Thus, water flow does not appear to be linked to the biodiversity of an area per se, but to the land-cover type and the non-biological elements of the local context.

For carbon sequestration, planting fast-growing exotic tree species generally results in greater carbon uptake than planting native tree species. Eucalyptus plantations, for example, are fast growing but have among the lowest biodiversity of any primary, secondary, or plantation forest (Barlow et al. 2007) and have been shown to be detrimental to local biodiversity (Zhao et al. 2007). Thus, the regulating services of carbon sequestration via reforestation or afforestation do not necessarily have a connection to biodiversity conservation.

Soil retention is another ecosystem service that is vital to human wellbeing because it helps maintain soil fertility. Yet soil retention has minimal links with biodiversity. A study in China, for example, found that planting an area with tea resulted in a greater reduction in soil erosion than reforesting with pine, fir, or natural secondary forest (Zheng et al. 2008). From terracing to contour ploughing, soil retention services can be provided with nominal biodiversity.

Unfortunately for conservation and conservation organizations, the level of biodiversity required to provide a number of essential ecosystem services is often low. The functional diversity of organisms rather than the biological diversity may be the most significant feature explaining the optimal provisioning of an ecosystem service (Cimon-Morin et al. 2013).

8.3 SOCIAL CONSIDERATIONS

Ecosystem services are fundamentally about people (core element 1). It can be argued that where there are no people, there are no ecosystem services because there is no one to 'serve'. Biodiversity conservation is increasingly about people, and

Conservation International, The Nature Conservancy, and several other conservation organizations have changed their mission statements in recent years to explicitly include people. This change is a tacit acknowledgement that social considerations are vital to biodiversity conservation success (Mascia 2003). Hence, it is no surprise that conservation organizations have adopted ecosystem service-based approaches.

Yet, for at least one common tool for protecting ecosystem services, social considerations appear to limit benefits to biodiversity. Payments for ecosystem services schemes have been widely used in Latin America, China, and Vietnam (Martin-Ortega et al. 2013; Wunder 2013), and many of the larger conservation organizations have payment for ecosystem services projects ongoing. In a payment for ecosystem services scheme, participation is voluntary (Wunder 2005). This allows, in theory, for sellers to opt in or opt out, depending on their preferences.

For a payment for ecosystem services arrangement protecting a watershed, a withdrawal of one or more service sellers may compromise the overall ability to provide the service. A seller, for instance, withdrawing a large area of land and changing the land use practices to something detrimental to clean-water provisioning could negate the service provisioning of the other sellers. Moreover, opportunity costs may shift suddenly as new economic opportunities arrive. If the price of milk or beef increases, for example, payments for not grazing riparian areas may no longer be enough, and sellers may opt out.

A payment for ecosystem services also requires long-term trust between buyers and sellers (Wunder 2013) that can be easily lost. If the buyer, for instance, is a government agency, funding priorities can shift with a change of government, and if payments stop, long-term trust is lost.

The voluntary nature of a payment for ecosystem services arrangement, the changes in opportunity costs over time, and the need for long-term trust between buyers and sellers, combine to make a payment for ecosystem services scheme unstable economically and socially. Hence, one of the primary tools of ecosystem services-based approaches may be less useful to biodiversity conservation than it seemed at first.

8.4 FEW GOVERNMENT DECISION-MAKERS WILLING AND ABLE TO PRIORITIZE ECOSYSTEM SERVICE PROTECTION

In relation to core element 4 (assessment for decision-making), while businesses and individuals are important (see Houdet et al. and Corral-Verdugo et al. in this book), governments play perhaps the key role in protecting ecosystem services. Ecosystem services often come from a large area (e.g., a watershed) and require long-term, proactive protection of the service source. As a result, it largely falls to governments to protect ecosystem services directly or create the regulatory structure and policy incentives to protect ecosystem services indirectly.

An assumption implicit in ecosystem services is that the identification and valuation of the services will result in better decision-making by governments about how ecosystems are used and managed (Martin-Ortega et al., this book; TEEB 2010). Yet this assumption may not always hold; there is a history of government decision-makers willingly trading one ecosystem service for another. Hydropower dams, for example, often trade the ecosystem service of power production for the ecosystem service of migratory fish catches. During the Cold War, the Soviets abstracted large volumes of water from the rivers that fed the Aral Sea in order to boost cotton production, and decision-makers knowingly traded the fish catches in the Aral Sea for self-sufficiency in cotton (Edelstein 2012). Better information on the value of ecosystem services may not necessarily result in greater protection of the services, and government decision-makers are not always willing to prioritize the protection of an ecosystem service over economic or strategic interests.

Inertia plays a role as well. People may be unwilling to change how they manage a provisioning or regulating service so long as the service is continuing to provide benefits, even though these are much-diminished. There are often multiple possible causes for a decline in a provisioning or regulating ecosystem service, and there may be little agreement on how to address the issue. Policies to address climate change, for example, have been hampered by polarized public opinions about the status quo (Kahan et al. 2012). In fisheries, there is evidence that local fishers will change their practices only once there is a widely perceived crisis in the fishery (Pollnac et al. 2001). This suggests that the status quo for an ecosystem service may be 'sticky' unless there is a crisis.

Among lists of ecosystem services (e.g. Costanza et al. 1997; Daily 1997), most services provide wellbeing benefits to people indirectly rather than directly. Where time-horizons are short and discount rates high – as they are in many developing countries – decision-makers may undervalue these indirect benefits because the cause and effect is less evident. From the clearance of mangroves to the logging of forests, there is a history of indirect ecosystem services being appreciated only when they are lost. During China's Great Leap Forward, for example, up to 50% of the trees were cut down in parts of China in a policy to create more arable land (Dikötter 2010). This led to severe soil erosion, a drop in crop production, and eventually government incentives to replant the trees (Yu et al. 2011).

Perhaps better upfront information about the value of indirect ecosystem services would result in fewer decision-makers overlooking them, but this has not been the case in China; the ecosystem services literature on China is extensive, and a

centralized governance structure facilitates rapid policy change, yet China remains one of the more environmentally blighted nations of the world (Watts 2010).

In short, the identification and valuation of ecosystem services (core element 4) may be insufficient to protect the services due to overriding economic and strategic interests, inertia to maintain the status quo, and myopia towards indirect benefits (see also Febria et al., this book). Thus, there may be fewer government decision-makers willing to pay for the protection of an ecosystem service than advocates might hope.

There may also be a limited ability to pay. The provisioning of clean water is one of the most important ecosystem services for most people's lives. More than half the people in the world live in urban areas (UN 2012), and in 2005 there were 590 cities in the world with more than 750,000 inhabitants (Ahlenius 2010). A recent study of the water sources for cities with greater than 750,000 inhabitants found that 80% of them depend primarily on surface water for their municipal water supply (McDonald et al. 2014). This creates the potential for a large number of watershed-focused payments for ecosystem services initiatives. Yet most urban water systems in Africa, Latin America, and South Asia do not recover even their operation and maintenance costs (Foster & Yepes 2006; Ying et al. 2010; Water Sanitation Program 2011). The cost of urban households' water supply is often subsidized by municipal or national governments (Ying et al. 2010), so unless subsidy transfers are increased or water tariffs raised, many municipal water authorities do not have the ability to pay for the ecosystem service of watershed protection. As discussed in detail by Turner in this book, water is also viewed by many governments as a basic right. (In 2010, 122 countries adopted UN General Assembly Resolution A/64/292 making water a basic human right.) In developing countries with high levels of poverty – where most biodiversity is located – increases in water tariffs to fund ecosystem service protection may be difficult to square with water as a basic right.

The benefit to biodiversity conservation from more governments prioritizing the protection of ecosystem services is likely to be underwhelming because too few government decision-makers are willing and able to prioritize ecosystem service protection. This calls into question the premise that a focus on ecosystem services will result in greater political will or greater resources for reducing the decline in biodiversity.

8.5 CONCLUSIONS

The value to biodiversity conservation of ecosystem services-based approaches may have been oversold. The food, fibre, fodder, and fuel provisioning services upon which people depend most directly are only weakly correlated with biodiversity. Voluntary payments for ecosystem services may be hampered

Box 8.2 Key messages

- Agencies and organizations focused on biodiversity conservation should use ecosystem services-based approaches primarily as an information and advocacy tool.

- Agencies and organizations focused on biodiversity conservation should avoid a focus on the provisioning and regulating ecosystem services that have weak correlations with biodiversity.

- Payments for ecosystem service initiatives should build-in frequent checks of sellers' perceptions of net benefits. Where possible, long-term, binding agreements between buyers and sellers should be used, and where the local legal system allows, consider the use of permanent conservation easements.

- Capitalize on a crisis to change how a natural resource is managed.

- For payments for watershed services initiatives, limit investments to where water treatment costs can be reduced substantially by better watershed management, and focus on the worst pollution sources first.

by inherent instability. The willingness and ability of governments to prioritize the protection of ecosystem services may be limited. Overall, the logic chain of how ecosystem services-based approaches could lead to greater biodiversity conservation is weak.

With the global population expected to grow by approximately 50% in the next 40 years (UN 2012), the imperative for biodiversity conservation is to provide people with compelling reasons for protecting nature. Ecosystem services-based approaches contribute towards this imperative. Such approaches make people more aware of the goods and services that nature provides; they give a framework for thinking about people's dependence on nature and to a lesser degree biodiversity; they help decision-makers to make better-informed decisions about trade-offs among ecosystem services. In short, they are not without merit. But ecosystem services-based approaches are unlikely to generate substantial new political will or resources for biodiversity conservation.

References

Ahlenius, H. (2010). Nordpil and UN Population Division and World Urbanization Prospects, 2007 Revision. http://nordpil.com/go/resources/world-database-of-large-cities (last accessed May 2014).

Andréassian, V. (2004). Waters and forests: from historical controversy to scientific debate. *Journal of Hydrology* **291**, 1–27.

Barlow, J., Gardner, T. A., Araujo, I. S., *et al.* (2007). Quantifying the biodiversity value of tropical primary, secondary, and plantation forests. *Proceedings of the National Academy of Sciences* **104**, 18555–18560.

Boulton, A. J., Fenwick, G. D., Hancock, P. J., *et al.* (2008). Biodiversity, functional roles and ecosystem services of groundwater invertebrates. *Invertebrate Systematics* **22**, 103–116.

Brisson, J. & Chazarenc, F. (2009). Maximizing pollutant removal in constructed wetlands: should we pay more attention to macrophyte species selection? *Science of the Total Environment* **407**, 3923–3930.

Brockington, D. (2002). *Fortress Conservation: The Preservation of the Mkomazi Game Reserve, Tanzania.* Indiana University Press, Bloomington, IN.

Cardinale, B. J., Duffy, J. E., Gonzalez, A., *et al.* (2012). Biodiversity loss and its impact on humanity. *Nature* **486**, 59–67.

Cimon-Morin, J., Darveau, M., & Poulin, M. (2013). Fostering synergies between ecosystem services and biodiversity in conservation planning: a review. *Biological Conservation* **166**, 144–154.

Convention on Biological Diversity (1992). Montreal: UN Secretariat of the Convention on Biological Diversity. www.cbd.int/doc/legal/cbd-en.pdf (last accessed May 2014).

Costanza, R., d'Arge, R., De Groot, R., *et al.* (1997). The value of the world's ecosystem services and natural capital. *Nature* **387**, 253–260.

Daily, G. C. (ed.) (1997). *Nature's Services: Societal Dependence on Natural Ecosystems.* Island Press, Washington, DC.

De Groot, R. D. (1992). *Functions of Nature: Evaluation of Nature in Environmental Planning, Management and Decision Making.* Wolters-Noordhoff BV, Wageningen.

Dikötter, F. (2010). *Mao's Great Famine: The History of China's Most Devastating Catastrophe, 1958–1962.* Bloomsbury Publishing, New York.

Edelstein, M. R. (2012). Disaster by design: the multiple caused catastrophes of the Aral Sea. *Research in Social Problems and Public Policy* **20**, 105–151.

FAO (1995). *Dimensions of Need: An Atlas of Food and Agriculture.* FAO, Rome.

FAO (2012). *The State of the World Fisheries and Aquaculture.* FAO, Rome.

Farber, S. C., Costanza, R., & Wilson, M. A. (2002). Economic and ecological concepts for valuing ecosystem services. *Ecological Economics* **41**, 375–392.

Farley, K. A., Jobbágy, E. G., & Jackson, R. B. (2005). Effects of afforestation on water yield: a global synthesis with implications for policy. *Global Change Biology* **11**, 1565–1576.

Fisher, B., Turner, R. K., & Morling, P. (2009). Defining and classifying ecosystem services for decision making. *Ecological Economics* **68**, 643–653.

Foster, V. & Yepes, T. (2006). *Is Cost Recovery a Feasible Objective for Water and Electricity? The Latin American Experience.* World Bank, Washington, DC.

Gaodi, X., Yu, X., & Chunxia, L. (2006). Study on ecosystem services: progress, limitation and basic paradigm. *Acta Phytoecological Sinica* **30**, 191.

Gómez-Baggethun, E., De Groot, R., Lomas, P. L., *et al.* (2010). The history of ecosystem services in economic theory and practice: from early notions to markets and payment schemes. *Ecological Economics* **69**, 1209–1218.

Kahan, D. M., Peters, E., Wittlin, M., *et al.* (2012). The polarizing impact of science literacy and numeracy on perceived climate change risks. *Nature Climate Change* **2**, 732–735.

Kareiva, P. & Marvier, M. (2003). Conserving biodiversity coldspots. *American Scientist* **91**, 344–351.

Kosoy, N. & Corbera, E. (2010). Payments for ecosystem services as commodity fetishism. *Ecological Economics* **69**, 1228–1236.

Leisher, C. & Sanjayan, M. (2013). Conservation and the world's poorest of the poor. In: *Encyclopaedia of Biodiversity*, 2nd edn, Vol. **2**, ed. S. A. Levin. Academic Press, Waltham, MA.

Martin-Ortega, J., Ojea, E., & Roux, C. (2013). Payments for water ecosystem services in Latin America: a literature review and conceptual model. *Ecosystem Services* **6**, 122–132.

Mascia, M. B., Brosius, J. P., Dobson, T. A., *et al.* (2003). Conservation and the social sciences. *Conservation Biology* **17**, 649–650.

McDonald, R., Weber, K., Padowski, J., *et al.* (2014). Water on an urban planet: urbanization and the reach of urban water infrastructure. *Global Environmental Change* **27**, 96–105.

McShane, T. O. & Wells, M. P. (eds) (2004). *Getting Biodiversity Projects to Work: Towards More Effective Conservation and Development.* Columbia University Press, New York.

Millennium Ecosystem Assessment (2005). *Ecosystems and Human Well-Being: Synthesis.* Island Press, Washington, DC.

Myers, N., Mittermeier, R. A., Mittermeier, C. G., Da Fonseca, G. A., & Kent, J. (2000). Biodiversity hotspots for conservation priorities. *Nature* **403**, 853–858.

Noss, R. F. (1983). A regional landscape approach to maintain diversity. *BioScience* **33**, 700–706.

Ojea, E., Martin-Ortega, J., & Chiabai, A. (2012). Defining and classifying ecosystem services for economic valuation: the case of forest water services. *Environmental Science & Policy* **19**, 1–15.

Opdam, P. (2013). Using ecosystem services in community-based landscape planning: science is not ready to deliver. In: B. Fu and K. B. Jones (eds), *Landscape Ecology for Sustainable Environment and Culture*. Springer, Dordrecht.

Pollnac, R. B., Crawford, B. R., & Gorospe, M. L. (2001). Discovering factors that influence the success of community-based marine protected areas in the Visayas, Philippines. *Ocean and Coastal Management* **44**, 683–710.

Redford, K. H. & Adams, W. M. (2009). Payment for ecosystem services and the challenge of saving nature. *Conservation Biology* **23**, 785–787.

Redford, K. H., Coppolillo, P., Sanderson, E. W., *et al.* (2003). Mapping the conservation landscape. *Conservation Biology* **17**, 116–131.

Sanjayan, M. & Shen, S. (1997). *Experiences with Integrated Conservation Development Projects in Asia*. World Bank, Washington, DC.

Sayer, J., Sunderland, T., Ghazoul, J., *et al.* (2013). Ten principles for a landscape approach to reconciling agriculture, conservation, and other competing land uses. *Proceedings of the National Academy of Sciences* **110**, 8349–8356.

Tallis, H., Kareiva, P., Marvier, M., *et al.* (2008). An ecosystem services framework to support both practical conservation and economic development. *Proceedings of the National Academy of Sciences* **105**, 9457–9464.

The Economics of Ecosystems and Biodiversity (TEEB) (2010). *Mainstreaming the Economics of Nature: A Synthesis of the Approach, Conclusions and Recommendations of TEEB*. TEEB, Nagoya.

Turner, R. K. & Daily, G. C. (2008). The ecosystem services framework and natural capital conservation. *Environmental and Resource Economics* **39**, 25–35.

UN (2012). *World Urbanization Prospects, the 2011 Revision: Highlights*. UN Department of Economic and Social Affairs, Population Division, New York.

Water Sanitation Program (2011). *Cost Recovery in Urban Water Services: Select Experiences in Indian Cities*. World Bank, Washington, DC.

Watts, J. (2010). *When a Billion Chinese Jump: How China Will Save Mankind – or Destroy it*. Scribner, New York.

Wunder, S. (2005). *Payments for Environmental Services: Some Nuts and Bolts*. CIFOR, Jakarta.

Wunder, S. (2013). When payments for environmental services will work for conservation. *Conservation Letters* **6**, 230–237.

Ying, Y., Skilling, H., Banerjee, S., *et al.* (2010). *Cost Recovery, Equity and Efficiency in Water Tariffs*. World Bank, Washington, DC.

Yu, D. Y., Shi, P. J., Han, G. Y., *et al.* (2011). Forest ecosystem restoration due to a national conservation plan in China. *Ecological Engineering* **37**, 1387–1397.

Zhao, Y. H., Yang, Y. M., Yang, S. Y., *et al.* (2007). A review of the biodiversity in *Eucalyptus* plantation. *Journal of Yunnan Agricultural University* **22**, 741.

Zheng, H., Chen, F., Ouyang, Z., *et al.* (2008). Impacts of reforestation approaches on runoff control in the hilly red soil region of Southern China. *Journal of Hydrology* **356**, 174–184.

Part III
Assessing water ecosystem services

9 The first United Kingdom's National Ecosystem Assessment and beyond

Marije Schaafsma, Silvia Ferrini, Amii R. Harwood, and Ian J. Bateman

9.1 INTRODUCTION

As water and land resources become scarcer, further conflicting demands of different uses and users will arise (Vörösmarty *et al.* 2000). Sustainable management is required to secure water resources for future generations. Ecosystem services-based approaches aim to ensure that the values of a broad range of benefits to humanity that are provided by our natural environment are accounted for in policy making, in order to foster sustainable development (Chapter 2). National-level incorporation of sustainable development goals has propelled interest in large-scale assessments of ecosystem services which can help address complex problems of ecosystem change (Bateman *et al.* 2013).

The central question of this chapter is whether large-scale ecosystem services-based approaches provide an opportunity for improving water management. The UK National Ecosystem Assessment (UK-NEA) was the first analysis of the societal benefits of the UK natural environment (UK-NEA 2011a). Moreover, it was one of the leading initiatives worldwide to assess ecosystem services at national level after the Millennium Ecosystem Assessment (Millennium Ecosystem Assessment 2005) produced a global assessment. The first phase of the UK-NEA provided a wealth of policy-relevant information, and we use it here as a case study.

UK rivers, lakes, and ponds make up around 250 000 hectares (1.1%) of the UK total surface area. These surface waters, together with unseen groundwater systems, contribute significant ecosystem services and goods to human wellbeing in the UK. The quality of UK freshwaters has improved over the last 50 years following direct regulatory interventions in rural and agricultural practices and EU Directives, such as the Water Framework Directive (Watson 2012). These policies have led to a reduction of point and diffuse chemical pollution and improved ecological conditions. Nonetheless, pressures from agricultural, industrial, and domestic use on water resources remains high, both in terms of quality and quantity (Watson 2012). Agricultural practices and landscape modifications, such

as use of fertilisers, habitat fragmentation, and degradation, reduce the ecosystem service provision and resulting human benefits. Under the Water Framework Directive, which commits member states to acquire good ecological status of water bodies by 2015, only 26% of rivers and 36% of lakes in England and Wales presently meet or exceed this target status. The supply of water from most natural habitats is decreasing (e.g. driven by urban expansion) and continued population growth will put increasing pressure on these water resources (UK-NEA 2011a).

We first introduce the conceptual framework underpinning the UK-NEA and highlight differences between ecosystem services-based water management approaches and traditional ones. Next, we provide an application of the UK-NEA framework to water-related ecosystem services. As an example of the UK-NEA 'at work', an assessment is included of the non-market values of recreation to water bodies under two contrasting scenarios. We discuss the impact of the UK-NEA on UK water-related policies and we end the chapter shortlisting some of the main challenges for integral management of water and other ecosystem services.

9.2 ECOSYSTEM SERVICE ASSESSMENT UNDER THE MILLENNIUM ASSESSMENT AND THE UK-NEA

9.2.1 UK-NEA conceptual framework

The conceptual framework of the UK-NEA builds upon the circular relationship between human societies (their actions and their wellbeing) and the environment and its ecosystem services provision, in line with core element 1 outlined in Chapter 2. The UK-NEA makes an explicit distinction between ecosystems processes and functions, and intermediate services which underpin the final ecosystem goods that are of human benefit (e.g. Fisher & Turner 2008). In turn, the wellbeing we derive from ecosystems, together with drivers such as changes in policy regimes, social institutions, and demographics, affects

---

Box 9.1 Key facts on UK-NEA

Start: mid May 2009

Findings published: June 2011

Impact: profound influence on 'The natural choice: securing the value of nature', the most fundamental overhaul of government policy regarding the English natural environment for 20 years.

Researchers: more than 500 natural, economic, and social scientists. 'The UK NEA was an inclusive process; many government, academic, NGO and private sector institutions helped to design the assessment, contribute information and analyses, review the preliminary findings, and promote the results' (UK-NEA website, http://uknea.unep-wcmc.org).

Core funders: The UK Department for Environment, Food and Rural Affairs (Defra), the devolved administrations of Scotland, Wales, and Northern Ireland, the Natural Environment Research Council (NERC), and the Economic and Social Research Council (ESRC). The economic analysis was part-funded by the Social and Environmental Economic Research (SEER) project (ESRC Funder Ref: RES-060-25-0063).

---

the way we manage the environment and its potential to deliver valuable services.

Figure 9.1 shows how the UK-NEA framework can be applied to water-related ecosystem services. Physical conditions affect the core ecosystem processes, such as nutrient and water cycling, or soil formation. These ecosystem processes drive various ecosystem functions, such as nutrient uptake by plants, but also aquifer recharge and storage of flood water that regulate water levels and thereby, among others, carbon storage in soils. The final ecosystem services that are of human benefit include better water quality and quantity, but also reductions in flood risk and climate regulation. The 'good(s)' that people derive from these services range from basic needs, e.g. drinking water and food, to safety in terms of flood control and longer-term effects of climatic conditions. In many cases, final ecosystem services have to be combined with other resource inputs, such as manufactured or human capital, to generate valuable goods. While some of the water-related final services are incorporated in commercial goods or services (e.g. the whisky industry heavily relies on water), other important benefits are non-marketed, with recreation and tourism benefits as an important example.

The new classification into final goods and intermediate services is also adopted in the UK-NEA. The classification

recognises the complexity of ecosystems as it highlights the interactions and dependencies between ecosystem structures, processes, functions, and services. The main advantage of a focus on final outputs of ecosystem services is that it avoids double counting of benefits of ecosystem services that have both intermediate and final states, and thereby helps to avoid excessively high costs or benefits. Attention to double counting is of particular relevance for the valuation of water-related services (Fisher *et al.* 2008). A typical example of double counting is when nutrient retention (to improve water quality) integrally supports biodiversity conditions. Including both the value of biodiversity and of nutrient retention in benefit estimation would lead to overestimation of the welfare impact.

The UK-NEA is oriented around eight different habitats, including freshwaters. These broad habitat types capture the thematic diversity of the UK's natural environment (Jackson 2000). Mapping enables the spatial diversity of these habitats to be captured. A spatially explicit approach of the analysis of ecosystem services and benefits is one of the key characteristics of the UK-NEA (Bateman *et al.* 2011).

The spatial aspects are reflected in biophysical models of ecosystem stocks and service provision, as well as in the economic models that underpin the benefits attached to these services. Moreover, the scenario analysis, which outlines the outcomes related to different policy interventions (Haines-Young *et al.* 2010), is subsequently related to corresponding maps of land use changes with associated welfare changes (Bateman *et al.* 2013). This spatially explicit approach demonstrates where costs and benefits of interventions are expected to occur, where policies can achieve trade-offs and synergies between ecosystem services and may help to define areas where improving ecosystem conditions would have the highest net benefits for society.

The UK-NEA framework aims to raise awareness of the relationship between habitats, water quality and quantity, and goods; it stresses the necessity of better understanding of the biophysical underpinnings of ecosystem functions and service delivery (core element 2, Chapter 2). However, the understanding of the links between ecosystem structure, functioning, habitat type, location, and size (and related issues of fragmentation) is far from complete (Maltby *et al.* 2011). Figure 9.1 is only a first attempt to sketch ecosystem links and interactions, and the arrows are by no means intended to indicate linear relationships. Links may be non-linear or bi-directional, with final ecosystem services influencing ecosystem functions. Interactions of freshwater characteristics, land types, and temporal hydrological dynamics define ecosystem services that are highly spatially heterogeneous.

The UK-NEA identified many knowledge gaps and highlighted the uncertainty about how changes in ecosystems affect

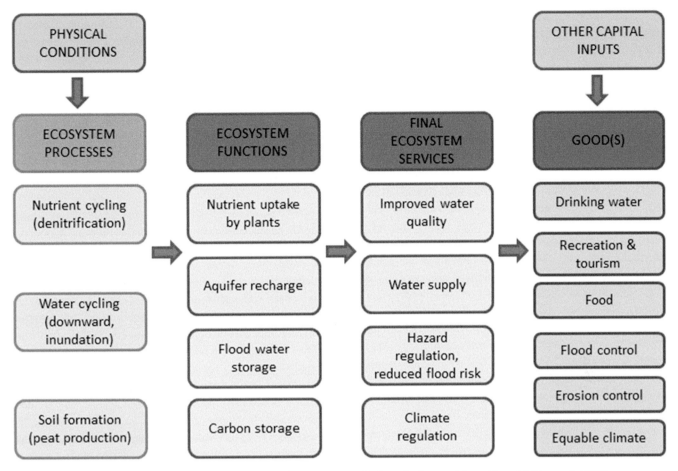

Figure 9.1 Schematic diagram of ecosystem assessment, linking processes to final goods. Source: Adapted from Maltby *et al.* (2011).

wellbeing, which makes it difficult to operationalise ecosystem services-based approaches. For example, there are no hydrological models for the quantification of ecosystem service delivery that fully capture these relationships. Similarly, the value of ecosystems is not fixed, and benefits depend on timing and location of ecosystem service delivery, on the relationship between water quality and quantity, on other ecosystem services related to water and finally on stakeholders' preferences. Moreover, ecosystems and service provision may be vulnerable to 'regime' shifts (Maltby *et al.* 2011). Once a shift occurs, large losses of ecosystem services may occur which may be irreversible or difficult to restore.

9.2.2 Water management and ecosystem services-based approaches

Water management has traditionally focused on the key task of water supply to industry, households, and agriculture, while at the same time managing water quality in watersheds as well as sewage treatment. However, the true societal cost of water is not reflected in water pricing mechanisms and decisions on water allocation. For example, the negative effects of depletion of groundwater resources on future water supply are not reflected in water prices.

Establishment of the Water Framework Directive was a first step into managing water bodies at the integral level of river basins. Moreover, the Water Framework Directive made an explicit attempt to integrate economic values into water-related policies and adopted economic criteria to decide on the efficiency and disproportionality of costs versus the economic benefits of its implementation. This ignited a series of valuation studies to assess the non-market benefits of water quality and quantity improvements, including recreation, biodiversity conservation, and habitat improvements (e.g. Hanley *et al.* 2006; Bateman *et al.* 2011; Metcalfe *et al.* 2012).

Ecosystem services-based approaches will require responsible agencies to broaden their scope even more, towards management of the habitats and associated ecosystem services of which the water bodies form an integral part. This has several opportunities for improved water management, which are discussed in depth by Blackstock *et al.* in this book. Exploring the biophysical underpinnings of ecosystem service delivery in ecosystem services-based approaches (core element 2) may better reveal the trade-offs and synergy effects between water supply and other ecosystem services and help to address unintended negative consequences (Martin-Ortega 2012). On the positive side, the quantification of these services may reveal potential co-benefits of achieving improved water quality in terms of other ecosystem services, including effects on terrestrial ecosystems. Such co-benefits might justify investments in actions of which the costs would otherwise be deemed disproportionate, thereby changing policy outcomes (core element 4) (see Brils *et al.*, this book, for the flagship example in New York City). It may also support decisions on water distribution among different stakeholders, ensuring that human needs as well as environmental demands are met. Last but not least, water and land managers may seek cooperation to strike a balance between on-site ecosystem service delivery and off-site water resources. For example, improved peatland management to reduce carbon emissions and conserve biodiversity could have positive effects on water quality (Martin-Ortega *et al.* 2014). Essentially, a more holistic approach may provide better understanding of the effects of ongoing land use changes on water resources (and vice versa) and subsequently on final ecosystem services and associated benefits.

Ecosystem services-based approaches have the potential to improve decision-making, but an inevitable consequence of broadening the scope of management is the increased complexity introduced in analysis as well as policy making. The considerable knowledge gaps with respect to the effect of ecosystem management on water-related ecosystem services introduce high uncertainty in decision-making. Different disciplines, stakeholders, and policy targets with conflicting needs and different nomenclature have to come together and cooperate: the

trans-disciplinarity of ecosystem services-based approaches (core element 3) is also of high relevance in water-related governance and decision-making. The risk is that this may lead to delay in implementation, and working towards ecosystem assessments should not cause inertia.

9.3 ECONOMIC ASSESSMENT OF THE CURRENT STATUS OF WATER ECOSYSTEM SERVICES IN THE UK-NEA

The first phase of the UK-NEA gave an overview of the wide range of water-related benefits by summarising the existing information on economic values of ecosystem services related to water, including a wide range of non-market ecosystem service values. We include some of the main findings and studies that provide aggregated values. Although these studies differ in accuracy, the results provide evidence that the UK population derives considerable wellbeing from water ecosystem services (see Table 9.1).

Mourato *et al.* (2010) use residential property transaction data to analyse how environmental characteristics influence house prices. They demonstrate that freshwater sites along with other natural characteristics (woodlands, green spaces, etc.) within a 1 km range from the property attract a significant price premium of around 1% of the value of average house market prices, reflecting the positive value that society attaches to living closer to environmental assets. In comparison, other services such as schools or rail stations provide a 2% increase in property values, but only if they are within a 200 m range.

Morris and Camino (2010) provide estimates of ecosystem services for inland wetlands based on a global meta-analysis of wetland valuation studies by Brander *et al.* (2008). Globally, flood control protection by wetlands is estimated to be worth

Table 9.1 *Summary of water ecosystem services studies in the UK-NEA**

Habitat	Ecosystem service	Aggregate value (10^6 per year) (£)	Reference
Wetland inland	Water supply	2	Morris and Camino (2010)
Wetland inland	Flood protection	366	Morris and Camino (2010)
Wetland inland	Water quality improvement through nutrient recycling	263	Morris and Camino (2010)
Rivers, lakes	Biodiversity, amenity, recreation	1140 (England & Wales)	NERA (2007); Morris and Camino (2010)
Inland surface water	Recreation	603	Sen *et al.* (2014)

* We only report studies that are related to specific habitats and are aggregated at national level. Note that the values provided by Sen *et al.* (2014), NERA (2007), and the water quality values by Morris and Camino (2010) reflect overlapping ecosystem services and, in order to avoid double counting, these estimates should not be summed. Source: Abstracted from UK-NEA (2011b) and follow-on work.

approximately £336 $\times$ 10^6 per year. The value of water supply by wetlands is very small compared to wetland water quality improvement and flood control protection (Table 9.1), also because wetlands mainly function as water flow regulators rather than suppliers. However, these estimates may not accurately reflect the values of water supply in the UK, and are based on the assumption that all wetlands deliver these functions (irrespective of their location relative to the population that would benefit from the wetland ecosystem services). Another study which attempted to provide national estimates of water ecosystem services is NERA (2007). This study used stated preference surveys to assess the value that households attached to water quality improvements in rivers and lakes that would affect biodiversity, aesthetic, and recreational quality. It showed that, on average, households are willing to pay £40 per year for a nation-wide improvement of water quality. Although the willingness-to-pay values are sensitive to the elicitation of formal and statistical models, the study clearly shows the importance that society attaches to water ecosystem services.

The spatially explicit approach of the UK-NEA is clearly demonstrated by Sen *et al.* (2014), who modelled the non-market value of open-air recreation. In this chapter we focus on the results of this study related to freshwater benefits. The model is based on a large survey about recreational behaviour among households in England (Natural England 2010). The model predicting annual visitor numbers takes into account a wide range of spatial characteristics, including habitats, population, and accessibility. One of the findings is that the number of trips to freshwater sites is higher than for most other types of habitat, including grasslands, mountains, or woodlands. This visitor number model is combined with a meta-analysis on the value per recreational trip across different types of habitats. The results show that the value per trip for freshwater areas is higher than that for grassland and farmland, but lower than for most other habitats (mountains, moors, heaths, woodlands, and marine and coastal areas). The rather low value of freshwater may reflect the abundance of the different ecosystem services in the UK. By multiplying the estimated number of visits by the value per trip, an estimate of the total annual value of visits to freshwater sites is obtained of approximately £603 $\times$ 10^6 per year (Table 9.1). This value is smaller than the estimates reported by NERA (2007), which also reflect non-use values.

9.4 SCENARIO ANALYSIS FOR WATER RECREATIONAL SERVICES

9.4.1 The importance and complications of scenario analysis

One of the key objectives of the Water Framework Directive is to foster sustainable water management and secure water resources into the future. Scenario analysis is a useful tool to evaluate current levels of ecosystem use, and support decision-making by examining the trade-offs implied by each of a set of feasible policy options. Scenario analysis aims to ensure that ecosystem services are incorporated into decision-making and policy prioritisation (core element 4). Economic valuation contributes to this by estimating the societal costs and benefits when moving from a baseline scenario to an alternative state, and helps to identify options with positive net benefits. Scenario analysis for ecosystem service assessments hence requires that services are not considered in isolation, but in combination, showing where trade-offs have to be made or synergies can be achieved in ecosystem management.

In the UK-NEA, a scenario analysis was undertaken to compare ecosystem services in the 2010 baseline with various possible future states in 2060. The UK-NEA scenarios team (Haines-Young *et al.* 2010) generated a number of plausible scenarios that are likely to arise under different policy formulations. Moreover, the conceptual framework of the UK-NEA helped to explore the effects on future water security and wellbeing of climate change exacerbation and human demand pressures.

While this scenarios analysis provides interesting insights for policy makers, it also requires a deep and flexible understanding of the impacts of many indirect and direct drivers (e.g. policy, technology, freshwater pollution) on ecosystem services. The impact of each driver varies over space and time and the UK-NEA scenarios analysis struggles to capture this dynamic. Furthermore, the understanding of the interaction of multiple drivers on a specific ecosystem service (e.g. nutrient cycles) is less well known and represents a major research challenge for the future. As a result, only a few of the water-related ecosystem goods and benefits were assessed in biophysical and economic terms in the UK-NEA.

9.4.2 UK-NEA scenario analysis for water-related recreation

To demonstrate the scenario analysis of the UK-NEA, we provide an example of two scenarios (UK-NEA 2011a) and then describe how these impact freshwater recreational services.

- 'Green and Pleasant Land'. Here, economic growth is mainly driven by secondary and tertiary sectors as opposed to intensive primary land uses. Pressures on rural areas are assumed to be declining as a result of increased concern for the conservation of biodiversity and landscape. A key objective for policy makers is biodiversity preservation, and aesthetic values of landscapes are enhanced by increases in improved grassland (temporary or permanent grassland with reduced fertiliser), semi-natural grassland, and conifer woodland. This implies a decrease in food production which is compensated for by increased imports to offset the demands of a larger population.

Table 9.2 *Changes in recreational values for sites (1 × 1 km cells) with ≥1 ha freshwater*

	Visits (10^3)	Total value (10^3 per year) (£)*	Change in values ($\Delta10^3$ per year) (£)*	
	Baseline	Baseline 2010	Green and Pleasant Land 2060	World Market 2060
England	257 347	414 393	327 448	–54 164
Scotland	71 915	157 904	142 214	–78 794
Wales	14 344	30 563	22 249	–15 386
GB	**343 606**	**602 861**	**491 911**	**–148 343**

* All values are converted to 2010 prices.
Source: abstracted after UK-NEA (2011b).

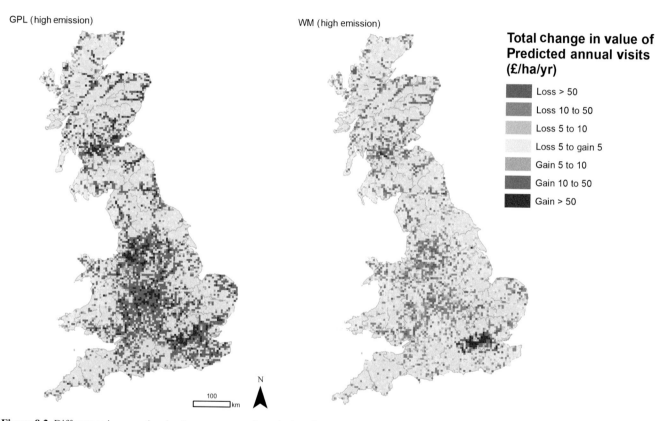

Figure 9.2 Differences in recreational value as we move from the baseline (2010) to two different policy options for 2060: GPL = Green and Pleasant Land (left) and WM = World Markets (right). A black and white version of this figure will appear in some formats. For the colour version, please refer to the plate section.
Contains Ordnance Survey data © Crown copyright and database right 2014.

- 'World Market'. In this scenario the area of arable land increases and improved grassland and semi-natural areas decrease to accommodate population-driven urban growth, which in turn drives further biodiversity declines.

The storylines of the Green and Pleasant Land and World Market scenarios were translated into alternative habitat maps for 2060 (UK-NEA 2011a). Both scenarios are analysed under the high CO_2 emissions trajectory for climate variables (see Murphy *et al.* 2009) and this will contribute for both to a modest increase in the percentage of freshwater land (currently 0.80%, in 2060 1.60%).

The visitor number model of Sen *et al.* (2014) was applied to these new habitat maps to estimate the welfare changes of outdoor recreation. As population and habitats change under the scenarios, the predicted numbers of visits change, and so do the aggregate values of visits to sites.

Figure 9.2 presents two maps with the spatial changes in the value of recreation under these two scenarios for sites with freshwater habitats. This comparison shows that there is a stark contrast between the recreational benefits under both scenarios relative to the baseline, with much higher recreational values under the Green and Pleasant Land scenario.

The sum of country-level changes for recreational benefits is reported in Table 9.2 (alongside baseline estimates for reference). Note that these values do not reflect the financial value of the tourism sector (which would be captured by gross domestic product changes), and excludes any benefits that international visitors may attach to these freshwater areas. Also, freshwater sites may contain other broad habitats where they are spatially indistinguishable, e.g. a mountain stream.

Under the Green and Pleasant Land scenario, preservation of biodiversity and aesthetic quality of landscapes results in major increases in the benefits over the baseline, especially around urban areas, of almost £500 × 10^6 per year. This effect comes at the expense of a decrease in primary sector production, substituted by imports.

In the World Market scenario, major recreational losses are found around small urban centres and in remote areas. There are still considerable benefits enjoyed by the population in and around large urban centres where a substantial reduction of urban and peri-urban recreational areas (including urban greenspaces) is envisioned under the scenario and therefore an increase in water recreation values is expected. However, overall the World Market scenario results in substantial losses of recreational values, mainly for Wales, while for England and Scotland the impact is less severe.

9.5 THE IMPACT OF THE UK-NEA FINDINGS IN WATER-RELATED DECISION-MAKING

The results of the first phase of the UK-NEA were published in 2011 (UK-NEA 2011a) and the ultimate impact on policy making will only be apparent in years to come. Nevertheless, there have already been some important achievements and shifts in policy making, where the UK-NEA has contributed. The UK-NEA has strongly influenced the development of the Natural Environment White Paper (H.M. Government 2011a), described as the most important change in UK policy for the past 20 years (Watson 2012). This policy argues for an adoption of the ecosystem services-based approach at the national scale across the UK. Subsequent water-related policies such as the Water White Paper (H.M. Government 2011b), the Water For Life policy (Defra 2011), and the National Policy Statement for Waste Water (Defra 2012) also build upon the findings of the UK-NEA and seek to reform the water industry in ways which sustain and improve ecosystem services.

Change in decision-making is also being driven through private-sector initiatives. Rather than their traditional focus upon end-of-pipe, treatment-oriented approaches to delivering water supplies, the private sector is getting involved in joint initiatives with environmental organisations and statutory bodies to seek out solutions for using better environmental management of ecosystems as mechanisms for delivering improved water-related services (BES & UK BRAG 2011). In north-west England, United Utilities has developed a sustainable catchment management programme in collaboration with the Royal Society for the Protection of Birds to both improve water supplies and reduce carbon emissions.[1] This wider scope is also reflected in the assessment of the potential of woodland management to contribute to the achievement of the objectives of the Water Framework Directive (Nisbet et al. 2011). Another example is provided by the South West Water 'Upstream Thinking' initiative in Cornwall and Devon. This seeks to work with farmers to improve the quantity and quality of water through land use change as an alternative to engineering and chemical treatment options.[2]

These private-sector initiatives are being actively encouraged through public-sector changes in the rules governing water company operations and through an extension of Payments for Environmental Services schemes (Defra 2010). Water-related Payments for Environmental Services schemes are an example of the potential of capturing benefits when downstream beneficiaries pay for the benefits they derive from better land management by land users upstream, but well-working schemes can be hard to define (Muradian et al. 2010).

9.6 DISCUSSION AND CONCLUSION

One of the core findings of the UK-NEA was that many of the ecosystem services provided by natural habitats, including freshwater habitats, remain poorly identified and under-valued in policy-making, resulting in ongoing habitat loss, degradation, and modification. This argues for a number of developments in both academic research and policy development.

A key issue for assessments of water-related ecosystem service using land cover or habitat maps is that it does not reflect the complexity of ecosystems and it is often unclear what the implications of changes in the extent of natural habitats on changes in ecosystem services such as water provisioning will be. There are still important knowledge gaps and methodological issues related to quantitative analysis of ecological and economic linkages, and their relation to water values. This includes temporal and spatial effects: it is not yet well understood what the longer-term effects of current water uses are and how resilient and resistant freshwater ecosystems are, and what the interactions of various freshwater and land types and their effects on spatial ecosystem service delivery are. With better understanding of dynamic and spatial effects and interactions, the transition paths in scenario analysis could be explored.

[1] http://corporate.unitedutilities.com/scamp-index.aspx
[2] http://www.financeforthefuture.co.uk/Upload/PageAttachments/page1577/files/South_West_Water_case_study_final.pdf.

Furthermore, the non-financial nature of many water-related ecosystem goods and benefits and the absence of relevant economic value estimates increase the risk that these benefits are ignored in policy making. The UK-NEA framework, such as presented in Figure 9.1, provides insight into the changes in human welfare that may result from environmental changes and understanding of the service delivery process. It may enhance communication and integration between natural scientists and economists, with several potential improvements, e.g. (1) including a wide(r) range of goods and services in economic assessment as a result of the evaluation of a range of impacts of environmental change; (2) building valuation scenarios on adequate ecological knowledge. However, additional funding for primary research and valuation studies may be required for a reliable assessment of the full set of water-related goods and services.

The UK-NEA framework provides guidance on the economic assessment and mapping of ecosystem goods and benefits, and spatially explicit scenarios can be used to inform efficient land use policies. However, the importance of services such as water and flood protection may override considerations of cost–benefit ratios. When, besides efficiency, equity criteria play a role, further insight in the distribution of benefits and costs of changes in water-related goods and services among stakeholder groups in society will be required. Economic assessments can be extended by, for example, the disaggregation approach by Krutilla (2005) to address equity considerations. We have provided a range of examples of recent policy initiatives that have adopted the UK-NEA approach, although there is still considerable room for enhancement of the institutional engagement with ecosystem services-based approaches. Particular challenges include:

- robustly valuing ecosystem services in ways which reflect the inherent variability in those services across locations;
- incorporating dynamic effects such that decisions become more robust across time;
- engaging with and generating enhancements to existing legislation;
- drawing in and incentivising the various actors necessary to ensure that decisions are effectively and efficiently turned into actions.

This is, we recognise, a substantial research agenda and one which we expect to provide a major focus for both researchers, public and private institutions, and indeed society for many years to come.

References

Bateman, I. J., Mace, G. M., Fezzi, C., *et al.* (2011). Economic analysis for ecosystem service assessments. *Environmental and Resource Economics* **48**, 177–218.
Bateman, I. J., Harwood, A., Mace, G. M., *et al.* (2013). Bringing ecosystem services into economic decision making: land use in the UK. *Science* **341**, 45–50.
BES and UK BRAG (2011). Where next for the UK National Ecosystem Assessment and IPBES? Report of a joint session between the British Ecological Society (BES) and the UK Biodiversity Research Advisory Group (UK BRAG), 13 September, Council Chamber, Octagon Centre, University of Sheffield.
Brander, L. M., Ghermandi, A., Kuik, O., *et al.* (2008). Scaling up ecosystem services values: methodology, applicability and a case study. Final Report, EEA. Fondazione Eni Enrico Mattei. Available at: www.feem.it/userfiles/attach/2010471736364 NDL2010-041.pdf (last accessed 21 October 2014)
Defra (2010). *Payments for Ecosystem Services: A Short Introduction.* Defra, London.
Defra (2011). *Water for Life: Market Reform Proposals.* Defra, London.
Defra (2012). *National Policy Statement for Waste Water.* Defra, London.
Fisher, B. & Turner, R. K. (2008). Ecosystem services: classification for valuation. *Biological Conservation* **141**, 1167–1169.
Fisher, B., Turner, R. K., Zylstra, M., *et al.* (2008). Ecosystem services and economic theory: integration for policy-relevant research. *Ecological Applications* **18**, 2050–2067.
Haines-Young, R., Paterson, J., Potschin, M., *et al.* (2010). *The UK NEA Scenarios: Development of Storylines and Analysis of Outcomes.* UNEP-WCMC, Cambridge.
Hanley, N., Wright, R. E., & Alvarez-Farizo, B. (2006). Estimating the economic value of improvements in river ecology using choice experiments: an application to the Water Framework Directive. *Journal of Environmental Management* **78**, 183–193.
H.M. Government (2011a). *The Natural Choice: Securing the Value of Nature (the Natural Environment White Paper).* The Stationery Office, London.
H.M. Government (2011b). *The Water White Paper.* The Stationery Office, London.
Jackson, D. L. (2000). *Guidance on the Interpretation of Biodiversity Broad Habitat Classification (Terrestrial and Freshwater Types): Definitions and the Relationship with Other Habitat Classifications.* Joint Nature Conservation Committee, Peterborough.
Krutilla, K. (2005). Using the Kaldor–Hicks Tableau format for cost–benefit analysis. *Public Policy Analysis and Management* **24**, 864–875.
Maltby, E., Omerod, S., Acreman, M., *et al.* (2011). *Freshwater: Open Waters, Wetlands and Floodplains.* UNEP-WCMC, Cambridge.
Martin-Ortega, J. (2012). Economic prescriptions and policy applications in the implementation of the European Water Framework Directive. *Environmental Science and Policy* **24**, 83–91.
Martin-Ortega, J., Allot, T. E., Glenk, K., & Schaafsma, M. (2014). Valuing water quality improvements from peatland restoration: evidence and challenges. *Ecosystem Services* **9**, 34–43.

Box 9.2 Key messages

- The UK-NEA provides a useful, national-level model for the assessment of changes to ecosystem services.
- As such it has generated substantial policy impact both in terms of general natural capital decision-making and for specific issues such as those relating to the water environment.
- Addressing water policy through ecosystem services-based assessment may help reveal the importance of water and ecosystems to human wellbeing.
- Many of the water-related services provided by ecosystems are not traded in markets, yet they are of significant value to society.
- The full assessment of water-related ecosystem services will require more knowledge development to better understand how ecosystems contribute to human wellbeing.

Metcalfe, P. J., Baker, W., Andrews, K., *et al.* (2012). An assessment of the nonmarket benefits of the Water Framework Directive for households in England and Wales. *Water Resources Research* **48**, W03526.

Millennium Ecosystem Assessment (2005). *Ecosystems and Human Well-being: A Framework for Assessment.* Island Press, Washington, DC.

Morris, J. & Camino, M. (2010). *Economic Assessment of Freshwater, Wetland and Floodplain Ecosystem Services.* UK-NEA, Cranfield.

Mourato, S., Atkinson, G., Collins, M., *et al.* (2010). *Economic Assessment of Ecosystem Related UK Cultural Services.* The Economics Team of the UK-NEA, London.

Muradian, R., Corbera, E., Pascual, U., *et al.* (2010). Reconciling theory and practice: an alternative conceptual framework for understanding payments for environmental services. *Ecological Economics* **69**, 1202–1208.

Murphy, J. M., Sexton, D. M. H., Jenkins, G. J., *et al.* (2009). *UK Climate Projections Science Report: Climate Change Projections.* Met Office Hadley Centre, Exeter.

Natural England (2010). *Monitor of Engagement with the Natural Environment: The National Survey on People and the Natural Environment.* Natural England, Sheffield.

NERA (2007). *The Benefits of Water Framework Directive Programmes of Measures in England and Wales.* NERA, London.

Nisbet, T., Silgram, M., Shah, N., *et al.* (2011). Woodland for water: summary report. Environment Agency and Forestry Commission.

Sen, A., Harwood, A. R., Bateman, I. J., *et al.* (2014). Economic assessment of the recreational value of ecosystems: methodological development and national and local application. *Environmental and Resource Economics* **57**, 233–249.

UK-NEA (2011a). *The UK National Ecosystem Assessment: Synthesis of the Key Findings.* UNEP-WCMC, Cambridge.

UK-NEA (2011b). *The UK National Ecosystem Assessment: Technical Report.* UNEP-WCMC, Cambridge.

Vörösmarty, C. J., Green, P., Salisbury, J., & Lammers, R. B. (2000). Global water resources: vulnerability from climate change and population growth. *Science* **289**(5477), 284–288.

Vörösmarty, C. J., McIntyre, P. B., Gessner, M. O., *et al.* (2010). Global threats to human water security and river biodiversity. *Nature* **467**, 555–561.

Watson, R. T. (2012). The science–policy interface: the role of scientific assessments – UK National Ecosystem Assessment. *Proceedings of the Royal Society A* **468**, 3265–3281.

10 Using an ecosystem services-based approach to measure the benefits of reducing diversions of freshwater

A case study in the Murray-Darling Basin, Australia

Neville D. Crossman, Rosalind H. Bark, Matthew J. Colloff, Darla Hatton MacDonald, and Carmel A. Pollino

10.1 INTRODUCTION

Ecosystem services-based approaches have been applied to decisions about trade-offs between alternative uses of land (Raudsepp-Hearne *et al.* 2010; Maes *et al.* 2012; Bryan & Crossman 2013; Geneletti 2013; Seppelt *et al.* 2013), but have been used less commonly to assess trade-offs in alternative uses of water (Schlu-ter *et al.* 2009; Rouquette *et al.* 2011; Liu *et al.* 2013). In this chapter we provide an overview of a case study into quantifying the ecosystem services and associated benefits (and their monetary values) of a new water-sharing plan that will return water to the environment in the Murray-Darling Basin, Australia. This serves as an illustration of how to operationalize an ecosystem services-based approach, as defined in this book. Chapter 2 in this book emphasizes that there is a gap between the conceptualization and endorsement of ecosystem services by both researchers and policy makers and the incorporation of ecosystem services-based approaches into natural resources management practice. The present chapter demonstrates the operationalization of an ecosystem services-based approach in the context of water resource planning and management. We estimate the changes to a range of final ecosystem services (Boyd & Banzhaf 2007; Kumar 2010) that result from the implementation of a discrete policy scenario, and provide economic estimates for the associated benefits. Our work contributes to the still scarce literature on real-world examples of integrating empirical data on the biophysical supply of ecosystem services with their socio-cultural context and monetary valuation to inform investment decisions (Martín-López *et al.* 2014; see also Mulligan *et al.*, this book).

The Murray-Darling Basin contains iconic and internationally important wetlands and is Australia's major food-producing area. In terms of gross value, about 40% of Australia's agriculture and 50% of irrigated agriculture is produced in the Basin (Australian Bureau of Statistics 2013). However, the dominance of food production has come at the expense of other ecosystem services

provided by land and water resources in the Basin, primarily due to the decline in health of river, wetland, and floodplain ecosystems (Kingsford 2000; Kingsford *et al.* 2011). Here we summarize a project (CSIRO 2012) commissioned by the Murray-Darling Basin Authority, an Australian Federal Government Agency, to support decision-making on water allocations associated with the development of policy guiding the re-allocation of water resources under a new government policy and legislative framework, the Murray-Darling Basin Plan. A detailed report of the research is presented in CSIRO (2012). The objective of that project was to quantify the benefits, and where possible the monetary values, of returning water to the environment to improve the supply of the other, non-provisioning ecosystem services. We use the 'cascade diagram' conceptual framework from Kumar (2010) to structure the analysis because this framework clearly shows the links between biophysical changes in ecosystems, to changes in ecosystem services, through to changes in benefits and then monetary values. Specifically, we: (1) modelled increases in river flows in each of the catchments of the Murray-Darling Basin; (2) related additional flows to predicted ecological responses at important wetland indicator sites; (3) identified the ecosystem services associated with those predicted ecological responses; (4) assessed the marginal change in supply of selected regulating, habitat, and cultural ecosystem services under the Basin Plan scenario compared with a baseline or 'do nothing' scenario; and (5) undertook monetary valuation, where possible, of marginal changes in supply of ecosystem services for use by the Australian government in cost–benefit analysis of the impact of the proposed regulations.

10.2 THE MURRAY-DARLING BASIN

The Murray-Darling Basin covers one-seventh of the land area of Australia (Figure 10.1) and contains the only major permanently

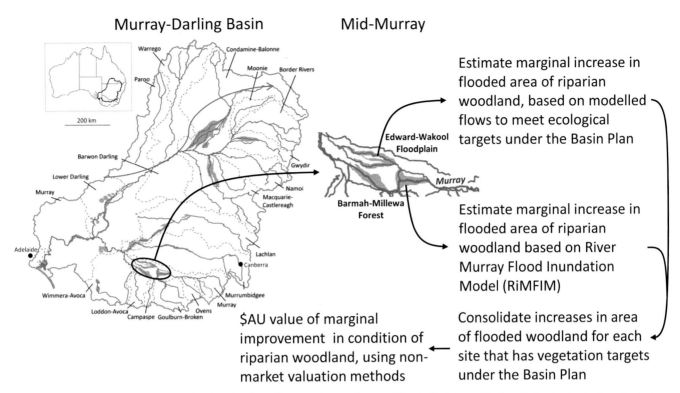

Figure 10.1 Steps in the methodology for the valuation of improved vegetation condition in the Murray-Darling Basin, Australia. Hydrological modelling provided an estimate of marginal change in areas of riparian woodland likely to be inundated under the Basin Plan and maintained in good ecological condition. The monetary value of the increase in areas of woodland in good condition was then estimated. Shadowed areas = hydrological indicator sites subject to ecological targets for vegetation.

Box 10.1 The Murray-Darling Basin

The Murray-Darling Basin covers one-seventh of the land area of Australia (Figure 10.1) and contains the only major permanently flowing river systems on the continent, including the Murray, Murrumbidgee, Barwon–Darling, Condamine–Balonne and Macquarie–Castlereagh River systems and tributaries. Many of the catchments contain nationally and internationally significant wetlands (including 16 Ramsar wetlands) that provide foci for aquatic biodiversity, recreational activities, and spiritual values. Gross value of agricultural production was AU\$18.6 billion in 2012, of which 36% was from irrigation (Australian Bureau of Statistics 2013). However, many of the rivers, wetlands, floodplains, and the Murray estuary are in poor ecological condition and have been for some time, in part as a result of changes in flood and flow regimes due to increased water diversions from the rivers (Sims & Colloff 2012). Poor ecological condition has been exacerbated by five severe, widespread droughts since 1940, of which the Millennium Drought (1997–2010) was the most severe in recorded history.

flowing river systems on the continent (see Box 10.1). Environmental degradation has prompted a series of water reforms by the Australian government. Introduction of the Water Act (Commonwealth of Australia 2007) provided the legislative mechanism to reduce the volume of water that can be diverted for irrigated agriculture. In support of the Water Act is a planning process (the Basin Plan) that stipulates the volume of water that would need to be returned to the environment in order to meet a set of hydrological targets that, if achieved, will match the water requirements of aquatic ecosystems to maintain them in good condition. While not explicit in the draft Basin Plan, there is an assumption that maintaining the aquatic ecosystems in good condition will ensure the continued supply of ecosystem services from those ecosystems, especially the non-provisioning services which have been compromised by increased water diversions for irrigation. The draft Basin Plan contained a proposed reduction of 2800 gigalitres[1] (GL) per year from the 2009 average irrigation diversions of 13 623 GL per year (a 21% reduction).[2]

---

[1] 1 gigalitre is equal to 1 billion (10^9) litres, or approximately 810 acre feet.
[2] In the final Basin Plan, an annual average of 2750 GL of water will be recovered for the environment by 2019, with an additional 450 GL to be recovered by 2024.

During the latter stages of the development of the draft Basin Plan in 2011, attention focused on the economic costs of reduced irrigation diversions. These were estimated by the Australian Bureau of Agricultural and Resource Economics and Sciences as an average annual reduction of AU$542 million in the gross value of irrigation production (Murray-Darling Basin Authority 2012b). But there was no detailed assessment and valuation of the social, economic, and environmental benefits of the Basin Plan. In response to this knowledge gap, the Murray–Darling Basin Authority, the agency responsible for developing and implementing the Basin Plan, commissioned us to identify and quantify the ecological and economic improvements that were likely to eventuate from returning 2800 GL per year of water to the environment. We used an ecosystem services-based approach as a framework and reporting tool to quantify the benefits, and when possible the monetary values, of reduced diversions. Monetary estimates of the values were an important input into the cost–benefit analysis of the proposed Basin Plan used by the Australian government to assess the potential impacts of new policy regulations (Murray-Darling Basin Authority 2012a).

10.3 ASSESSING THE ECOSYSTEM SERVICES BENEFITS AND MONETARY VALUES OF WATER RE-ALLOCATION

People depend on potable freshwater for drinking and domestic supply, and indirectly via production of food and energy, industry, and transportation (Grey & Sadoff 2007). Flow-dependent ecosystems also provide other important hydrologically mediated ecosystem services that support human wellbeing, including recreation and amenity value, habitat for biodiversity, and spiritual and cultural values (Brauman *et al.* 2007; Maltby & Acreman 2011; Keeler *et al.* 2012). One of the greatest challenges in natural resource management is to implement equitable sharing of finite water resources between consumptive uses and the environment in order to maintain condition and function of these flow-dependent ecosystems and maintaining the services these ecosystems provide (Gordon *et al.* 2010; Grafton *et al.* 2013). In the Murray-Darling Basin, equitable sharing involves reducing water diverted for irrigation, thereby re-balancing supply of ecosystem services from the provisioning services to the other non-provisioning services (Gordon *et al.* 2010).

10.3.1 Changes in the biophysical conditions underpinning ecosystem services

Following the 'cascade diagram' (De Groot *et al.* 2010; Haines-Young & Potschin 2010; Kumar 2010) that shows the link

between changes in ecological processes, functions, services, human wellbeing, and benefits and their values, the marginal change in ecosystems, ecosystem services, and subsequent benefits and their (monetary) values that result from reducing the water that is diverted for irrigation was calculated by comparing a baseline scenario (the current level of diversions) to a future scenario (the Basin Plan scenario of reducing diversions by 2800 GL per year, hereafter the '2800 GL/year scenario'). The baseline for ecological condition was established using a modelled hydrologic flow sequence of 114 years (1895–2009), assuming historic climate, current river operation rules and basin infrastructure (includes dams, infrastructure for moving water to key environmental assets, and diversions for consumptive use). The 2800 GL/year scenario was also based on the 114-year flow sequence, assuming historic climate, current infrastructure, and Basin Plan operating rules, including the new river flow regimes resulting from the 2800 GL/yr reduction in water diverted for irrigation. This data was provided by the Murray-Darling Basin Authority. The models, scales, spatial extents, and sources of information used in the analyses are summarized in Table 10.1; Figure 10.1 demonstrates one of the methods, in this case to estimate changes to extent of inundation of mapped floodplain vegetation. To undertake the biophysical analyses, the following steps were used:

(1) Calculate flow metrics for the baseline hydrologic scenario (i.e. current flow) and for the 2800 GL/year scenario. Flow metrics were used to calculate frequency of exceeding known thresholds of salinity, bank erosion, and sedimentation.

(2) Using the hydrologic model scenarios as inputs, ecological response models were used to predict likely changes to ecosystem condition for the 2800 GL/year scenario. Model predictors were the frequency of waterbird breeding events, habitat condition for native fish, the extent of inundation of mapped floodplain vegetation, and the condition of the Coorong Lakes. Water quality models were used to predict the likelihood of blackwater events[3] and blue-green algal blooms, and the potential for acidification of the Lower Lakes. Carbon sequestration measures were derived from a floodplain vegetation model.

ECOLOGICAL IMPROVEMENTS FROM REDUCING DIVERSIONS

All ecological response variables were modelled as improved under the 2800 GL/year scenario relative to the 'do nothing'

---

[3] Blackwater can be a natural feature of lowland river systems and occurs during flooding when organic material is washed into waterways and consumed by bacteria, leading to a sudden depletion of dissolved oxygen in water.

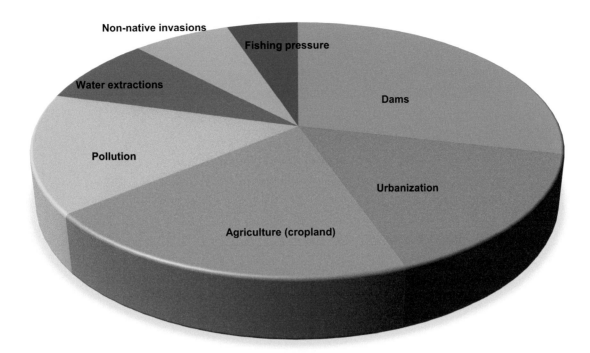

Figure 4.4 Most important threats to river biodiversity based on global-scale data from Vörösmarty *et al.* (2010). At local scales, impacts from various sources will vary as a function of land use, population, status of development, and lifestyles, and will influence stakeholder prioritization of freshwater goods and services.

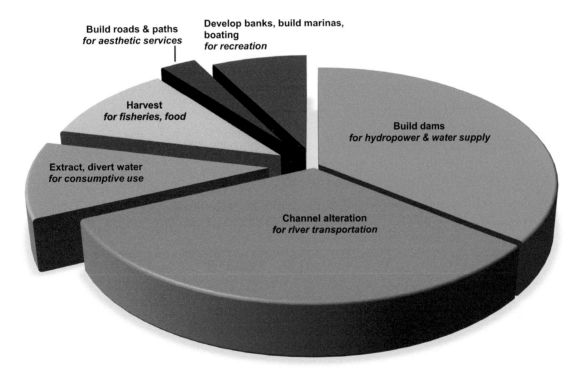

Figure 4.5 Impacts of human actions to river ecosystem services. Accessing river ecosystem services other than biodiversity can have unintended negative impacts on freshwater biodiversity. Here the impacts of individual human actions (or stressors) are reported by individual stressor (shown here as a percentage of all stressors; data from Vörösmarty *et al.* 2010. In order to benefit from or gain access to certain river ecosystem services, humans have altered fundamental bio-physical processes and ecosystem attributes. For example, the use of rivers for transportation by large ships or barges has involved extensive alteration of the channel to ensure it is wide and deep enough for passage. To ensure sufficient water for consumptive use, humans have extracted large amounts from rivers, diverted water flows to agricultural fields, and built dams to store water. Even cultural ecosystem services such as those associated with aesthetics, spiritual values, and recreation may require the building of roads near waterways or the construction of marinas; however, those impacts on biodiversity are modest relative to the other categories. All of these actions have negative consequences for biodiversity. At least one ecosystem service provided by rivers – flood protection – requires no action unless the region is developing. In that case, actions that support or enhance freshwater biodiversity (i.e. preservation of floodplains and riparian corridors) may be necessary.

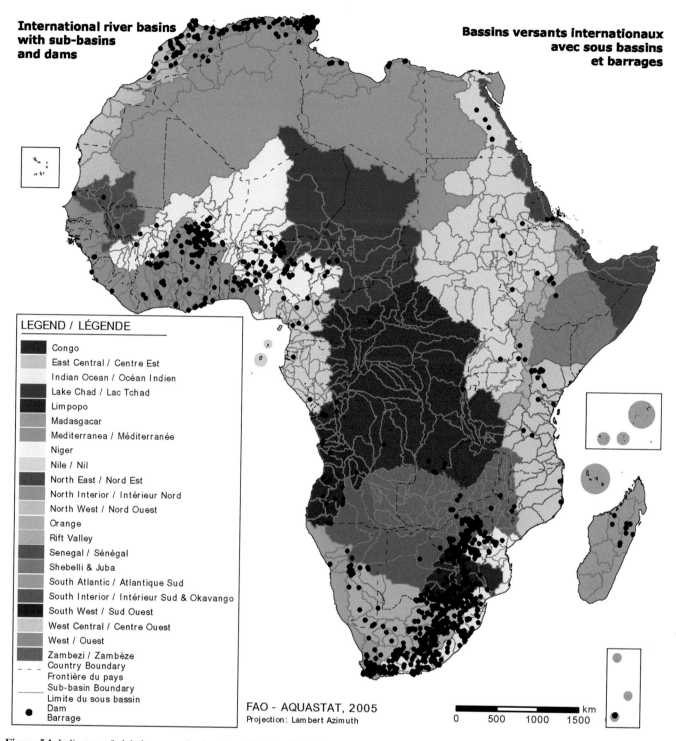

International river basins with sub-basins and dams

Bassins versants internationaux avec sous bassins et barrages

LEGEND / LÉGENDE

- Congo
- East Central / Centre Est
- Indian Ocean / Océan Indien
- Lake Chad / Lac Tchad
- Limpopo
- Madasgacar
- Mediterranea / Méditerranée
- Niger
- Nile / Nil
- North East / Nord Est
- North Interior / Intérieur Nord
- North West / Nord Ouest
- Orange
- Rift Valley
- Senegal / Sénégal
- Shebelli & Juba
- South Atlantic / Atlantique Sud
- South Interior / Intérieur Sud & Okavango
- South West / Sud Ouest
- West Central / Centre Ouest
- West / Ouest
- Zambezi / Zambèze
- - - Country Boundary / Frontière du pays
- —— Sub-basin Boundary / Limite du sous bassin
- • Dam / Barrage

FAO - AQUASTAT, 2005
Projection: Lambert Azimuth

0   500   1000   1500   km

Figure 5.1 Indicators of global crop production intensification, 1961–2007. Index (1961 = 100) Source: FAO (2011b).

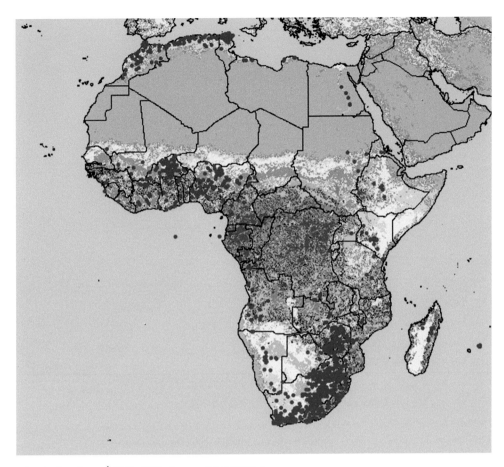

Figure 5.2 Increase of agricultural area,[1] 1961–2011. Source: FAO (2011c).

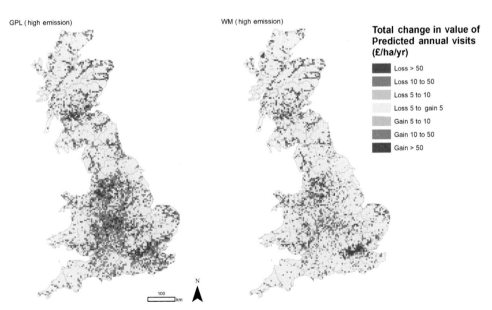

Figure 9.2 Differences in recreational value as we move from the baseline (2010) to two different policy options for 2060: GPL = Green and Pleasant Land (left) and WM = World Markets (right). Contains Ordnance Survey data © Crown copyright and database right 2014.

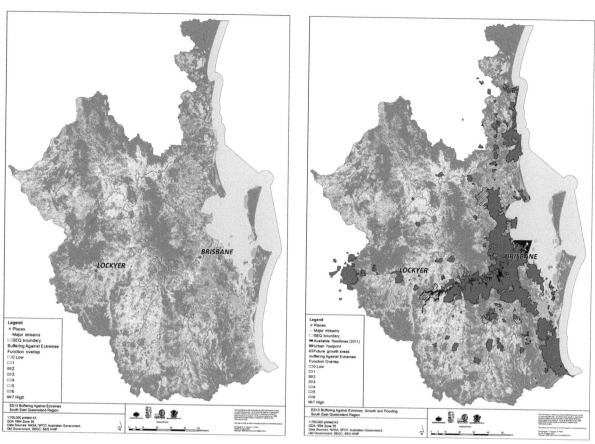

Figure 11.2 Both maps show areas of ecosystem function with potential to contribute to buffering against extremes. Low ecosystem function (white) = 0 functions occurring. High ecosystem function (green) = 7 function occurring. Additionally, the map on the right shows current (grey) and proposed (hatched) urbanisation in South East Queensland. Areas in red show 2011 floodlines.

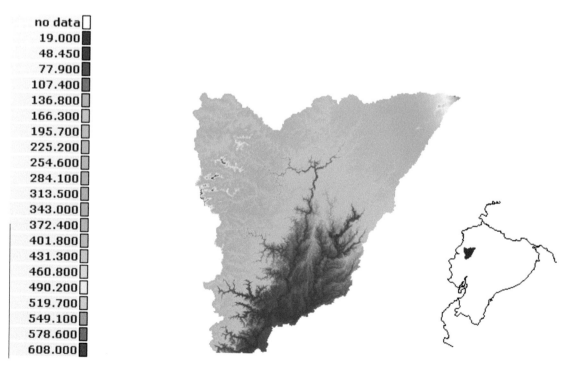

Figure 12.1 Digital elevation model for the upper Daule watershed, Ecuador set within the context of Ecuador. Source: WaterWorld, based on SRTM HydroSHEDS.

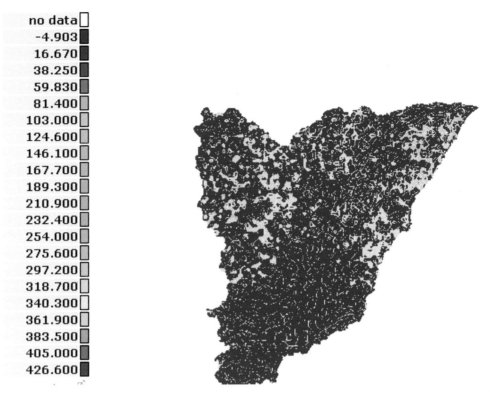

Value	
no data	
−4.903	
16.670	
38.250	
59.830	
81.400	
103.000	
124.600	
146.100	
167.700	
189.300	
210.900	
232.400	
254.000	
275.600	
297.200	
318.700	
340.300	
361.900	
383.500	
405.000	
426.600	

Figure 12.2 Change in annual total gross hillslope soil erosion (%).

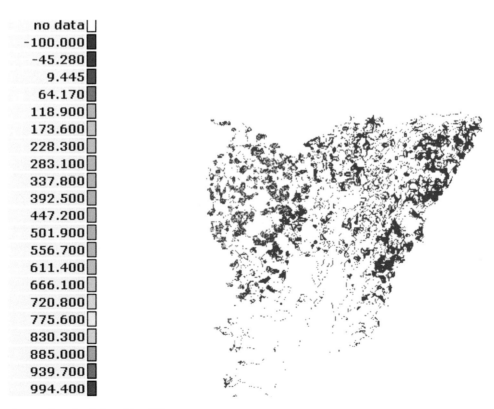

Value	
no data	
−100.000	
−45.280	
9.445	
64.170	
118.900	
173.600	
228.300	
283.100	
337.800	
392.500	
447.200	
501.900	
556.700	
611.400	
666.100	
720.800	
775.600	
830.300	
885.000	
939.700	
994.400	

Figure 12.3 Change in annual total soil deposition (%).

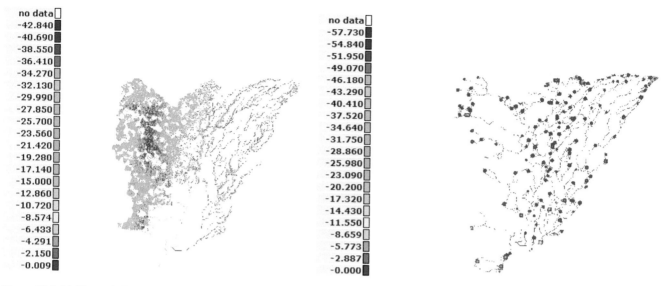

no data ☐
-42.840 ■
-40.690 ■
-38.550 ■
-36.410 ☐
-34.270 ☐
-32.130 ☐
-29.990 ☐
-27.850 ☐
-25.700 ☐
-23.560 ☐
-21.420 ☐
-19.280 ☐
-17.140 ☐
-15.000 ☐
-12.860 ☐
-10.720 ☐
-8.574 ☐
-6.433 ☐
-4.291 ☐
-2.150 ☐
-0.009 ■

no data ☐
-57.730 ■
-54.840 ■
-51.950 ■
-49.070 ■
-46.180 ☐
-43.290 ☐
-40.410 ☐
-37.520 ☐
-34.640 ☐
-31.750 ☐
-28.860 ☐
-25.980 ☐
-23.090 ☐
-20.200 ☐
-17.320 ☐
-14.430 ☐
-11.550 ☐
-8.659 ☐
-5.773 ☐
-2.887 ☐
-0.000 ■

Figure 12.4 (a) Change in human footprint on water quality for ecoefficient agriculture scenario. (b) Change in human footprint on water quality for sanitation scenario.

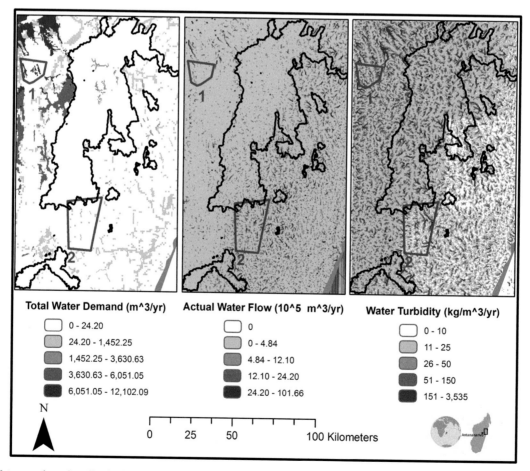

Total Water Demand (m^3/yr)

☐ 0 - 24.20
☐ 24.20 - 1,452.25
☐ 1,452.25 - 3,630.63
☐ 3,630.63 - 6,051.05
■ 6,051.05 - 12,102.09

N

Actual Water Flow (10^5 m^3/yr)

☐ 0
☐ 0 - 4.84
☐ 4.84 - 12.10
☐ 12.10 - 24.20
■ 24.20 - 101.66

0   25   50   100 Kilometers

Water Turbidity (kg/m^3/yr)

☐ 0 - 10
☐ 11 - 25
☐ 26 - 50
■ 51 - 150
■ 151 - 3,535

Antananarivo

Figure 13.2 Water supply and quality in the Ankeniheny–Zahamena Forest Corridor area of Madagascar. From the left: total water demand across sectors, surface-water flow that is used by beneficiaries, and amount of sediment that is transported by hydrologic flows. Regions 1 and 2 show the areas selected for comparison.

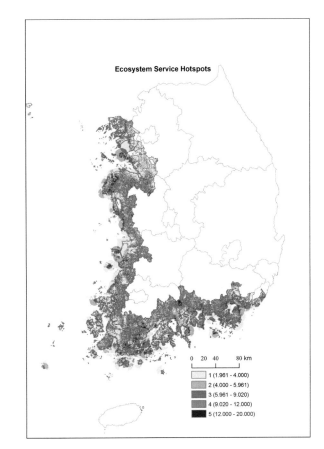

Figure 14.2 Hotspots of ecosystem services based on six components of ecosystem services in coastal areas of Korea (source: Chung 2013).

Figure 16.1 New York City's water supply system.

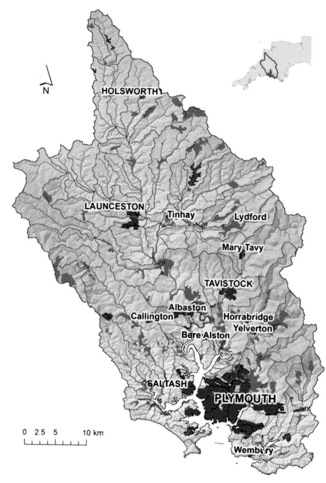

Figure 16.2 River Tamar catchment. (Source and more information: http://river-gateway.org.uk/catchments/tamar.html).

Figure 16.3 Nature friendly banks created voluntarily by farmers at their farmland. Photograph by N. van Everdingen.

Table 10.1 *Monetary values of benefits under the 2800 GL/year scenario compared with the 'do nothing' scenario*

Ecosystem service	Biophysical metrics	Economic modelling	AU$ million
Regulating			
Carbon sequestration	Hectares of native vegetation in good condition and woody carbon potential	Based on different carbon prices	120.0–1000.0
Moderation of acid sulphate soils	Lower Lakes height threshold	Avoided costs	9.2
Moderation of sedimentation	End-of-system flows and Mouth Opening Index	Avoided costs	17.8
Maintenance of bank stability	River in-channel height and threshold	Avoided costs	23.8
Provisioning			
Floodplain (grazing)	Hectares	Transfer from another study	32.2
Freshwater quality	Salinity concentrations	Avoided salinity productivity losses and costs to utilities and users	1.1
	Cyanobacterial bloom risk	Avoided treatment costs	0.9
Fish	Commercial catch, Coorong and Lower Lakes Fishery	Regression estimates	0.2 (annual)
Cultural			
Aesthetic appreciation	House prices in basin 2003–2010, historic and modelled river flows and lake level height	Hedonic models	337.0
Indigenous values	Geocoded cultural and bush tucker sites for Wamba Wamba of the Werai Forest		+
Tourism	Swimmable, fishable, boatable water quality days	Benefit transfer values	161.4 10.3–20.6
Native species diversity			
Native vegetation	Inundation model and floodplain vegetation mapping	Choice modelling	2303.9
Native fish	Response relationships derived from the Murray Flow Assessment Tool	Choice modelling	339.9
Colonial waterbird breeding	Environmental Water Requirements; Ecological Response Models	Choice modelling	693.1
Coorong, Lower Lakes	Ecosystem states model	Choice modelling	480.0/ 4000.0/ 4300.0*

* Depends on assumptions.

scenario (Table 10.1). Ecological responses tended to be greater for those response variables that depend on flooding (e.g. waterbird breeding) than for those that depend on in-channel flows (fish groups). Increased floodplain inundation under the 2800 GL/year relative to the 'do nothing' scenario benefited the lignum shrubland and river red gum forest and woodland vegetation communities on the lower- and mid-level floodplains along the Murray River. Higher elevation floodplains along the Murray River are likely to remain vulnerable under the 2800 GL/year scenario and their capacity to continue to support river red gum and black box communities could be compromised. There are important ecological benefits for the Coorong, Lower Lakes, and Murray Mouth under the 2800 GL/year scenario, including reduced occurrence of time when the Coorong is in an ecologically unhealthy state.

WATER QUALITY IMPROVEMENTS FROM REDUCING DIVERSIONS

Reducing by 2800 GL per year the amount of water diverted in the Basin resulted in improved water quality (Table 10.1). First, through reduced numbers of days of low flow when cyanobacterial blooms could develop. Second, through less frequent periods of low water levels in the Lower Lakes, when acidification could occur. Third, through more frequent inundation of vegetated floodplains, which reduces the number of days of high oxygen demand due to oxidation of floodplain carbon sources, which in turn reduces the number of blackwater events and fish kills.

10.3.2 Assessing marginal changes in ecosystem service supply and value

Building on the modelled improvements to biophysical conditions, the following steps were used to undertake the analyses of change in ecosystem service supply and value:

(1) Flow metrics (salinity, bank erosion, sedimentation) and ecological response models were used as inputs to the ecosystem services assessment. The incremental changes in the supply of ecosystem services were predicted under the 2800 GL/year scenario relative to the baseline scenario. The major ecosystem services modelled are listed in Table 10.1.

(2) The monetary value of the incremental changes in the supply of ecosystem services was calculated for different scenarios. Standard economic valuation techniques were used to value services listed in Table 10.1. These included benefit transfer methods, using values obtained from previous studies inside and outside the basin and hedonic methods, which were used to estimate aesthetic appreciation. Improved quality of freshwater sourced from the Murray-Darling Basin was modelled to reduce treatment costs and costs associated with lost recreation and tourism opportunities. Further detail on the valuation is provided below.

REGULATING ECOSYSTEM SERVICES

We valued a subset of the regulating ecosystem services for which we were able to quantify marginal change in supply from reducing irrigation diversions, namely climate regulation (through carbon sequestration), water purification (through moderation of acid sulphate soils), and erosion prevention (through moderation of sedimentation and maintenance of river bank stability). Carbon market prices were used to value climate regulation, and avoided cost was used to value the other regulation services based on methods described by Banerjee *et al.* (2013). Carbon sequestration was estimated as the incremental increase in standing carbon between the two scenarios as a result of changes in inundation. The changes in the supply of the other regulating services were valued using data on remediation costs incurred by governments and individuals during the 1997–2010 Millennium Drought (Banerjee *et al.* 2013) combined with the probability of exceeding hydrologic and ecological response thresholds between the two scenarios. Despite their theoretical limitations (National Research Council 2005), damage cost avoidance methods provided a reasonable proxy estimate of value because remediation costs have previously been incurred, demonstrating demand for the services.

PROVISIONING ECOSYSTEM SERVICES

We valued the marginal change in supply of freshwater, livestock production from floodplains, and fish production provisioning ecosystem services using available data and models.

Improvements in freshwater quality between the scenarios were estimated by modelling changes in likelihood of algal blooms and the subsequent reduced treatment costs for water utilities. Increases in livestock production on floodplains were estimated in floodplains where grazing already occurs and where floodplains are expected to receive more frequent inundation under the 2800 GL/year scenario. Commercial fishery outcomes in the delta region were estimated using regression relationships between river flows and historic catch.

CULTURAL ECOSYSTEM SERVICES

Noting that many cultural ecosystem services exist (see Church *et al.*, this book), a number of which are very difficult to value (Chan *et al.* 2012), we selected those services which have pedigree in economic valuation studies, namely recreation and tourism (Rolfe & Dyack 2011) and aesthetics (Tapsuwan *et al.* 2012). Tourism and recreation values were estimated using benefit transfer of recreation estimates from the Basin (Morrison & Hatton MacDonald 2010), threshold water quality indicators and historic visitation data. Following methods in Tapsuwan *et al.* (2012), an original hedonic study modelled the relationship between river flow and lake level height and nearby house sale prices. These results were then used with modelled changes in river flow and lake level height between the two scenarios to estimate aesthetic values.

HABITAT ECOSYSTEM SERVICES

The Kumar (2010) ecosystem service framework identifies a group of habitat services as a discrete fourth category of final services that can be sensibly valued. We valued enhanced native species diversity from improved health of floodplain vegetation, increased waterbird breeding, and increased stocks of native fish; healthier Coorong and Lower Lakes ecosystems were valued using monetary estimates from an earlier study undertaken in the Basin (Hatton MacDonald *et al.* 2011), a benefit transfer approach (Morrison & Hatton MacDonald 2010), combined with incremental ecological outcomes between the two scenarios.

ECONOMIC VALUES OF REDUCING DIVERSIONS

In general, a healthy and functioning environment will provide positive economic value to society through enhanced supply of ecosystem services. Table 10.1 lists our monetary estimates for the benefits in terms of ecosystem services from increasing the volume of water in the Basin by an annual average 2800 GL per year. The values are dominated by habitat ecosystem services, specifically improved health of floodplain vegetation, increased waterbird breeding, increased stocks of native fish, and a healthier Coorong. There are other large benefits from carbon sequestration, aesthetics, recreation, and enhanced provision of regulating ecosystem services. In total the estimated monetary value of the marginal change in ecosystem services between the two scenarios is around AU$4 billion to AU$9 billion in current prices, depending on assumptions.

10.4 WHAT DOES AN ECOSYSTEM SERVICES-BASED APPROACH BRING?

This case study provides an opportunity to reflect on advances made in the integration of biophysical quantification of ecosystem services delivery with valuation techniques. Here we discuss how our operationalization of the ecosystem services-based approach can help support improved decision-making in the context of the four core elements outlined in Chapter 2 of this book.

The integration of natural and social sciences and other strands of knowledge for a comprehensive understanding of the service delivery process (core element 3).

Absent in the drafting of the Basin Plan was a clear description of the social and economic benefits that would accrue to Australian communities from reducing by 2800 GL per year the volume of water diverted to agriculture. The draft Basin Plan was focused mainly on the volumes of water required to reach hydrological targets that if achieved would maintain or improve the ecological health of important wetlands. The narrow disciplinary focus made it more difficult for the Australian government to counter the concerns of those potentially impacted by the reduction in diversions for irrigation. Ecosystem services-based approaches provide an analytical framework for interdisciplinary integration between biophysical and socio-economic sciences. Integration allows for improved quantification and explanation of the benefits to human wellbeing and the economy flowing from sustainable resource management and policy.

The understanding of the bio-physical underpinning of ecosystem functions in terms of service delivery (core element 2).

An important step to take advantage of the integration potential offered by an ecosystem services approach is to co-develop ecosystem services endpoint models. We relied primarily on existing biophysical models and valuation studies to apply the ecosystem services approach to assess the benefits of reducing the amount of water diverted for irrigation. A note of caution in such circumstances is that biophysical models and valuation studies not developed for the purpose of an ecosystem services assessment may face scale mismatches between models and assessment, and model outputs may not be entirely fit for purpose. The outcome for the ecosystem services assessment is that confidence in the underlying biophysical models and valuation techniques varies with each ecosystem service assessed.

Nonetheless, the ecosystem services-based approach applied here was found to be a useful tool to improve understanding of how changes to biophysical processes in freshwater ecosystems lead to multiple benefits that arise from reduced irrigation diversions in the Murray-Darling Basin. In our case study we found that existing biophysical models and valuation studies linking changes in flow and inundation to regulating and cultural ecosystem services were particularly lacking. In this way ecosystem services-based approaches can identify research gaps in how biodiversity and ecosystem functioning supplies ecosystem services that contribute to human wellbeing, and provide opportunity for an integrated future research agenda. A prominent example is the need to better understand how improvements in wetland health and functioning impact (positively or otherwise) on spiritual and cultural values held by Australia's indigenous people.

The assessment of the services provided by ecosystems for its incorporation into decision-making (core element 4).

Integrating biophysical modelling and valuation within a single ecosystem services framework provides a means for decision-makers to trace the connection between policy reform (reductions in irrigation diversions), to hydrologic outcomes in terms of changed flow and inundation timing, extent, and patterns, to incremental changes in ecosystem outcomes and the flow of ecosystem services, to monetarily valuing the ecosystem services benefits. The integration of biophysical quantification of ecosystem services delivery with valuation methodology provided valuable information for the Australian government's cost–benefit analysis that is required for any major policy implementation. Our study was also useful as a communication tool, particularly through conceptual maps of the connections between biophysical changes and human wellbeing. Nevertheless, the acceptance of ecosystem services-based approaches by decision-makers and stakeholders in the Murray-Darling Basin is mixed: survey results from Hatton MacDonald *et al.* (2014) report that the approach is considered experimental and is not well understood, particularly outside of the Australian government. Studies that investigate science impact have the potential to provide lessons on how to improve the relevance of ecosystem services-based approaches, or science more generally, in policy decision-making.

The recognition that the status of ecosystems has an effect on human wellbeing (core element 1).

Using an ecosystem services-based approach is challenging because the science of quantifying and valuing the contribution of ecosystem services to human wellbeing, while conceptually powerful, is relatively novel and experimental. While the contribution to wellbeing is recognized, we found, in our case study, a relative dearth in the indicators and data available at appropriate spatial and temporal scales to describe the full suite of ecosystem service benefits that may arise from recovering more water for the environment. More difficult to identify with any precision were the contributions to wellbeing and subsequent benefits of the regulating services, including wastewater treatment, erosion prevention, maintenance of soil fertility, and moderation of extreme events. Also, more research needs to be done to place (monetary and non-monetary) value on cultural ecosystem services, such as spiritual and sense of place, and mental health. In some cases improvements might be best captured with indices, or mapped using participatory approaches (Plieninger *et al.* 2013).

However, monetary measurements of improvements to wellbeing and the subsequent benefits from reducing water diverted for irrigation in the Murray-Darling Basin are amenable to cost–benefit analysis which places in context any costs arising from reduced irrigated agricultural production. Valuing ecosystem service benefits is fraught with difficulties because many of the ecosystem services provided by the wetlands and floodplains in the Murray-Darling Basin are public goods for which there are no indicators of market value (Boyd 2007). While there is a growing body of non-market valuation techniques and an increasing acceptance of placing a monetary value on the environment (Atkinson *et al.* 2012), the commoditization of nature comes with deep ethical and moral challenges (Chan *et al.* 2012). Other ways to measure ecosystem service contributions to human wellbeing may be more acceptable to different people, for example by using non-monetary measurements and indicators, and ranking these in participatory process such as multi-criteria analyses (Liu *et al.* 2013).

10.5 CONCLUSION

The objectives of the Water Act (Commonwealth of Australia 2007), among others, are to 'protect, restore and provide for the ecological values and ecosystem services of the MurrayDarling Basin' and to 'maximise the net economic returns to the Australian community from the use and management of the Basin water resources'. Yet in Australian (and international) water management and planning there are few examples where an ecosystem services approach helps determine how much water is delivered where, and when to maintain or improve freshwater ecosystem health. In Australia, water-sharing plans typically contain targets for water flow volumes and timing based on relatively simple relationships between flow and ecology, with the assumption that achieving particular flow targets at key locations along the river will achieve ecological goals in the river and on floodplains, and then presumably achieve ecosystem outcomes.

We demonstrate that reducing by 2800 GL per year the water that can be diverted for irrigation in the Murray-Darling Basin, and thereby leaving this water in the river system, can offer significant improvements to ecosystems, which translate to

---

Box 10.2 Key messages

- The Australian government has proposed a 21% reduction in the volume of water that can be diverted for irrigation in the Murray-Darling Basin, Australia.
- The reduced irrigation diversions may lead to economic costs through lower crop production.
- Using ecosystem services-based approaches, we estimate that the benefits, and where possible the monetary values, of reduced diversions are in the same order of magnitude as the costs.
- Our case study demonstrates that ecosystem services-based approaches are very useful for supporting decision-making and the cost–benefit analyses often required under policy implementation.

---

improved flows of ecosystem services and the potential for significant economic benefits. At the risk of double counting, the assessment of the benefits may be worth up to AU$9 billion. Reduced diversions, principally for irrigation, will have an economic cost, estimated at approximately AU$550 million annually (Murray-Darling Basin Authority 2012b), or in present value terms approximately AU$7 billion (7% discount rate over 30 years). The cost of returning water to the environment may not be as high as projected if the economic value of improved ecosystem services is considered. Research into restoration of dryland systems is relatively well advanced in the investigation of spatially explicit land management strategies that maximize ecological, ecosystem service and therefore benefits and associated values (Crossman & Bryan 2009; Bryan *et al.* 2011). We suggest ecosystem services-based approaches offer a new way to manage water resources for maximum economic benefits to all water users, including the environment.

ACKNOWLEDGEMENTS

The authors thank all the other team members involved in the CSIRO Multiple Benefits of the Basin Plan Project. Funding for this research was provided by the Murray-Darling Basin Authority and the CSIRO Water for a Healthy Country National Research Flagship.

References

Atkinson, G., Bateman, I., & Mourato, S. (2012). Recent advances in the valuation of ecosystem services and biodiversity. *Oxford Review of Economic Policy* **28**, 22–47.

Australian Bureau of Statistics (2013). *Gross Value of Irrigated Agricultural Production, 2011–12*. Australian Bureau of Statistics, Canberra.

Banerjee, O., Bark, R., Connor, J., & Crossman, N. D. (2013). An ecosystem services approach to estimating economic losses associated with drought. *Ecological Economics* **91**, 19–27.

Boyd, J. (2007). Nonmarket benefits of nature: what should be counted in green GDP? *Ecological Economics* **61**, 716–723.

Boyd, J. & Banzhaf, S. (2007). What are ecosystem services? The need for standardized environmental accounting units. *Ecological Economics* **63**, 616–626.

Brauman, K. A., Daily, G. C., Duarte, T. K. E., & Mooney, H. A. (2007). The nature and value of ecosystem services: an overview highlighting hydrologic services. *Annual Review of Environment and Resources* **32**, 67–98.

Bryan, B. A. & Crossman, N. D. (2013). Impact of multiple interacting financial incentives on land use change and the supply of ecosystem services. *Ecosystem Services* **4**, 60–72.

Bryan, B. A., Crossman, N. D., King, D., & Meyer, W. S. (2011). Landscape futures analysis: assessing the impacts of environmental targets under alternative spatial policy options and future scenarios. *Environmental Modelling & Software* **26**, 83–91.

Chan, K. M. A., Satterfield, T., & Goldstein, J. (2012). Rethinking ecosystem services to better address and navigate cultural values. *Ecological Economics* **74**, 8–18.

Commonwealth of Australia (2007) *Water Act 2007: Basin Plan 2012*. Commonwealth of Australia, Canberra.

Crossman, N. D. & Bryan, B. A. (2009). Identifying cost-effective hotspots for restoring natural capital and enhancing landscape multifunctionality. *Ecological Economics* **68**, 654–668.

CSIRO (2012). Assessment of the ecological and economic benefits of environmental water in the Murray-Darling Basin. In: *CSIRO Water for a Healthy Country*. National Research Flagship, Canberra.

De Groot, R. S., Alkemade, R., Braat, L., Hein, L., & Willemen, L. (2010). Challenges in integrating the concept of ecosystem services and values in landscape planning, management and decision making. *Ecological Complexity* **7**, 260–272.

Geneletti, D. (2013). Assessing the impact of alternative land-use zoning policies on future ecosystem services. *Environmental Impact Assessment Review* **40**, 25–35.

Gordon, L. J., Finlayson, C. M., & Falkenmark, M. (2010). Managing water in agriculture for food production and other ecosystem services. *Agricultural Water Management* **97**, 512–519.

Grafton, R. Q., Pittock, J., Davis, R., *et al.* (2013). Global insights into water resources, climate change and governance. *Nature Climate Change* **3**, 315–321.

Grey, D. & Sadoff, C. W. (2007). Sink or swim? Water security for growth and development. *Water Policy* **9**, 545–571.

Haines-Young, R. & Potschin, M. (2010). The links between biodiversity, ecosystem services and human well-being. In D. G. Raffaelli & C. L. J. Frid (eds), *Ecosystem Ecology: A New Synthesis*. Cambridge University Press, Cambridge.

Hatton MacDonald, D., Morrison, M. D., Rose, J. M., & Boyle, K. J. (2011). Valuing a multistate river: the case of the River Murray. *Australian Journal of Agricultural and Resource Economics* **55**, 373–391.

Hatton MacDonald, D., Bark, R. H., & Coggan, A. (2014). Is ecosystem service research used by decision-makers? A case study of the Murray-Darling Basin, Australia. *Landscape Ecology* **29**, 1447–1460.

Keeler, B. L., Polasky, S., Brauman, K. A., *et al.* (2012). Linking water quality and well-being for improved assessment and valuation of ecosystem services. *Proceedings of the National Academy of Sciences* **109**, 18619–18624.

Kingsford, R. T. (2000). Ecological impacts of dams, water diversions and river management on floodplain wetlands in Australia. *Austral Ecology* **25**, 109–127.

Kingsford, R. T., Walker, K. F., Lester, R. E., *et al.* (2011). A Ramsar wetland in crisis: the Coorong, Lower Lakes and Murray Mouth, Australia. *Marine and Freshwater Research* **62**, 255–265.

Kumar, P.(2010). *The Economics of Ecosystems and Biodiversity: Ecological and Economic Foundations*. Earthscan, London and Washington, DC.

Liu, S., Crossman, N. D., Nolan, M., & Ghirmay, H. (2013). Bringing ecosystem services into integrated water resources management. *Journal of Environmental Management* **129**, 92–102.

Maes, J., Paracchini, M. L., Zulian, G., Dunbar, M. B., & Alkemade, R. (2012). Synergies and trade-offs between ecosystem service supply, biodiversity, and habitat conservation status in Europe. *Biological Conservation* **155**, 1–12.

Maltby, E. & Acreman, M. C. (2011). Ecosystem services of wetlands: pathfinder for a new paradigm. *Hydrological Sciences Journal* **56**, 1341–1359.

Martín-López, B., Gómez-Baggethun, E., García-Llorente, M., & Montes, C. (2014). Trade-offs across value-domains in ecosystem services assessment. *Ecological Indicators* **37**(Part A), 220–228.

Morrison, M. & Hatton MacDonald, D. (2010). *Economic Valuation of Environmental Benefits in the Murray-Darling Basin. Report Prepared for the Murray–Darling Basin Authority*. Murray–Darling Basin Authority, Canberra.

Murray-Darling Basin Authority (2012a). *Regulation Impact Statement: Basin Plan*. Murray-Darling Basin Authority, Canberra.

Murray-Darling Basin Authority (2012b). *The Socio-economic Implications of the Proposed Basin Plan*. Murray-Darling Basin Authority, Canberra.

National Research Council (2005). *Valuing Ecosystem Services: Toward Better Environmental Decision-Making*. National Academies Press, Washington, DC.

Plieninger, T., Dijks, S., Oteros-Rozas, E., & Bieling, C. (2013). Assessing, mapping, and quantifying cultural ecosystem services at community level. *Land Use Policy* **33**, 118–129.

Raudsepp-Hearne, C., Peterson, G. D., & Bennett, E. M. (2010). Ecosystem service bundles for analyzing tradeoffs in diverse landscapes. *Proceedings of the National Academy of Sciences* **107**, 5242–5247.

Rolfe, J. & Dyack, B. (2011). Valuing recreation in the Coorong, Australia, with travel cost and contingent behaviour models. *Economic Record* **87**, 282–293.

Rouquette, J. R., Posthumus, H., Morris, J., Hess, T. M., Dawson, Q. L., & Gowing, D. J. G. (2011). Synergies and trade-offs in the management of lowland rural floodplains: an ecosystem services approach. *Hydrological Sciences Journal* **56**, 1566–1581.

Schluter, M., Leslie, H., & Levin, S. (2009). Managing water-use trade-offs in a semi-arid river delta to sustain multiple ecosystem services: a modeling approach. *Ecological Research* **24**, 491–503.

Seppelt, R., Lautenbach, S., & Volk, M. (2013). Identifying trade-offs between ecosystem services, land use, and biodiversity: a plea for combining scenario analysis and optimization on different spatial scales. *Current Opinion in Environmental Sustainability* **5**, 458–463.

Sims, N. C. & Colloff, M. J. (2012). Remote sensing of vegetation responses to flooding of a semi-arid floodplain: implications for monitoring ecological effects of environmental flows. *Ecological Indicators* **18**, 387–391.

Tapsuwan, S., MacDonald, D. H., King, D., & Poudyal, N. (2012). A combined site proximity and recreation index approach to value natural amenities: an example from a natural resource management region of Murray-Darling Basin. *Journal of Environmental Management* **94**, 69–77.

11 An ecosystem services-based approach to integrated regional catchment management

The South East Queensland experience

Simone Maynard, David James, Stuart Hoverman, Andrew Davidson, and Shannon Mooney

11.1 AN ECOSYSTEM SERVICES-BASED APPROACH TO INTEGRATED REGIONAL CATCHMENT MANAGEMENT

The importance of ecosystem services to sustainable development and the wellbeing of communities is well recognised and has been emphasised in many international policies and programmes (World Commission on Environment and Development 1987; Millennium Ecosystem Assessment 2005; European Commission *et al.* 2012). In recognition of this importance, stakeholders (e.g. government, non-government, business, industry, community, Traditional Owners, researchers) across South East Queensland (Australia) came together to develop the South East Queensland Ecosystem Services Framework, a tool to identify, measure, and value the ecosystem services provided by the region (Maynard *et al.* 2010, 2012). The Framework, which operationalises an ecosystem services-based approach, is

---

Box 11.1 Evolution of integrated catchment management in South East Queensland, Australia

- Community-led watershed management was underway in the Lockyer catchment, the 'food bowl' of the South East Queensland region, in the early 1980s.
- Integrated Catchment Management was officially adopted as a natural resource management programme in the state of Queensland in the early 1990s.
- Integrated Catchment Management emerged from concerns relating to the degradation of natural resources; conflicting government policies; and increasing public expectations for involvement in decision-making.
- The national Regional Natural Resource Management Body network established by the Australian government evolved from these initiatives.
- Each region has a Regional Body whose primary role is to work with stakeholders to better manage the natural resources of the region.

---

described briefly in this chapter; also given is an example of the Framework's application to better align land use planning and on-ground catchment management.

'Integrated Catchment Management' takes a catchment-scale approach to the management of natural resources (Murray Darling Basin Commission 2001; Falkenmark 2004). Catchments are explicitly defined structures; they represent the natural order of landscape processes and provide spatial gradients and thus water pathways and soil generation profiles. These gradients determine the vegetation cover and ecological structure of the catchment, which in turn influence the spatial organisation of human settlements (Priscoli 1999). In essence, a catchment is the matrix in which socio-ecological systems organise themselves (Priscoli 1999; Ostrom 2009).

Integrated Catchment Management is an overarching approach that is related to many other community-led environmental activities which share the message 'what we do on the land is reflected in the water' (Condamine Alliance 2010; SEQ Catchments 2013a). The 'integrated' component of Integrated Catchment Management suggests that to manage natural resources sustainably we must understand the individual parts of the catchment (i.e. its land systems and hydrologic and biotic resources), as well as their interconnections (Mitchell & Hollick 1993; Falkenmark 2004).

At the heart of Integrated Catchment Management is its engagement process – the catchment's social component. Integrated Catchment Management provides a collaborative tool for a diverse range of stakeholders to share goals and aspirations within the context of the catchment (Murray Darling Basin Commission 2001; Binney & James 2011). Supportive, adaptive, and coordinated 'management' and governance is necessary to achieve desired outcomes (Mitchell & Hollick 1993; Murray Darling Basin Commission 2001; Falkenmark 2004). Common to both Integrated Catchment Management and ecosystem services-based approaches is recognition that different people hold different types of knowledge, manage different parts of the socio-ecological system, and have different values (Murray Darling Basin Commission 2001). Participatory and multidisciplinary approaches are required to extract and understand this knowledge, to improve the credibility, legitimacy, and saliency of information, and bridge the often competing epistemologies of stakeholders (core element 3 of ecosystem services-based approaches as defined in this book, Martin-Ortega *et al.*).

The challenge of applying an ecosystem services-based approach to Integrated Catchment Management increases when multiple agencies and institutional arrangements for managing catchments overlap at the regional scale. This requires assessing, managing, and planning for ecological processes providing multiple ecosystem services across adjacent ecosystems and catchments. For example, hydrological cycles typically cannot be considered or managed within the confines of a single agency, ecosystem, or catchment.

Land use planning is one area of decision-making where multiple responsibilities and assessment needs coincide. Spatial information associated with land use planning can strongly support Integrated Catchment Management, including clear images (e.g. maps) of the relationships between stakeholders and the benefits they receive from ecosystems. An ecosystem services-based approach to Integrated Catchment Management can enable on-ground management activities to be aligned with measureable spatial land use planning objectives. As background to the example described in this chapter, the following section discusses Integrated Catchment Management in the context of the South East Queensland region.

11.2 INTEGRATED CATCHMENT MANAGEMENT IN THE SOUTH EAST QUEENSLAND REGION

South East Queensland is one of the fastest growing metropolitan regions in Australia, with an area of $23\,000$ km^2 (Queensland Government 2009b). The natural systems are being subjected to considerable pressures resulting from population growth and the expansion of urban, commercial, and industrial activities (Queensland Government 2009a, 2009b). An adequate quantity and quality of natural resources is required to support the well-being of the 3.4 million people residing in South East Queensland, as well as supplying inputs to local industries such as agriculture, mining, manufacturing and tourism (Queensland Government 2009a, 2009b).

Integrated Catchment Management was officially adopted as a natural resource management programme in Queensland in the early 1990s when the state government established three pilot projects in heterogeneous catchments (McDonald *et al.* 1999). However, community-led watershed management was underway in the Lockyer catchment, the 'food bowl' of the South East Queensland region, in the early 1980s (McDonald *et al.* 1999). Integrated Catchment Management primarily emerged from concerns relating to the continuing and increasing degradation of natural resources; conflicting government policies and programmes; and increasing public expectations for involvement in decision-making (McDonald *et al.* 1999). In South East Queensland, the long-term uptake of Integrated Catchment Management by governments and landholders has proved challenging and in many respects Lockyer is still one of the more difficult areas for stakeholder engagement and land use practice change.

Present-day Landcare and Catchment Management Authorities have evolved from these initiatives, providing the foundations for the national regional natural resource management body network established by the Australian government (McDonald *et al.* 1999; Australian Government 2013b; Landcare Australia 2013). The relevant Regional Body for the South East Queensland region is SEQ Catchments, a non-government, not-for-profit organisation. South East Queensland Catchment's role is to work with stakeholders to better manage the natural resources of the region (Australian Government 2013a). Funding for natural resource management is generally delivered to Regional Bodies through national and state government programmes, primarily on an asset basis. An asset approach to natural resource management focuses investment on highly valued parts of the landscape (e.g. water quality, soil fertility, threatened species), rather than investing in large areas against broad-scale threats to the environment (Curtis & Lefroy 2010).

Land use planning in Australia remains mostly a state and local government issue. In Queensland, local government planning schemes are required to meet obligations under state planning policies and as determined in statutory regional plans for managing growth (Queensland Government 2009b). These regional plans are informed by non-statutory regional natural resource management plans (Queensland Government 2009a). Importantly, although land use and natural resource management plans are delivered at the regional scale, there are no regulatory agencies for either at this scale.

The 2011 and 2013 extreme flooding events in South East Queensland led to loss of life and catastrophic property and costly infrastructure damage greater than previously recorded in Australia's history (Urich *et al.* 2011). These events have once again turned the attention of stakeholders to improving their understanding of the capacity of catchments to mitigate the effects of such events on property, life, and other essential services to the large urban population (Urich *et al.* 2011). A spatially integrated multi-functional landscape can potentially buffer communities from floods and drought, as well as regulate and store water important for supporting other species and ecosystem services (SEQ Catchments 2013b). Tools are clearly required that encapsulate information on local ecosystems and ecological processes, and integrate stakeholder understanding and values of these with regional-scale land use planning.

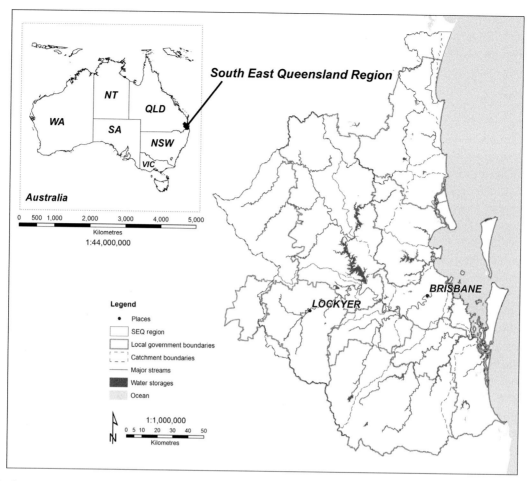

Figure 11.1 The location of South East Queensland in the context of Queensland, Australia, showing major rivers and catchments under the jurisdiction of 11 local governments. Coordinated natural resource management and land use planning is crucial to the health of waterways feeding into the receiving coastal areas, which often suffer from the cumulative impacts of land use practices and landscape change in the catchments.

11.3 THE SOUTH EAST QUEENSLAND ECOSYSTEM SERVICES FRAMEWORK

The South East Queensland Ecosystem Services Framework (the Framework) identifies the key ecosystems and ecological processes in South East Queensland, their spatial distribution, and the extent to which they potentially can contribute to ecosystem services important to the wellbeing of the South East Queensland community. The primary objective behind the Framework's development was to better plan for and manage ecosystem service delivery within the region (core element 4). The development of the Framework was coordinated by SEQ Catchments through an extensive stakeholder process involving over 190 professionals with knowledge, skills, and experience in the natural and social sciences and the region (core element 3) (Maynard *et al.* 2010; SEQ Catchments 2013b). Underpinning the decision to adopt this process was: (1) the complexity and multidisciplinary nature of ecosystem services; (2) the need for local knowledge; (3) the idea that those responsible for planning and

managing ecosystem services are more likely to understand the concept and apply the Framework if they were involved in its development; and (4) the limited resources available to construct and apply the Framework (Maynard *et al.* 2010 2012; Petter *et al.* 2012).

11.3.1 The Framework's structure

The Framework is best described as a participatory systems analysis model using data and evaluation techniques similar to those applied in multi-objective decision support systems (Janssen 1993). By making use of qualitative and often quantitative data these models determine interconnections between model components in the simplest possible way, additionally creating opportunities for stakeholders and experts to be actively involved in the processes of model construction and application. More detailed description of the Framework and its development can be found in Maynard *et al.* (2010, 2012) and SEQ Catchments (2013b).

Table 11.1 *The four Components for Assessment under the South East Queensland Ecosystem Services Framework.*

Ecosystem reporting categories	Ecosystem functions	Ecosystem services	Constituents of wellbeing
Deep ocean	Gas regulation	Food products	Breathing
Open water – pelagic	Climate regulation	Water for consumption	Drinking
Open water – benthic	Disturbance regulation	Building and fibre products	Nutrition
Coral reefs	Water regulation	Fuel resources	Shelter
Seagrass	Soil retention	Genetic resources for cultivated products	Physical health
Rocky shores	Nutrient regulation	Biochemicals, medicines, and pharmaceuticals	Mental health
Beaches	Waste treatment and assimilation	Ornamental resources	Secure and continuous supply of services
Dunes	Pollination	Transport infrastructure	Security of health
Coastal zone wetlands	Biological control	Air quality	Security of person
Palustrine wetlands	Barrier effect of vegetation	Habitable climate	Community and social cohesion
Lacustrine wetlands	Supporting habitats	Water quality	Secure access to services
Riverine wetlands	Soil formation	Arable land	Family cohesion
Rainforests	Food	Buffering against extremes	Security of property
Sclerophyll forests	Raw materials	Pollination	Social and economic freedom
Native plantations	Water supply	Reduce pests and diseases	Self-actualisation
Exotic plantations	Genetic resources	Productive soils	
Regrowth	Provision of shade and shelter	Noise abatement	
Native and improved grasslands	Pharmacological resources	Iconic species	
Shrublands and woodlands	Landscape opportunity	Cultural diversity	
Moreton Island		Spiritual and religious values	
Bribie Island		Knowledge systems	
North Stradbroke Island		Inspiration	
South Stradbroke Island & other Bay islands		Aesthetic values	
Montane		Effect on social interactions	
Sugarcane		Sense of place	
Horticulture – small crops		Iconic landscapes	
Horticulture – tree crops		Recreational opportunities	
Other irrigated crops		Therapeutic landscapes	
Dams			
Hard surfaces			
Parks and gardens			
Residential gardens			

Source: Adapted from Maynard *et al.* (2010).

The Framework itself consists of:

(1) *Lists and descriptions of four 'Components for Assessment'* (see Table 11.1), including 32 groups of ecosystems categorised as Ecosystem Reporting Categories; 19 broad groups of ecological processes termed Ecosystem Functions; 28 Ecosystem Services derived from these Ecosystem Reporting Categories and Ecosystem Functions; and 15 Constituents of Wellbeing dependent on Ecosystem Services.

(2) Interconnections between these four Components for Assessment are represented by *scores arranged in matrix form*. This innovative approach relied on the judgement of experts in the biological and social sciences (including economics) to determine the relative magnitude of the interconnections (on a score of 0–5).

(3) *A series of maps* spatially identifying areas of the region with potential to contribute to ecosystem service provision. This includes one map for each of the 32 Ecosystem Reporting Categories and one map for each of the 19 Ecosystem Functions.

The simple structure of the Framework provides a flexible approach to its application. Not all the information or tools supporting the Framework are needed in any one application. Rather, the information and tools required are determined by the decision-making context (e.g. local government planning schemes; nature conservation strategies; flood impact assessments) and the decision-maker's capacity to apply information (e.g. funding; project personnel; political will). Although there are multiple current and potential applications of the Framework, given South East Queensland's recent experience with serious floods, it is instructive to consider what the Framework can tell us about the ecosystem service 'Buffering Against Extremes' in the context of Integrated Catchment Management in South East Queensland.

11.3.2 An example application: Buffering Against Extremes

In the South East Queensland Framework the ecosystem service 'Buffering Against Extremes' is described as *the role of ecosystems in maintaining normal situations (e.g. buffering against extreme natural events such as droughts, floods, storms, tsunamis), including providing natural irrigation and drainage (e.g. water table regulation)* (SEQ Catchments 2013b). The relevant expert scores for Buffering Against Extremes indicate that this service contributes strongly (i.e. with a score of 4 or 5) to the following Constituents of Wellbeing important to the South East Queensland community: Shelter, Security of Person, Security of Property, Security of Health, Secure and Continuous Supply of Services, and Secure Access to Services (SEQ Catchments 2013b).

The scores provided by experts also indicate that the following ecosystem functions contribute most to Buffering Against Extremes: Climate Regulation, Disturbance Regulation, Water Regulation, Soil Retention, the Barrier Effect of Vegetation, Water Supply, and the Provision of Shade and Shelter (SEQ Catchments 2013b). Overlaying maps developed for each of these functions identifies key areas with low to high potential (0–7 functions) to help buffer the South East Queensland community from the effects of extreme events.[1]

The important role of aquatic ecosystems and vegetation in maintaining processes that contribute to this ecosystem service are highlighted through primary data sets underpinning this map. These include: Good Ground Cover; Woody Vegetation (on

streams); Flood Plains and Coastal Deposits; South East Queensland Water Bodies; and Wetlands. An explanation of these data sets is beyond the scope of this chapter, but full references and the rationale for the data sets used can be found in Petter *et al.* (2012). The map on the right of Figure 11.2 shows in grey the current South East Queensland urban footprint. Comparing these maps, it is evident that areas of current development dominated by hard surfaces and limited in vegetation cover have only low to medium potential to buffer against extreme events. Areas in red on the right-hand map show floodlines for the 2011 flood event.

The ability of catchments to actually perform these functions and to buffer communities in the South East Queensland region was demonstrated during the 2011 and 2013 floods. The main creeks of the Lockyer catchment did not have sufficient capacity to absorb the water discharged from the upper catchments when the 2011 floods occurred (van den Honert & McAneney 2011). The hydrological regime pre-European settlement (in the late eighteenth century) would have involved similar events; however, the extensive clearing of upper slopes and riparian zones (loss of ecological infrastructure) and increased impermeable surfaces has increased both the volume and velocity of these episodic events (van den Honert & McAneney 2011).

Consequently, low-lying areas of South East Queensland such as Queensland's capital city, Brisbane, were subject to flood flows which caused major damage (van den Honert & McAneney 2011; SEQ Catchments 2013a). Vital productive top soils and bed and bank sediment from non-vegetated and non-stabilised areas of the Lockyer travelled far through the basin, over-running and smothering hard infrastructure, as well as fragile coastal ecosystems in the receiving waters (i.e. wetlands and seagrass) (Maxwell *et al.* 2013; SEQ Catchments 2013a). Although significant attempts were made to repair catchments and communities after 2011, the 2013 floods gave little time to recover from the previous event and damage was exacerbated in both the upper and lower parts of the region (SEQ Catchments 2013a).

While the majority of catchments in the Lockyer have been subject to the intervention described above, a number of largely unmodified catchments still exist, with headwaters primarily in national parks. These unmodified catchments possess the characteristics of Integrated Catchment Management with well-vegetated headwaters and land management within the catchment in tune with the hydrological regime (Mitchell & Hollick 1993).

Overlaying areas of proposed urban growth (hatched areas in the right-side map in Figure 11.2) with areas of ecosystem function identifies where land use planning is compatible with, or in conflict with, effective Integrated Catchment Management. The maps suggest that much urban growth is still proposed in areas with medium to high potential to contribute to buffering against extreme events (i.e. areas containing 5–7 ecosystem functions).

---

[1] By following the same process, these types of maps, based on relevant suites of ecosystem functions identified by experts as important to delivering specific ecosystem services listed in the Framework, can be developed for each ecosystem service of interest or bundled to assess the potential for multiple service provision.

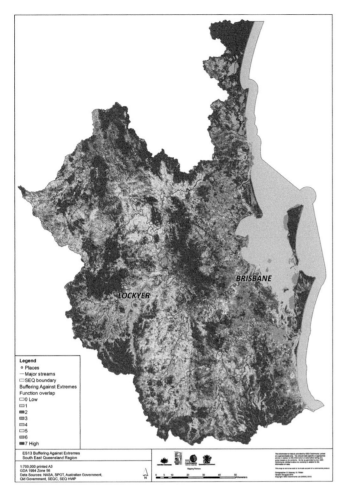
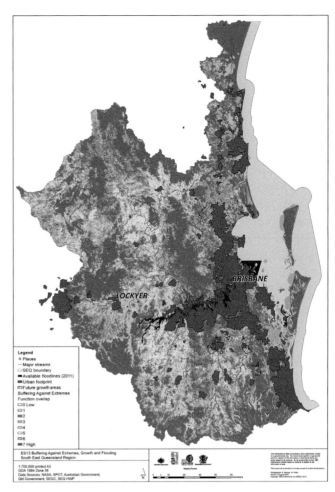

Figure 11.2 Both maps show areas of ecosystem function with potential to contribute to buffering against extremes. Low ecosystem function (white) = 0 functions occurring. High ecosystem function (green) = 7 functions occurring. Additionally, the map on the right shows current (grey) and proposed (hatched) urbanisation in South East Queensland. Areas in red show 2011 floodlines. A black and white version of this figure will appear in some formats. For the colour version, please refer to the plate section.

11.4 THE SOUTH EAST QUEENSLAND EXPERIENCE

Chapter 2 described the core elements of an ecosystem services-based approach and these are strongly reflected in the South East Queensland experience. The Framework developed by integrating natural and social science knowledge (core element 3) defines the different ecosystems in South East Queensland and recognises their interconnectedness through ecosystem functions (core element 2). The community is an integral and dominant component of ecosystems, depending greatly on the ecosystem services within the region for their wellbeing (core elements 1 and 4). The South East Queensland experience provides a valuable opportunity to explore impediments to, and strengths of, applying an ecosystem services-based approach to Regional Integrated Catchment Management.

11.4.1 Impediments to an ecosystem services-based approach

Institutional, governance, and funding arrangements to support an ecosystem services-based approach are yet to be developed for the South East Queensland region. This lack of supportive governance required South East Queensland Catchments (with no regulatory authority) to apply a bottom-up approach, exerting much effort with limited resources for engaging stakeholders and developing and implementing ecosystem service information. Supportive, adaptive, and coordinated governance that allows for the management of resources in an integrated manner (rather than in silos) would improve time frames from Framework development to on-ground implementation.

The limitations of traditional asset approaches to natural resource management and funding were evident when developing the Framework and attempting to apply a more systems-based approach to natural resource management and land use planning. The primary limitation inherited was data and information gaps. For example, information was limited on the location of groundwater ecosystems, thresholds of ecosystems and maps of specific ecosystems in South East Queensland. Empirical data on causal relationships between ecosystem service provision and wellbeing as it relates to the South East Queensland community were almost negligible. Curtis and Lefroy (2010) suggest that asset approaches fail to sufficiently acknowledge the importance of engaging and building the human, social and cultural capital required to underpin long-term environmental management.

As well, the strategic nature of asset approaches has the potential to sacrifice effectiveness for efficiency by overlooking large-scale biophysical and social processes that underpin the viability of the discrete assets (Curtis & Lefroy 2010). This mis-match of scale between landscape processes, natural resource management, and land use planning combined with an asset approach to funding often results in place-based issues being divorced from the complex management, planning, and policy development occurring at broader scales; and vice versa. Symptoms of issues (e.g. sediment transport) rather than their causes (e.g. change in vegetation cover) then become the focus of management efforts. This mis-alignment of objectives is the antithesis of Integrated Catchment Management motivations.

The Framework relies heavily on qualitative information obtained from experts rather than quantitative data. An important element of this process was determining and then engaging the appropriate people, then placing them within the process of developing the Framework in a manner that extracts the required information to support it. The use of simple scoring methods, value weights (rather than monetary values), and linear systems mean that only broad-brush assessments can be made with the Framework. Despite its limitations, the Framework does have the advantage of indicating the relative importance of particular ecosystems and therefore managing resources and catchments more generally, in terms of the functions they perform, the ecosystem services they provide, and their contribution to the wellbeing of the region's population.

The size, location, and arrangement of ecosystems are important to determining how much ecosystem service is being provided; who benefits from ecosystem services; any synergistic benefits or impacts; priority areas for conservation; and maintaining the health and resilience of ecosystems (Millennium Ecosystem Assessment 2005). Generally speaking, functions occurring on small spatial and temporal scales (e.g. the Provision of Shade and Shelter on a property) are easier to manage. However, large spatial and temporal scales of consideration (e.g. the scale of climate regulation) are essential to understanding the cumulative and/or future impacts of individual management

decisions. Available resources constrained the Framework to assess only ecosystem services derived from within the South East Queensland region. Information on ecosystem function and service flows in and out of the region has yet to be incorporated.

It is important to note that ecosystem services are derived from complex interactions between biotic and abiotic components of an ecosystem and the size, distribution, and diversity of ecosystems occurring across the landscape (SEQ Catchments 2013b). No individual species, group of species or individual ecosystem can provide the full suite of ecosystem services on which the community depends (SEQ Catchments 2013b). An ecosystem services-based approach is therefore not the panacea for rare, endangered, threatened species or ecosystem conservation. Rather, ecosystem services-based approaches are better suited to landscape-scale conservation efforts and less suitable for site scale or asset-based conservation initiatives (South East Queensland Catchments 2013b).

11.4.2 Strengths of an ecosystem services-based approach

The ecosystem services-based approach encouraged stakeholders to look beyond traditional silo approaches to resource management, and develop a shared conceptual understanding of how the South East Queensland system operates, represented by the Framework. The classification and systemisation of the socio-ecological system developed by stakeholders offers a simple structure and language for all, which translates scientific information into an understanding of how individuals and communities across the region benefit from ecosystem service provision.

Ecosystem services-based approaches provide an opportunity to change how decision-makers communicate with their audiences. Through current and potential evolutionary applications the language of 'benefits' provides a basis for building social capacity and informed decision-making through positive communication of Integrated Catchment Management, rather than the negatives of poor catchment management. An ecosystem services-based approach to Integrated Catchment Management has the potential to strengthen links between community education and engagement in on-ground catchment management and land use planning at local, regional, and state scales.

The qualitative and participatory process to developing the Framework produced a transparent and repeatable approach to assessing ecosystem services by stakeholders. Over time, the outcomes of stakeholder assessments will build the capacity in South East Queensland to monitor, detect, and predict change in ecosystem service provision. The scores supporting the Framework provide the basis to identify and rank catchments in terms of their potential community benefits. Potential beneficial or adverse impacts may be incorporated, including those resulting from exogenous drivers such as climate change. These types of models can improve the identification, use, and management practices of catchment resources within a time-specific manner by including information on larger controlling factors of weather, geology, biology, and hydrologic regimes. The South East Queensland experience also reveals that an ecosystem function approach to ecosystem service assessments can identify key areas where few functions occur that might be suitable for rehabilitation or restoration to improve ecosystem service provision; or they may be areas suitable for future development or offsets.

The participatory process through which the Framework was developed has brought numerous advantages in terms of its own application. Expert participatory mapping can help to identify existing common information and key data sets of importance to integrating ecosystem services-based approaches and current local and state government planning and natural resource management. Overlays of ecosystem function maps with local government planning schemes and nature conservation strategies reveal the limitations of these planning instruments to capture and protect landscape processes, specifically floodplains in South East Queensland (S. Maynard and S. Mooney, personal communication 2013). Social learning ('learning while doing') has improved stakeholder understanding of ecosystem service concepts, providing ownership and empowerment of the Framework as is evident in the uptake of ecosystem services in the regional statutory planning document and natural resource management plan for South East Queensland (Queensland Government 2009a, 2009b; Maynard *et al.* 2012).

11.5 CONCLUSION

The South East Queensland experience shows how an ecosystem services-based approach can add effectiveness to Integrated Catchment Management by integrating human wellbeing considerations into natural resource management decisions. The Framework's development is premised on the idea that different stakeholders are responsible for planning and managing different parts of a catchment, and that different people benefit from the derived ecosystem services in different ways. Consequently, to sustainably manage natural resources for the provision of ecosystem services a consistent approach to assessments is required across stakeholders.

The complexity and multidisciplinary nature of ecosystem services and the lack of data and information available on the link between ecosystems and community wellbeing provided limitations. To overcome these limitations the Framework relied on qualitative information obtained from local experts, and developed simple scoring methods, value weights, and linear systems to support the Framework. The Framework's common language, its tested knowledge, and tools that resource managers

---

Box 11.2 Key messages

- To sustainably manage natural resources for the provision of ecosystem services a framework is required to consistently conduct assessments across stakeholders.
- Integrated Catchment Management must underpin land use planning to work within the natural capacity of catchments to provide ecosystem services.
- An ecosystem services-based approach adds effectiveness to traditional Integrated Catchment Management and land use planning by incorporating human wellbeing considerations into decisions.
- Supportive, adaptive, and coordinated governance is required to effectively align Integrated Catchment Management and land use planning.
- The use of simple scoring methods, value weights (rather than monetary values), and linear systems mean that only broad-brush assessments can be made.
- An ecosystem services-based approach can help build bridges across sectors, organisations, and disciplines.

---

can apply, provides a platform that combines the physical, social, and management objectives of the region.

Although catchments are generally managed for 'normal' events (e.g. average rainfall), the South East Queensland experience identifies the importance of managing catchments to protect communities against extreme events. Studies conducted in the aftermath of the 2011 and 2013 floods have called for an Integrated Catchment Management approach to maintain and enhance landscape functions that contribute to ecosystem service provision (SEQ Catchments 2013a). Integrated Catchment Management must underpin land use planning to work within the natural capacity of catchments to provide ecosystem services (SEQ Catchments 2011). For example, the repair of river bank vegetation would do much to regulate water flows; mitigate the transportation of sediments; and contribute to the provision of multiple ecosystem services such as providing water quality improvements and regularising water supplies (South East Queensland Catchments 2011).

The process to develop the Framework and its application has proved useful to resource management in South East Queensland as it continues to build bridges across sectors, organisations, and disciplines. Collaboration has improved across stakeholders based on a common understanding that a catchment operates as a system with a multiplicity of interconnecting parts. The health and utility of the catchment depends on the decisions made by people using and managing the land, water, and other natural resources. Catchment health is a responsibility shared by all South East Queensland stakeholders.

Society's ability to manage the resilience in our catchments, so both the catchment and stakeholders can adapt to pressures such as irregular rainfall, rising temperatures, and increasing populations, requires a new social narrative that allows stakeholders to better understand the consequences of their demands on the landscape to continue delivering the goods and services of Nature on which they ultimately depend.

References

Australian Government (2013a). Australia's bioregional framework. Available at: www.environment.gov.au/parks/nrs/science/bioregion-framework (last accessed 12 June 2013).

Australian Government (2013b). Caring for our country: regional delivery. Available at: www.nrm.gov.au/funding/regional/index.html (last accessed 8 August 2013).

Binney, J. & James, D. (2011). Sharing the load: a collaborative approach to investing in South East Queensland's Waterways. Mainstream Economics.

Condamine Alliance (2010). *Natural Resource Management Plan.* Condamine Alliance, Toowoomba.

Curtis, A. & Lefroy, E. (2010). Beyond threat- and asset-based approaches to natural resource management in Australia. *Australasian Journal of Environmental Management* 17(3), 134–141.

European Commission, Food and Agriculture Organization, International Monetary Fund, *et al.* (2012). System of environmental-economic accounting: central framework. White Paper.

Falkenmark, M. (2004). Towards integrated catchment management: opening the paradigm locks between hydrology, ecology and policy-making. *International Journal of Water Resources Development* 20(30), 275–281.

Janssen, R. (1993). *Multi-objective Decision Support for Environmental Management.* Kluwer Academic Publishers, Dordrecht.

Landcare Australia (2013). Landcare: about. Available at: www.landcareonline.com.au/?page_id=2 (accessed 1 August 2013).

Maxwell, P., Burfeind, D., Pitt, K., *et al.* (2013). The effects of catchment run-off on seagrass in Moreton Bay. Griffith University, Report to Healthy Waterways Limited.

Maynard, S., James, D., & Davidson, A. (2010). The Development of an ecosystem services framework for South East Queensland. *Environmental Management* 45(5), 881–895.

Maynard, S., James, D., and Davidson, A. (2012). An adaptive participatory approach for developing an ecosystem services framework for South East Queensland, Australia. *International Journal of Biodiversity*, 7(3), 1–8.

McDonald, G., Bellamy, J., McDonald, K., *et al.* (1999). ICM in Queensland 1990–1999: an anthology. In: *Institutional Arrangements for ICM in Queensland.* CSIRO Sustainable Ecosystems, Brisbane.

Murray-Darling Basin Commission (2001). *Integrated Catchment Management in the Murray–Darling Basin 2001–2010: Delivering a Sustainable Future.* Murray-Darling Basin Commission, Canberra.

Millennium Ecosystem Assessment (2005). *Ecosystems and Human Wellbeing: General Synthesis.* Island Press, Washington, DC.

Mitchell, B. & Hollick, M. (1993). Integrated catchment management in Western Australia: transition from concept to implementation. *Environmental Management* 17(6), 735–743.

Ostrom, E. (2009). A general framework for analyzing sustainability of social-ecological systems. *Science* 24(325), 419–422.

Petter, M., Mooney, S., Maynard, S., *et al.* (2012). A methodology to map ecosystem functions to support ecosystem services assessments. *Ecology and Society*, 18(1), 31.

Priscoli, J. (1999). Water and civilization: using history to reframe water policy debates and to build a new ecological realism. *Water Policy* 1, 623–636.

Queensland Government (2009a). SEQ natural resource management plan 2009–2031. Queensland Government Department of Environment and Resource Management.

Queensland Government (2009b). SEQ regional plan 2009–2031. Queensland Government Department of Infrastructure and Planning.

SEQ Catchments (2011). *Submission to the Queensland Floods Commission of Inquiry Restoring Ecological Infrastructure for Flood Resilience: the 2011 Southeast Queensland Floods and Beyond.* SEQ Catchments Ltd, Brisbane.

SEQ Catchments (2013a). *Flood Impacts Report: February 2013.* SEQ Catchments Ltd, Brisbane.

SEQ Catchments (2013b). The SEQ Ecosystem Services Framework. Available at: www.ecosystemservicesseq.com.au (last accessed 11 February 2013).

Urich, P., Li, Y., Kouwenhoven, P., *et al.* (2011). Analysis of the January 2011 extreme precipitation event in the Brisbane River Basin. A CLIMsystems Technical Report.

van den Honert, R. & McAneney, J. (2011). The 2011 Brisbane floods: causes, impacts and implications. *Water* 3, 1149–1173.

World Commission on Environment and Development (1987). *Our Common Future.* Oxford University Press, Oxford.

12 Policy support systems for the development of benefit-sharing mechanisms for water-related ecosystem services

Mark Mulligan, Silvia Benítez-Ponce, Juan S. Lozano-V, and Jorge Leon Sarmiento

12.1 INTRODUCTION

There are a number of persistent oversimplifications in the lay understanding of water ecosystem services; for example: *forests generate more water, forests prevent floods, forests sustain dry season flows, forests improve water quality.* The reality is that the role of forests in water ecosystem services depends on (1) the landscape and climate context (terrain, rainfall, seasonality, storm characteristics, drought characteristics), (2) the type of forest, (3) the land cover and land use alternatives to forest cover and its management; and (4) the distribution of people locally and downstream of the site in question and their demand for ecosystem services. Another key control is the area of the forest in relation to other land uses and its location in relation to spatial heterogeneity of climate and other environmental properties and relative to downstream populations. Water ecosystem services are hence fundamentally a property of climate, but land cover and land use can have an impact on: water balance (through land cover effects on evapotranspiration and fog inputs); on runoff partitioning (through land cover and management effects on infiltration and runoff rates, on slope gradients and on subsurface flows); and through secondary impacts on water through agricultural water use and management infrastructure.[1] Water balance in turn impacts on the services of water provision through control of infiltration (soil water used in transpiration) and river runoff. Seasonality is also a strong control on water regulation and on water quality. The impact of

human-induced land cover and land use on water ecosystem services will depend on the magnitude of human intervention in relation to other cover types at the catchment scale and the location of human land cover and land use in relation to topographic, climatic, and soil factors in the catchment and in relation to beneficiaries downstream. The impact of a single farmer's actions on downstream water ecosystem services may be small, but the action of many farmers can produce non-linear cumulative downstream responses.

There is thus no simple 'rule of thumb' for understanding the impact of climate or land use change on water ecosystem services: local geography is critical and sophisticated analyses combining locally specific data and models of generic processes must be applied in each case to understand both the baseline and the likely impacts of management change (see also Capon *et al.* in this book in relation to climate change and water ecosystem services). This chapter contributes to the discussion on ecosystem services-based approaches as proposed in this book (Chapter 2) with a particular emphasis on core element 2, relating to the biophysical underpinning of the services delivery process and its linkage to economic information (core element 3), and how that can be used for the development of process-based policy support systems (core element 4). It is common that data on the biophysical underpinning of the services delivery process (for example, in relation to sediment dynamics, flow regulation, nutrient retention) are not available for watersheds. Acquiring such data from the field can be expensive and lengthy, and often not possible at all scales at which they are required. In this chapter we use the Guayaquil Water Fund in Ecuador as a case study for the application of an ecosystem services-based approach to water management. We examine the role of scientific data and tools in the effective investment of water fund payments to optimise the water ecosystem service outcomes of conservation and land management interventions. Water funds are innovative benefit-sharing mechanisms in which water users such as hydropower, municipal water companies, and private indus-

---

[1] It is important to establish a clear distinction between environmental and ecosystem services where the former are a function of the broader environment (including climate and terrain) and thus not manageable at the typically local to regional policy and land management scales. The latter are, however, a service provided by the ecosystem on the ground (vegetation, soil, wetlands) and thus can be manipulated by farmers or conservationists for both positive and negative ecosystem service delivery outcomes (Mulligan 2013), and hence fall within the definition of ecosystem services-based approaches as proposed in this book (Chapter 2).

Box 12.1 InVEST-RIOS

InVEST is a suite of modelling tools designed to inform decision-making about natural resource management. It can help answer questions related to where and how much of an ecosystem service is provided, and what is the value of that provision (Kareiva *et al.* 2011; Tallis *et al.* 2013). This focus on services and value is critical to an ecosystem services-based approach. RIOS was developed by the Natural Capital Project, a partnership between The Nature Conservancy, the World Wildlife Fund, Stanford University, and Minnesota University. The water-related models (e.g. sediment retention, water purification) have been used successfully to support the development of water funds. For the creation of water funds it has been useful to understand the change in the value of an ecosystem service, given some scenarios of change created under the assumption of investing or not in ecosystem conservation and other land management interventions. The estimated benefits of conservation allow the potential partners of the water fund to establish goals and improve the decision-making process on where to establish the conservation activities. Although the use of InVEST is helpful in the process of creating a water fund, the design of scenarios for intervention was always a big challenge for stakeholders and associates to the water fund. To help solve this issue, the Natural Capital Project with the Latin America Water Funds Partnership recently developed RIOS (Resource Investment Optimization System), a tool designed specifically to prioritise areas to implement land management interventions based on a set of objectives (ecosystem services improvements), in order to obtain the highest return on a financial investment (the greatest benefits for nature and people at the least cost; see Vogl *et al.* 2013). The scenarios created by RIOS are usually analysed with InVEST to estimate the changes in ecosystem service provision.

Box 12.2 WaterWorld

WaterWorld[a] is a spatially explicit, physically based globally applicable model for baseline and scenario water balance that is particularly well suited to heterogeneous environments with little locally available data (e.g. ungauged basins) and which is delivered through a simple web interface, requiring little local capacity for use. The model is 'self parameterising' in the sense that all data required for model application anywhere in the world are provided with the model. However, if users have better data than those provided with WaterWorld, it is possible to upload these as geographical information systems files. Results can be viewed visually within the web browser or downloaded as maps. The model's equations and processes are described in more detail by Mulligan and Burke (2005) and Mulligan (2013). WaterWorld is a grid-based water balance, water quality and soil erosion, transport and sedimentation model. Water balance is composed of wind-driven rainfall plus fog and snowmelt inputs, minus actual evapotranspiration calculated from the vegetation cover and type. The model can be applied at 1 ha and 1 km^2 spatial resolution with the available data for application to local and national scales, respectively. WaterWorld has been applied at sites throughout the world for estimating hydrological baselines and its inbuilt scenario generator has been used to estimate the impacts of changes in climate, land cover and use, and land and water management. In using a biophysical process model but connecting that to an understanding of projected changes in water ecosystem services at locations where services are provided to populations, we provide appropriate biophysical information underpinning ecosystem service assessments (core element 2).

[a] www.policysupport.org/waterworld

tries provide funding and payments to be invested in ecosystem service maintenance and improvement. Benefit-sharing mechanisms go beyond payments for ecosystem services since they can include any form of better sharing the benefits of ecosystem services, which may or may not include a return payment.

A variety of tools for mapping and modelling ecosystem services exist (Bagstad *et al.* 2013); Villa *et al.*, this book). We focus on two of these tools: InVEST-RIOS and WaterWorld (described in Boxes 12.1 and 12.2), which focus specifically on water ecosystem services and were developed for optimising water fund investments.

12.2 THE RIO DAULE: UNDERSTANDING WATER ECOSYSTEM SERVICES USING WATERWORLD

Coherently with core element 3 of an ecosystem services-based approach, during a series of consultations led by The Nature Conservancy in collaboration with members of the benefit sharing mechanisms project COMPANDES,[2] stakeholders in the basin identified a number of key issues that require improved land and water management, including (1) river navigation problems because of increased erosion and sedimentation as a result of deforestation in the basin; and (2) contamination of waters from agricultural pesticides and areas without sanitation. The proposed

[2] www.benefitsharing.net

Guayaquil Water Fund could support measures to reduce existing deforestation and to improve sanitation if socially and politically appropriate and biophysically effective intervention strategies can be identified. The questions asked of WaterWorld were thus:

(1) Where else will be deforested and converted to agriculture in the coming years and what will be the impacts on soil erosion and sedimentation in the river? How might such impacts be reduced by funding targeted at forest conservation and afforestation in the basin?

(2) Could investments in low-input agriculture or rural sanitation lead to water quality gains and what gains can be expected for particular scales of intervention?

In order to understand the optimal investments in forest cover, low-input agriculture, and sanitation for a given budget (a common question for water funds) we couple WaterWorld and RIOS using WaterWorld to understand the spatial variability in the impact of investments. We then use RIOS to allocate the resources invested to the areas in which the proposed intervention will have the greatest impact for the benefits in question. Next, we use WaterWorld again to apply these portfolios to understand the catchment-level impacts on benefits and beneficiaries. By coupling WaterWorld and RIOS we are able to provide a sound biophysical underpinning to our intervention impact analysis (WaterWorld) and a sound economic underpinning to our investment optimisation (RIOS).

We first ran a WaterWorld baseline to understand the current situation in the basin using land cover for the year 2000 and mean climate for 1950–2000. For question 1 we ran a business-as-usual land cover and land use change scenario using the WaterWorld model.[3] The resulting scenario converts forest cover from its baseline catchment mean value of 27% to 21% and increases cropland from 24% to 31%, herbaceous cover from 52% to 55%, and bare cover from 0.03% to 3.3%.

12.3 RESULTS

12.3.1 The hydrological baseline

The Rio Daule in Western Ecuador flows from Santo Domingo to the coastal city of Guayaquil (Ecuador's second city and its

industrial capital). The basin includes montane areas and the lowland Pacific plains of Ecuador. Here we use WaterWorld to examine the basin of the upper Daule, which has elevations between 19 and 630 m above sea level and covers some 506 065 ha (5061 km^2). The upper Daule feeds the large Velasco Ibarra reservoir which supplies 2.35 million people of metropolitan Guayaquil. Total annual precipitation within the basin varies from 1300 to 2900 mm yr^{-1} according to the WorldClim (Hijmans et al. 2005) and 990 to 3200 mm yr^{-1} according to WaterWorld's wind-driven rainfall metric, and mean annual temperature varies from 26 °C in the lowlands to 23 °C in the mountains. Seasonality (according to Walsh & Lawler 1981) of rainfall is 0.81 (markedly seasonal) on average, but varies from 0.69 (seasonal) in the eastern uplands to 0.97 (also markedly seasonal) in the southern lowlands. If we consider the growing season as the number of months with temperature greater than 6 °C and a positive local water balance, then the growing season varies from 12 months for much of the catchment to 1 month in the dry lowlands around Velasco Ibarra. The basin receives significant solar radiation despite the mean annual cloud frequency of 77%. According to GlobCover (2008), 21% of the catchment is rainfed cropland and there is no irrigated cropland.

WaterWorld's baseline water balance for the catchment varies from 210 to 3300 mm yr^{-1} with a mean of 1900 mm yr^{-1} and a gradient from the wettest areas in the north-west to the driest areas in the south-east. There are no spatial data on water quality so we use WaterWorld's Human Footprint on Water Quality index (Mulligan 2009), which examines the potential pollution based on the distribution of rainfall to human (polluting) land and natural (non-polluting) land covers. Human Footprint calculates the percentage of water in each pixel that fell as rain on potentially polluting land uses (cropland, pasture, urban, roads, mining, oil, and gas) upstream and thus the Human Footprint index varies from 0% to 100%. The Human Footprint index for the upper Daule is, on average, 40%, with the highest values in the west (associated with extensive agriculture in this area) and the lowest values in the east. WaterWorld calculates water resource stress as supply (using the water balance and thus including evapotranspiration from vegetation) minus domestic and industrial demand (estimated from the human population and a per-capita use of 47 m^3 yr^{-1} or 130 litres day^{-1}). Mean water resource stress is calculated as the percentage of demand not met by supply in months for which demand is greater than supply, averaged through the year. This is thus an index of seasonal and annual stress in the absence of water storage measures such as use of dams and groundwater. Mean water resource stress for the basin is zero in the uplands but up to 54% for areas away from rivers in the lowlands, and 14% as an average for the catchment. Around 240000 ha (48% of the upper basin) have rural populations receiving poor-quality water (defined as areas with a Human Footprint greater than 50%),

---

[3] The scenario used projects recent rates of land use change forward on the basis of tree cover change from terra-i (www.terra-i.org) and MODIS-VCF 2010–2000 (Hansen et al., 2006; Townsend et al., 2011) averaged over local administrative areas, with assignment of cells to be deforested dependent upon accessibility to population centres by road and river, as well as proximity to existing forest–agriculture boundaries. We used the model to convert current cover to a cropping land use at 2050, characterised by a per-pixel land cover of 10% trees, 70% herbaceous, and 20% bare ground. Assignment of land use and land cover change was made on the basis of the current road network, but also any known planned roads for the areas.

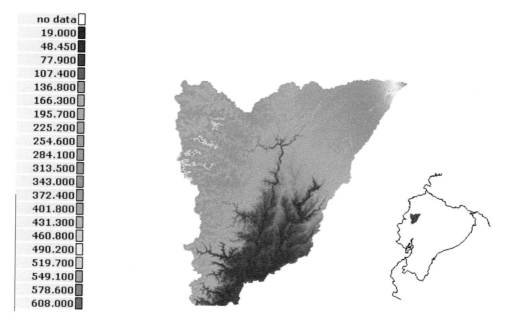

Figure 12.1 Digital elevation model for the upper Daule watershed, Ecuador set within the context of Ecuador. Source: WaterWorld, based on SRTM HydroSHEDS. A black and white version of this figure will appear in some formats. For the colour version, please refer to the plate section.

totalling some 1700 people according to the Landscan (2007) population data set.

12.3.2 Scenarios

The four different scenarios are presented here. Details of each of them can be found in Table 12.1.

SCENARIO 1: THE FUTURE OF DAULE'S FORESTS UNDER BUSINESS-AS-USUAL

The impacts of the business-as-usual scenario on soil erosion and sediment deposition in channels includes a catchment-average increase in annual total gross hillslope erosion of 0.14 mm yr^{-1} (270% of baseline erosion) over 19% of the basin, representing 870000 m^3 of extra sediment mobilised (see Figure 12.2). Over a significant area of the navigable channels, we can thus expect increases in sedimentation: where these are concentrated spatially in the channel we can expect continued problems for navigation.

SCENARIO 2: PROTECTING STEEP, WET SLOPES

We now examine the possibility of affording protection to the wettest and steepest lands. To achieve this we run another defor-estation scenario (protection), but this time considering the steepest and wettest slopes to be protected and thus preventing deforestation in those areas. The land use change model is run with the same settings as previously, but this time we use Water-World's zone of interest tool to define the wet and steep areas that are most prone to erosion and thus most in need of

protection. As well as a 3% lower loss of tree cover compared with business as usual, the distribution of converted areas also changes, with much less deforestation in the steeper, wetter western part of the catchment.

The protection scenario leads to greater sedimentation along the main channel than the business-as-usual scenario. This seems counter-intuitive but in fact is entirely logical: deforestation leads to increases in erosion but also to increases in runoff (and thus increases in the sediment transport capacity of rivers). In the business as usual scenario the runoff for the Strahler order 9 channels increases over 57% of the channel area and decreases over 43%, with a mean increase for the stream order 9 channels of 0.011% of the baseline. In the protection scenario, runoff over the same channels increases over 74% and decreases over 26% of the channel area, with a mean increase of 0.027% of the baseline. Thus the protection scenario leads to a greater increase in runoff *in the main channels*. This is because higher deforestation of cloud forest zones in the business-as-usual scenario leads to more areas of the basin in which water balance decreases (5.5% of the catchment under business as usual compared with 2.7% under protection) because of lower fog inputs, compared with areas in which water balance increases (11% of the catchment under business-as-usual compared with 7.4% under protection), because of reduced evapotranspiration under cropland.

Total change in fog inputs under business as usual is –50 mm yr^{-1} over 17% of the catchment, making –8.3 mm yr^{-1} overall compared with –56 mm yr^{-1} over 10% of the catchment, making –5.6 mm yr^{-1} overall for the protection scenario. Change

Table 12.1 *Details for each of the scenarios*

Scenario	Details
Scenario 1: the future of Daule's forests under business-as-usual	Much of the erosion is re-deposited on hillslopes so the annual total hillslope net erosion increases by only 0.13 mm yr^{-1} over 16% of the area. This represents 69 000 m^3 of extra sediment available to the channels. Sediment transportation increases by 11% along the main channels (representing 1% of the area) or 0.11% over the entire basin and sediment deposition increases by 77% in the areas downstream of converted areas and along the main channels (3% of the area) but decreases by 86% over the 6% of the area in which runoff increases after conversion to agriculture (see Figure 12.3). There is no change over 92% of the area so the catchment-wide response is a decrease in sediment deposition of 2.9%. If we examine channels only we also see that over 38% of the channels deposition increases by an average of 27%, but over 30% (the smaller channels draining the deforested areas) deposition decreases by 44% because of higher runoff. However, over the largest (and thus navigable channels of Strahler stream order 9, i.e. the Daule main stem), 62% of the channel shows an increase in deposition of 3%, 32% shows no change, and 6% of the channel shows a decrease in deposition of 25%, leading to an overall increase of 0.41%.
Scenario 2: protecting steep, wet slopes	The areas in which rainfall > 1500 mm yr^{-1} (wet) and slope gradient >5° (steep) represent 33% of the catchment. The resulting scenario of continued deforestation outside of these protected areas converts forest cover from its baseline catchment mean value of 27% to 24% and increases cropland from 24% to 29%, herbaceous cover from 52% to 54%, and bare cover from 0.03% to 2%. The protection scenario leads, for the Strahler stream order 9 channels, to 64% of the channel showing an increase in deposition of 3% relative to the baseline, 32% shows no change, and 4% of the channel showing a decrease of 41% relative to the baseline, leading to an overall increase in these channels of 0.49% for protection compared with 0.41% for business-as-usual.
Scenario 3: ecoefficient agriculture	This scenario is applied in all steep, wet areas, i.e. areas with precipitation >1500 mm yr^{-1} and slope gradient >5°, representing 33% of the catchment (170 000 ha).
Scenario 4: rural sanitation	Investment leads to an increase in the area with sanitation from 0.19% of the basin in the baseline to 6.2% in the scenario. This results in no change in human footprint over 90.2% of the area, but a decrease in the Human Footprint of –2.3% on average over the remaining 9.8% (an average of 0.2% decrease in Human Footprint overall for the basin), see Figure 12.4. Changes are particularly significant downstream of the most populated zone in the basin (Santo Domingo), with decreases of more than 10% observed as far as 10 km downstream.

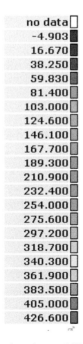

Figure 12.2 Change in annual total gross hillslope soil erosion (%). A black and white version of this figure will appear in some formats. For the colour version, please refer to the plate section.

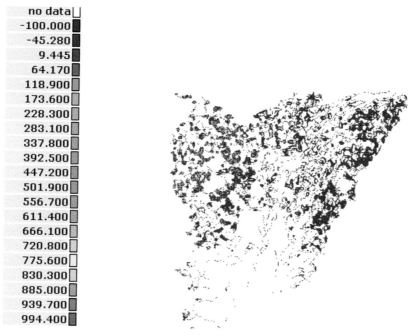

no data	⊔
−100.000	■
−45.280	■
9.445	■
64.170	■
118.900	□
173.600	□
228.300	□
283.100	□
337.800	□
392.500	□
447.200	□
501.900	□
556.700	□
611.400	□
666.100	□
720.800	□
775.600	□
830.300	□
885.000	■
939.700	■
994.400	■

Figure 12.3 Change in annual total soil deposition (%). A black and white version of this figure will appear in some formats. For the colour version, please refer to the plate section.

in evapotranspiration is –61 over 17% of the catchment, making –10 mm yr^{-1} overall for the business-as-usual scenario and –71 mm yr^{-1} over 10% of the catchment, making –7.1 mm yr^{-1} overall for the protection scenario. The outcome is a lower increase in catchment water balance (and thus runoff) because of greater cloud forest loss for pixels that have changed under business-as-usual compared with protection (11 mm yr^{-1} of business-as-usual compared with 15 mm yr^{-1} under protection). This leads to higher sediment transport capacity, so lower channel sedimentation, under the forest protection scenario. This kind of counter-intuitive result is precisely the rationale for using sophisticated spatial simulation models like WaterWorld to examine options *in silico* before they are trialled *in vivo*.

Though the protection scenario does not necessarily meet the objective of reducing sedimentation along the Daule main stem relative to the baseline, it does have a number of co-benefits including reducing forest loss (–6% loss for the business-as-usual scenario compared with –3% for protection) for biodiversity benefits, reducing gross hillslope soil erosion (50% increase for business-as-usual and only 29% increase for protection), meaning less soil degradation, reducing the increase in Human Footprint on water quality (9.6% increase for business-as-usual and only 8.9% for protection) and 12.5% fewer people affected by poor-quality water (Human Footprint greater than 50%).

For question 2 a further two scenarios were proposed to investigate the impacts on water quality of financing eco-efficient agriculture and rural sanitation through a water fund mechanism.

SCENARIO 3: ECO-EFFICIENT AGRICULTURE

In this scenario (ecoefficiency) the human footprint for all agricultural land (cropland and pasture) was reduced by 50% to reflect the investment of the water fund in farmers' use of low-input (ecoefficient) techniques. This results in a significant decrease in the Human Footprint on water quality of –23% in 28% of the basin and zero change in 72% of the basin, representing a fall of –6.5% in the Human Footprint over the catchment as a whole (see Figure 12.4). This leads to a decrease of 17% in the number of people exposed to poor-quality water. Water quality at the exit of the reservoir (affecting much more numerous urban populations supplied by it) improves by 7%.

SCENARIO 4: RURAL SANITATION

In this scenario (sanitation), investment in sanitation (treating 100% of effluent) for all non-urban areas in which population (persons km^{-2}) > 100 was adopted. This decreases the number of people affected by poor-quality water within the catchment by 35% and changes the area affected by poor-quality water from 48.0% of the area to 47.7%. Water quality at the exit of the reservoir improves by 0.35%. These figures may not be directly comparable with those of the ecoefficient agriculture scenario since 1% of contamination by animal waste or agrochemicals may not have the same impact on downstream human health as 1% contamination by unsanitised human waste.

These scenarios indicate that the significant continued losses of forest under the business-as-usual scenario will have spatially

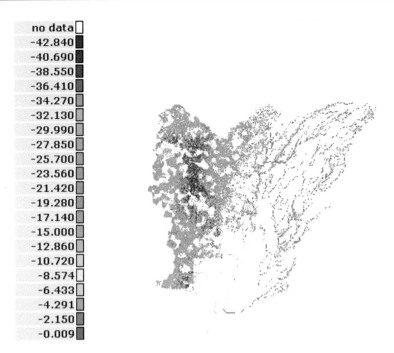

Figure 12.4 (a) Change in human footprint on water quality for ecoefficient agriculture scenario. A black and white version of this figure will appear in some formats. For the colour version, please refer to the plate section.

complex impacts on a number of water ecosystem services. Moreover, large-scale interventions such as forest protection, ecoefficient agriculture, and sanitation will also have significant benefits. However, comparison of the impacts of these scenarios is difficult since they would have very different costs of implementation, which would need to be considered in a cost–benefit analysis to understand and better optimise their implementation. Moreover, the typical budgets available from water funds would not allow very large-scale investments like this. To better understand the impacts of economically viable interventions we couple WaterWorld with RIOS, the Resource Investment Optimisation System.

12.3.3 Application of WaterWorld and RIOS to optimise water fund investments

Four further scenarios were applied. The business-as–usual scenario (–100% tree cover loss in all suitable, i.e. not roads or water pixels and land use converted to 'natural'), a new scenario called 'afforestation' (+100% afforestation in all suitable pixels and land use converted to 'natural'); 'ecoeffiency' and 'sanitation' were run against the same baseline as previously but with application over the entire catchment.

For the afforestation scenario, the combination of protecting natural ecosystems and afforesting non-natural areas was chosen as these would give a high benefit. The estimated cost of

'protection' in the study area is US$50/hectare and the afforestation costs are around US$1500 ha^{-1}. We assumed that US$100 000 would be invested on protection and the remaining US$900 000 on afforestation. The portfolio resulted in 2000 hectares of natural forests protected and 600 hectares of non-natural areas afforested (see Table 12.2, second column) with slight differences depending on whether reducing sediment deposition, Human Footprint on water quality, or soil erosion were the determinants for the spatial distribution of the intervention. The areas to be afforested and protected were set to +100% tree cover and 'natural' land use in WaterWorld.

For the ecoefficiency scenario, the RIOS activity 'sustainable agriculture' was chosen to allocate budget in order to meet the objectives. The activity costs around US$2000 ha^{-1} according to RIOS and involves many actions addressing good agricultural practices (green fences, allelopathy, and reducing contaminant inputs, among others). The resulting portfolio gave 500 hectares of crop areas converted to 'sustainable agriculture' (Table 12.2, third column). In WaterWorld the agricultural intensity (for cropland and pasture) in these areas was reduced from 1% to 0.5%. For the sanitation scenario runs in the portfolio generator, a set of assumptions were made. An average rural population of 200 persons km^{-2} was assumed, which represents 40 families km^{-2} if an average of five people per family is assumed. At a cost per hectare of implementing sanitation systems of US$300, the portfolio generator was able to provide 3334 ha of the study area to

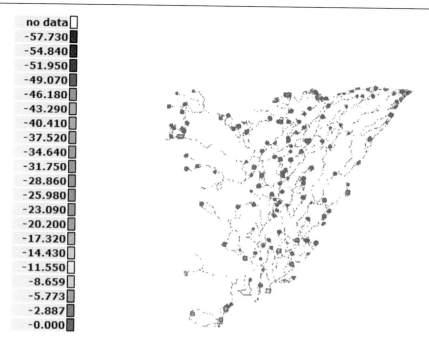

no data	
-57.730	
-54.840	
-51.950	
-49.070	
-46.180	
-43.290	
-40.410	
-37.520	
-34.640	
-31.750	
-28.860	
-25.980	
-23.090	
-20.200	
-17.320	
-14.430	
-11.550	
-8.659	
-5.773	
-2.887	
-0.000	

Figure 12.4 (b) Change in human footprint on water quality for sanitation scenario. A black and white version of this figure will appear in some formats. For the colour version, please refer to the plate section.

sanitation with the $1 000 000 available (Table 12.2, fourth column). In WaterWorld these areas were set to sanitation with 100% improvement in effluent. In each case it is clear that even $1 000 000 brings relatively small changes (at the catchment scale) in tree cover, cropland intensity, or area under sanitation (Table 12.2).

The resulting catchment-level changes in benefits of interest, area affected by benefits, and number of beneficiaries are shown in Table 12.3. For each investment scenario US$1 million is spent and even this budget does not buy very significant changes at the scale of large catchments like the Daule, which gives an idea of the reality of investments like this. Afforestation is applied three times according to the optimal configurations for the objectives of reducing sediment deposition, reducing human footprint, and reducing net hillslope erosion. In each case there are co-benefits for other objectives, but the US$1 million provides benefits to the most people (calculated as the population in areas where benefits are simulated) when applied to reducing sediment deposition, followed by hillslope erosion, and then water quality (Human Footprint). The area affected is also greatest where afforestation is applied to the sediment deposition reduction objective, though the catchment-scale magnitude of change is greatest for reducing net hillslope soil erosion and least for reduced sediment deposition

Ecoefficiency and sanitation are applied to the Human Footprint objective only since they have no role in soil erosion or deposition processes in WaterWorld. For the available budget,

while the ecoefficiency investment has the greatest impact on catchment-level water quality (an order of magnitude greater than sanitation), sanitation has benefits over a much larger area and affects a much greater population, again by an order of magnitude.

This analysis gives an indication of the level of information required for an assessment of the kinds of impacts associated with investments in water ecosystem services. WaterWorld provides an extensive and detailed spatial database for such analyses, but all of these data will be subject to unknown errors and uncertainties, as will our specification of water cycle processes. Moreover, the analysis makes a number of simplifications in outlining the available budget and the costs for specific interventions (which in reality will vary spatially according to landscape properties and accessibility). Despite these simplifications, it has been shown that it is possible to understand a spatial baseline (and WaterWorld can be applied in this way to any catchment globally, since all data are provided with the model). Moreover, it has shown that even complex interventions can be simulated and their sometimes counter-intuitive and always spatially complex impacts projected. The analysis provides analysis of different investments and provides maps of where these could be optimally spent on interventions with specific water ecosystem services improvement objectives (Table 12.3).

But how does one then measure and trade-off these impacts to come to a final decision? Three separate metrics are key: catchment average change in the metric; area in which the metric

Table 12.2 *Changes resulting from proposed intervention assigned by RIOS on the basis of available budget and areas creating the most positive change in the outcome of interest, from WaterWorld*

Objective	Scenario		
	afforestation (US$1 million)	Ecoefficiency (US$1 million)	Sanitation (US$1 million)
Reduce sediment deposition	+0.071% tree cover	N/A	N/A
Reduce Human Footprint	+0.066% tree cover	−0.00056 cropland intensity (fraction) −0.00056 pasture intensity (fraction)	+0.77% of area
Reduce soil erosion	+0.0701% tree cover	N/A	N/A

Table 12.3 *Impact of optimal resource investments on catchment level benefits.*

Objective	Scenario		
	Afforestation	Ecoefficiency	Sanitation
Reduce sediment deposition	−0.054% 2400 ha 93000 persons	N/A	N/A
Reduce Human Footprint	−0.068% 1700 ha 24000 persons	−0.039% 1800 ha 19000 persons	−0.0031% 17000 ha 510000 persons
Reduce net hillslope soil erosion	−0.087% 890 ha 36000 persons	N/A	N/A

changes positively (area of benefits); and number of people in this area (number of beneficiaries). Is a catchment-level reduction in sediment deposition of −0.054% better than a catchment-level decrease in Human Footprint of −0.068% or a catchment-level reduction in net hillslope soil erosion of −0.087%? Valuing these benefits economically is also fraught with complication and difficulty and thus prone to oversimplification. Through understanding the area affected we have an idea of the extent that benefits will occupy and by understanding the number of people in those areas we have an estimate of the beneficiaries. However, not all people in the areas with reduced sediment deposition will be affected by this because it depends upon their interaction with the rivers or land within which this change occurs, and their interaction will differ depending on their livelihood. Moreover, there will be many outside these areas affected by the benefits since the commodities (water, crops, energy) upon which they rely originate in these areas. These are some of the challenges for ecosystem services-based

approaches in general and policy support systems for the development of benefit-sharing mechanisms for water ecosystem services in particular.

12.4 CONCLUDING REMARKS

Water funds are benefit-sharing mechanisms that have much potential in Latin America and beyond, and may help manage for better food, water, and energy security for improved equity, justice, and poverty alleviation through livelihood diversification and positive impacts on ecosystem benefits received by the poor. Water funds can also help achieve environmental protection and conservation objectives with benefits for biodiversity and ecosystem services beyond water. In addition to the considerable challenges in engaging stakeholders, setting up, resourcing, and managing the fund, there are also significant challenges in providing the knowledge on what, where, and how to invest within the catchment (core element 3, Chapter 2). Key knowledge gaps could include: mis- or non-use of scientific input, (mis)understanding of processes by the scientists, mis-communication of outcomes between scientists and fund managers, insufficient quality of spatial data for reliable results, and insufficient quality and detail of temporal data to fully understand the impacts of climate variability on outcomes.

The long-term success of any benefit-sharing mechanism will depend upon the sustainability of the funding mechanism, the level of the transaction costs, and careful management of beneficiary–provider relationships and effective targeting of landscape interventions to achieve the desired outcomes. Some of the risks to success that are sometimes forgotten include the fact that many of the benefits of interventions can take years or decades to come to fruition (e.g. where reforestation benefits are not fully achieved until the forest reaches full majority) and in highly connected systems like hydrological ones, benefits produced by interventions in one part of a catchment can be degraded by dis-benefits produced by poor management elsewhere. To be effective, benefit-sharing mechanisms must be capable of influencing all parts of a catchment that might affect those receiving and paying for water ecosystem services.

Finally, benefits are always a function of climate as well as land cover and management. However effective a benefit-sharing mechanism is in bringing about benefits from better land management, these benefits will always be mediated by any variation or trend in climate over the same periods the scheme is implemented. Payments for ecosystem services schemes therefore need to be very careful to avoid overselling benefits achievable by managing ecosystem services where these benefits can be seriously affected by the essentially unmanageable environmental services on which ecosystems operate to generate ecosystem services.

Benefit-sharing mechanisms like this offer the potential to provide sustainable resources for watershed investments and to focus interventions on those which have proven, monitored outcomes for the beneficiaries who provide the investments. This differs from previous watershed management approaches in which management resources were not always sustainable and were much less conditional upon positive results. However, proven monitored outcomes can only be assessed *a posteriori* and even then are not guaranteed to continue being realised.

Research has come a long way in developing and coupling sophisticated spatial tools for optimising investments in water ecosystem services, but significant challenges remain. These challenges will only be addressed if these tools are used frequently and in many contexts by a range of users, in each case collaborating with the developers to ensure that the quality of output is sufficient for the purpose to which the tools are applied. In addition to their use in projecting the impacts of investments and thus optimising the spend of water funds, the long-term viability of water funds depends on their ability to show real impact on the ecosystem services being managed. There are thus considerable challenges for models and model–data hybrids as monitoring systems for understanding the impact of interventions made by water funds.

Spatial models have been very useful in the design of water funds, but it is important that a complementary monitoring system be implemented to measure the accomplishment or advance towards the goals of the water funds. It is important to remember that the ultimate goal is not the creation of the water fund, but the maintenance or improvement of ecosystem services provision for the long-term. The creation of a water fund is an important first step, but the fund will have to guarantee that it can: ensure financial sustainability (i.e. maintain permanent financial contributions from water users); maintain a good governance body where the different stakeholders have an appropriate balance of power in decision-making; and support a large constituency supporting the water fund (i.e. citizens of the basin). To obtain these, the water fund will need to implement good

> **Box 12.3** Key messages
>
> - Tools are available for the assessment of spatial ecosystem service baselines and the impacts of management interventions (see www.policysupport.org/waterworld).
> - These can be coupled with tools for the optimisation of investments, spatially and across multiple objectives (see www.naturalcapitalproject.org/RIOS.html).
> - Water funds are innovative benefit-sharing mechanisms that can provide multiple ecosystem services and other benefits if implemented properly.
> - The tools are often ahead of the available data for ecosystem services management decision-making within the context of water funds.
> - There remain a number of challenges in reducing uncertainties and in defining and trading-off benefits, but the tools are nevertheless readily available for application.

science to project and then measure the impacts of the water fund on the services it is aiming to maintain or restore. The ability to measure, report, and communicate the impact of the water fund will be the cornerstone of long-term success (The Nature Conservancy 2012).

ACKNOWLEDGEMENTS

Some of this work has been carried out within the context of the Consultative Group on International Agricultural Research (CGIAR) Challenge Programme on Water and Food (CPWF) project 'Mechanisms for benefit sharing to improve productivity and reduce conflicts for water in the Andes (COMPANDES)'. The support of the CPWF is gratefully acknowledged. Members of that project are gratefully acknowledged for their inputs, in particular Jorge Rubiano, Beth Sua Carvajal, and Leo Zurita.

References

Bagstad, K. J., Semmens, D. J., Waage, S., & Winthrop, R. (2013). A comparative assessment of decision-support tools for ecosystem services quantification and valuation. *Ecosystem Services* **5**, 27–49.

GlobCover (2008) Land Cover v2 database. European Space Agency, European Space Agency GlobCover Project, led by MEDIAS-France. http://ionia1.esrin.esa.int/index.asp (accessed June 2013).

Hansen, M., DeFries, R., Townshend, J. R., Carroll, M., Dimiceli, C., & Sohlberg, R. (2006). Vegetation Continuous Fields MOD44B, 2001 Percent Tree Cover, Collection 4, University of Maryland, College Park, MD.

Hijmans, R. J., Cameron, S. E., Parra, J. L., Jones, P. G., & Jarvis, A., (2005). Very high resolution interpolated climate surfaces for global land areas. *International Journal of Climatology* **25**: 1965–1978.

DEVELOPMENT OF BENEFIT-SHARING MECHANISMS

109

Kareiva, P., Tallis, H., Ricketts, T. H., Daily, G. C., & Polasky, S. (eds) (2011). *Natural Capital: Theory and Practice of Mapping Ecosystem Services*. Oxford University Press, Oxford.

LandScan (2007). *Global Population Database 2007*. Oak Ridge, TN: Oak Ridge National Laboratory. Available at www.ornl.gov/landscan (accessed June 2013).

Mulligan, M. (2009) The human water quality footprint: agricultural, industrial, and urban impacts on the quality of available water globally and in the Andean region. Proceedings of the International Conference on Integrated Water Resource Management and Climate Change, Cali, Colombia.

Mulligan, M. (2013) WaterWorld: a self-parameterising, physically-based model for application in data-poor but problem-rich environments globally. *Hydrology Research* **44**(5), 748–769

Mulligan, M. & Burke, S. M. (2005) FIESTA: Fog Interception for the Enhancement of Streamflow in Tropical Areas. Available at: www.ambiotek.com/fiesta (accessed June 2013).

Tallis, H. T., Ricketts, T., Guerry, A. D., et al. (2013). *InVEST 2.5.5 User's Guide*. The Natural Capital Project, Stanford, CA.

The Nature Conservancy (2012). *Water Funds: Conserving Green Infrastructure – A Guide for Design, Creation and Operation*. The Nature Conservancy, Bogota. Available at: www.femsafoundation.org/assets/003/21269.pdf (accessed June 2013).

Townshend, J. R. G., Carroll, M., Dimiceli, C., Sohlberg, R., Hansen, M., & DeFries, R. (2011). Vegetation Continuous Fields MOD44B, 2001 Percent Tree Cover, Collection 5, University of Maryland, College Park, MD.

Vogl, A. L., Tallis, H., Douglass, J., et al. (2013). *Resource Investment Optimization System (RIOS). Software and User's Guide, v1.0.0*. Available at: www.naturalcapitalproject.org/RIOS.html (accessed June 2013).

Walsh, P. D. & Lawler, D. M. (1981). Rainfall seasonality: description, spatial patterns and change through time. *Weather* **36**, 201–208.

13 Assessing biophysical and economic dimensions of societal value

An example for water ecosystem services in Madagascar

Ferdinando Villa, Rosimeiry Portela, Laura Onofri, Paulo A. L. D. Nunes, and Glenn-Marie Lange

13.1 INTRODUCTION

Policy decisions are often based on an assessment of value. However, the definition of value is highly context-dependent and policy decisions are ultimately a multiple-objectives problem, containing internal trade-offs that make alternative actions difficult to assess and rank. This chapter provides a demonstration of how to operationalize an ecosystem services-based approach, as defined in Chapter 2 of this book, to assess the social and economic value of water from a pluralistic viewpoint that can better support decision-making. Our approach to the biophysical analysis emphasizes explicitly identified beneficiaries (core element 1) along with an assessment and mapping of physical flows to them (core element 2). The biophysical interpretation is complemented by an analysis of economic productivity of water and by a policy analysis where the biophysical and economic analysis are integrated and discussed (core element 3), and implications of alternative management scenarios (protected versus non-protected areas) are addressed (core element 4). Notably, we quantify four key dimensions of ecosystem services, moving beyond the purely economic viewpoint that has dominated the policy translation of ecosystem services assessments so far. These are:

- Input productivity: the relationship between inputs and final output, estimated in this study as water productivity in four selected economic sectors.
- Economic value: the value of marginal productivity, in terms of the increases in economic productivity with the increase of an additional unit of input (in this case, per additional unit of water).
- Sustainability of supply: defined as the ratio between the amounts of ecosystem-provided benefit (here: water services) and the estimated maximum demand that can be met for it in the same conditions.
- Quality of supply: estimated by assessing the influence of the natural environment in preserving the quality of the ecosystem service.

Our study was performed in the context of a World Bank-led global initiative that aims to integrate natural capital values into national account systems. The Wealth Accounting and the Valuation of Ecosystem Services initiative[1] was launched in 2009 to promote sustainable development by mainstreaming natural capital accounting in development planning and national economic accounts and it currently supports programs in ten countries and will double that number by 2015. The Wealth Accounting and the Valuation of Ecosystem Services initiative promotes the development of methodologies for ecosystem services accounting through case studies, emphasizing (1) how to scale-up site-specific case studies to the national level; (2) handling thresholds and irreversibility in a valuation framework consistent with national economic accounts; and (3) valuation of assets in the face of uncertainties about future supply and demand for ecosystem services.

This chapter presents an in-depth assessment of the role of the largest remaining block of rainforest along the eastern escarpment of Madagascar, the Ankeniheny–Zahamena Forest Corridor (see Box 13.1). We focus on water supply, which is supported by the ecosystem in two primary ways: through the role of forests in retaining and recirculating precipitation (provisioning service) and in retaining sediment that would otherwise pollute the water supply (regulating service). The method used here emphasizes the spatial dynamics of ecosystem services flow and use by beneficiary groups, distinguishing potential ecosystem services value (water supply made available through ecosystem services but may or may not be used by beneficiaries) versus actually accrued ones. Such a beneficiary-based approach minimizes one of the risks commonly attributed to ecosystem services-based approaches: the potential for 'double counting' resulting from considering as ecosystem services all processes that may (directly or indirectly) produce benefits rather than only those processes that directly contribute value (Boyd & Banzhaf 2007, Wallace 2007, Fisher *et al.* 2008).

---

[1] (www.wavespartnership.org)

Box 13.1 Madagascar's water resources and the Ankeniheny–Zahamena Forest Corridor

Madagascar has relatively abundant water resources in the northern and central areas, but water is scarcer in the east and south of the country, where shortages and droughts occur regularly. Safe drinking water access is limited to 41% of the population nation-wide and largely concentrated in urban areas. Regional inequalities in water supply and sanitation, combined with growing population and rising demand for irrigated agriculture, are likely to increase pressure on water resources in Madagascar and pose serious threats to water security. Such impacts are of greater concern in the context of climate change and its expected impact on temperature and water availability. According to Hannah *et al.* (2008), projections of climate change for Madagascar indicate mean temperature increase of 1.1–2.6 °C throughout the island in this century, with the greatest warming in the south and the least along the coast and in the north. The Ankeniheny–Zahamena Forest Corridor is a region of rich biological diversity as well as an area of great poverty, where a large part of the area's 347 250 inhabitants is dependent on farming, practicing a mix of subsistence and cash crop production. Erosion and nutrient leaching are important problems in the region (Kull 2000), where sedimentation of channels leads to a lack of water in irrigated areas and to economic losses both due to lower productivity and needed repairs (Rakotoarison 2003). Significant deposits of sediment impact upon the turbines of the Andekaleka hydropower plant, the most important in the country.

13.2 METHODS: BIOPHYSICAL AND ECONOMIC ASSESSMENTS

In recent years, the first generation of integrated, multi-ecosystem services assessment methodologies and tools have been used to meet needs that cut across the academic, governmental, non-government organizations, and corporate sectors (Vigerstol & Aukema 2011; Waage *et al.* 2013). While rapid assessment and valuation methods have come to command wide interest from all these communities, it is generally recognized that systematic use of ecosystem services accounting in decision and policy making requires a degree of accuracy that is rarely met in practice (Eigenbrod *et al.* 2010; Bagstad *et al.* 2013). Most early assessment studies (Costanza *et al.* 1997; Troy *et al.* 2006) and some recent methods (Daily *et al.* 2009; Tallis *et al.* 2009) infer ecosystem services values through production functions whose driving input is land cover type, either alone or complemented by limited structural information (e.g. vegetation type). Other methods (Martinez-Harms *et al.* 2012) use models of a more functional nature to more accurately represent the mechanistic underpinning of ecosystem service dynamics (Johnson *et al.* 2010; Fisher *et al.* 2011; Kareiva *et al.* 2011; Johnson *et al.* 2012). Spatial maps of the biophysical distribution of ecosystem services have been used for many years (Kareiva *et al.* 2011) to provide static pictures of potential value to society resulting from the presence of natural features. More recently, ecosystem services-specific computing tools have emerged to systematize mapping and economic valuation with the aim of assisting policy decisions (e.g. Mulligan *et al.*, this book).

The methodological emphasis for this chapter is placed on new techniques for a biophysical analysis that progresses from the initial identification of beneficiaries, in contrast with most mainstream ecosystem services methodologies in use today, whose primary focus is on the ecosystem. This represents a step forwards in the development of core element 2 of ecosystem services-based approaches (as defined in this book), in the understanding of the biophysical underpinning of ecosystems in terms of service delivery, with the aim of assessing ecosystem services for its incorporation into decision-making (core element 4). The ARIES (ARtificial Intelligence for Ecosystem Services) methodology used here aims for a more realistic view of ecosystem services that accounts for some aspects of the complex dynamics of ecosystem services and enables spatially explicit quantification of the benefits provided. ARIES (Bagstad *et al.* 2013; Villa *et al.* 2014a) is already being used for a diverse set of ecosystem services applications, ranging from food security (Villa *et al.* 2014b) to integrated assessments, covering flood regulation, recreation, aesthetic services (Bagstad *et al.* 2011) and other services (Bagstad *et al.* 2014). Ecosystem services are seen in ARIES as the result of the flow of a beneficial or detrimental *carrier*, which may be physical (e.g. water, CO_2) or informational (e.g. culturally mediated services, aesthetic views). Using automated model composition driven by artificial intelligence algorithms (Villa 2007, 2009), ARIES chooses the most appropriate model taking into account the application context and the amount of available data; it then quantifies the benefits accrued by society by linking them to the spatial locations where the carrier has flowed or accumulated. The aim of this analysis is to more fully highlight the consequences of policy decisions on ecosystem services delivery than is usually achieved by mainstream methods.

Figure 13.1 depicts the various components of a benefit as seen in ARIES. In this conceptual model, the ecosystem acts as a source of a service carrier, which flows to societal beneficiaries (use) along *flow paths* that depend on the type of carrier and on landscape characteristics. During the process, the carrier may encounter *sinks* where the carrier is intercepted or depleted (in quantity and/or quality) so that it cannot reach the beneficiary.

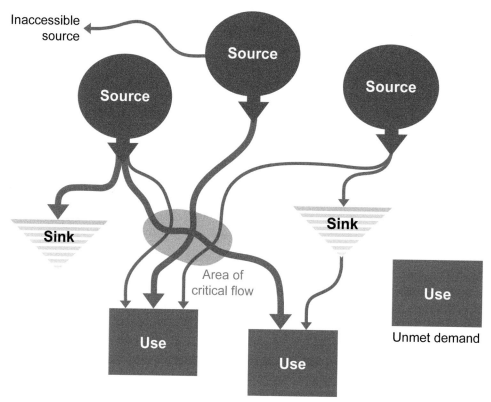

Figure 13.1 Conceptual model of the ecosystem services-based approach underlying an ARIES model.

The final amount of value that reaches the beneficiary is proportional to the amount of carrier obtained, directly for beneficial services (such as water for drinking or irrigation), or inversely for preventive services (such as floodwater or sediment involved in reservoir siltation). Rival users of ecosystem services compete for the benefit (e.g. the water that irrigates one crop is not available for others located downstream), while non-rival users do not (e.g. aesthetic views can be enjoyed regardless of how many people are there to watch).

ARIES can produce a full account of *winners* and *losers* and potential versus actual provision for each ecosystem service, allowing decision-makers to plan interventions and policy in a more precise way. Benefits are only accounted for when the beneficiaries are actually impacted, and each benefit is accounted for in proportion to what has actually been accrued by explicit beneficiaries. ARIES synthesizes the ecosystem services flow results into different groups of spatial maps, which can have different applications according to the policy context (Bagstad *et al.* 2013). In the case of water supply, water quantity is seen as a function of topographically based hydrologic simulation. Water supply is a complex ecosystem service to model spatially; the models presented here operate at an annual scale, using available spatial data to compute variables including precipitation, infiltration of water into the soil (depending on soil type and vegetation), and evapotranspiration (loss of water to the atmosphere from vegetation transpiration). While sediment

regulation has many dimensions, some of which can be classed as provision services (e.g. deposition of land for agriculture), here we have modeled only the regulation of sediment affecting water quality, where benefits provided by the Ankeniheny–Zahamena Forest Corridor natural features consist of protecting water sources from excessive turbidity due to dissolved sediment. Running the sediment flow model allows the mapping of spatial connections between sources of sediment, areas that promote sediment deposition, and users affected by sediment delivery. Most of the Ankeniheny–Zahamena Forest Corridor is not located in a floodplain, so the dissolved fraction of sediment is more of an issue than in other areas.

The flow paths of water are computed on an annual scale by a model that simulates surface water flow based on terrain and soil characteristics. This model computes all possible flow paths between water that reaches the soil as rainfall and each user of water, allowing us to estimate the spatial location and amounts of water that have flowed across the landscape over the course of a year, and how much of that water has been used by beneficiaries. At the same time, the sediment model simulates the erosion of sediment based again on slope, soil stability, and presence of vegetation. The output of the sediment model is intersected with that of the water model to estimate the contamination of the water flow in each point of the study area.

To understand the value of a protected area in providing benefits to downstream water users, we compared water budgets

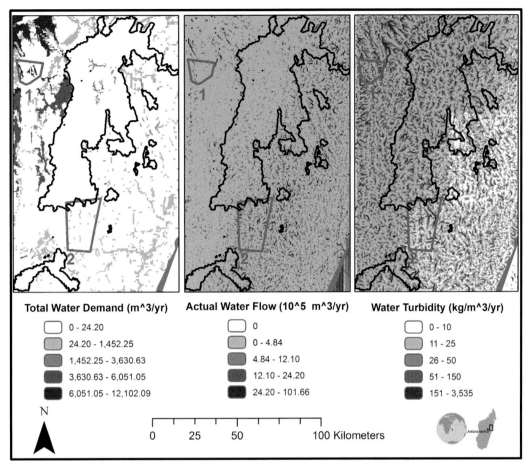

Figure 13.2 Water supply and quality in the Ankeniheny–Zahamena Forest Corridor area of Madagascar. From the left: total water demand across sectors, surface-water flow that is used by beneficiaries, and amount of sediment that is transported by hydrologic flows. Regions 1 and 2 show the areas selected for comparison. A black and white version of this figure will appear in some formats. For the color version, please refer to the plate section.

(supply and demand) and erosion for an area near the Ankeniheny–Zahamena Forest Corridor, but hydrologically unconnected to the protected area and with intensive agriculture (area 1, Figure 13.2) versus another adjacent and hydrologically connected to the protected area (area 2, Figure 13.2). In the rest of this chapter we refer to area 1 as 'outside forest corridor' and to area 2 as 'adjacent to forest corridor'.

The main source of water was assumed to be rainfall, estimated using WorldClim[2] data representing current rainfall conditions computed with data from 1950–2000. Water use was modeled separately for the agricultural, resident, and tourist sectors. Agricultural use was split into irrigation demand for rice, irrigation demand for other crops, and water demand for livestock (Portela *et al.* 2012). Irrigation demand was estimated using guidelines published by the Food Agricultural Organization, adopting the Blaney–Criddle method for evapotranspiration (Brouwer *et al.* 1985). Livestock demand was estimated using literature estimates of per capita consumption and Food Agricultural Organization

global livestock spatial data for sheep, goats, cattle, and pigs. Residential use was computed based on probability distributions of water use computed from available data in relatively comparable, developing countries (chiefly rural areas of Mexico), and 2006 population density data available globally. Water demand for tourism was estimated using data provided by the Vakona lodge in Andasibe-Mantadia National Park and extrapolated spatially to the other known lodges in the area. Models of tourism-related water use were run independently, considering the other water uses as sinks for the model. The water demand across all sectors is shown in Figure 13.2 (left).

From an economic point of view and in relation to the economic assessment of benefits, ecosystem services are seen as impacting the production of different goods and services which are traded in markets. A production function (Varian 1992) describes the change in the output of a process as a function of a unitary change in the amount of a number of inputs (marginal productivity). Such functions are used in computing economic value using inputs such as labor and capital, by multiplying the

[2] www.worldclim.org

marginal productivity obtained by the market price. We use production functions for ecosystem service economic assessment by considering natural capital as an input along with other conventional economic inputs. The general formulation we adopt is

$$Q = q(L; K; N, Z)$$

where Q is the productivity of some good (e.g. cobalt or rice), L is the input of human capital (e.g. labor); K is the input of financial capital (e.g. infrastructures and machineries); Z represents other inputs (e.g. land); and N denotes the input of natural capital (e.g. water) obtained from the ecosystem. We developed production functions for each of the principal economic sectors in the Ankeniheny-Zahamena Forest Corridor region (Box 13.1) and fitted them to data to obtain functional forms of the marginal productivity of water.

Two different production functions (Varian 1992; Portela *et al.* 2012) were estimated for the mining sector, using nickel and cobalt production as independent variables over a mine lifespan of 30 years. Ambatovy, the major mining enterprise located in the Ankeniheny–Zahamena Forest Corridor near the town of Moramanga, provided data on production, labor, machinery investments, water usage, land, raw materials (tons of limestone, sulfur, ammonia, and coal used), and energy. For the agriculture and farming sector we used data from the Ministry for Agriculture, Farming and Fishery, and surveys from the National Statistical Bureau, the national water company (JIRAMA), and the Rice Observatory to estimate quantities of produced rice and manioc as well as the number of households' farm animals in selected administrative areas within the Ankeniheny–Zahamena Forest Corridor in 2009. From the same sources we obtained estimates of labor (number of workers), water usage, land extent, and type of machinery used. Three production functions were estimated, with quantities of produced rice, manioc, and farm animals respectively as the dependent variables. For tourism we based our study on data from the main site in the Ankeniheny–Zahamena Forest Corridor area, Andasibe-Mantadia National Park, home of *indri* lemurs and many endemic species of flora and fauna. We integrated information from the Moramanga

Tourism Office, interviews with hotel managers and staff at Mantadia National Park and other parks with World Trade Organization data to obtain a tentative estimate of the number of beds in Ankeniheny–Zahamena Forest Corridor, from which the number of tourists per day was calculated.

13.3 RESULTS

ARIES models were run to produce result maps and aggregated water budgets for each area so that values could be compared. Table 13.1 shows aggregated water budgets, detailing the total demand for water per sector and the total supply of water from rainfall, for residential and agricultural users.

Each flow model run estimates spatially and quantitatively the amount of water actually flowed to the users; an example of the results of flow analysis is show in Figure 13.2 (center), where the amount of water that has flowed at each point in a year is shown. Flow analysis is, therefore, suitable to help understand whether water supply is at sustainable levels by comparing the actual supplied amounts to the demand. In order to obtain an approximate estimate of the sustainability of water supply, the flow models in each area were run repeatedly with proportional increases in demand, to understand the maximum *potential*, i.e. the water demand each area was able to meet. Sustainability of the water supply can then be estimated as the percentage ratio between the current demand and the maximum potential. As an illustration, the results of this analysis for the dominant water use (rice agriculture) are summarized in Table 13.2, which shows how levels of demand are essentially met in both areas, but while the area in the forest corridor has potential to sustain much greater demand, the area outside of its influence is already at critical levels (i.e. ratio of maximum potential and demand below 100%).

It is important to note that such estimates should only be considered as relative (to each other) and not as absolute values. Due to the many other factors that influence water supply which are not included in the model (not to mention the fact that the data are relatively old and global change has certainly impacted the area since), it is probably safe to consider such percentages as

Table 13.1 *Total estimated water budget (m^3 $year^{-1}$) for sample areas outside (1) and adjacent to (2) Ankeniheny–Zahamena Forest Corridor*

	Total in forest corridor	Area 1 (outside forest corridor)	Area 2 (adjacent to forest corridor)
Rice agriculture	512 187 528	15 943 889	5 958 885
Non-rice agriculture	31 718 842	444 689	6 512 517
Livestock	684 499	206 041	54 484
Residential	17 173 088	3 206 662	4 426 315
Annual precipitation	16 619 520 610	1 074 244 347	7 476 712 388

Table 13.2 *Water supply sustainability for rice agriculture in the two areas considered.*

	Sample area 1 (outside forest corridor)	Sample area 2 (adjacent to forest corridor)
Current water demand (m^3 year^{-1})	15 943 889	5 958 885
Maximum potential (m^3 year^{-1})	15 443 129	304 155 269
Ratio potential/demand	97%	5104%

overestimates. But we believe this still provides a useful representation to reflect on the usefulness of the approach.

Sediment release and build-up models were run in the two areas to compare the role of the Ankeniheny–Zahamena Forest Corridor natural features in influencing water quality in the form of dissolved sediment. The flow model runs eroded sediment through the water transport system and establishes the total amount of sediment that is likely to contaminate the water flow in each point. The model run, summarized in the map of Figure 13.2 (right), produces an estimated contamination of freshwater by sediment that is approximately six times higher outside the protected area (11.3 kg m^{-3} yr^{-1}) than adjacent to it (1.9 kg m^{-3} yr^{-1}). While estimating specific effects on the water quality and economics of the region requires more accurate data, it is well known that sediment contamination is detrimental to different degrees for all water uses considered. As average values of >10 kg m^{-3} are quite high, the analysis shows how sedimentation is a major factor in determining loss of productivity for the water supply and that the role of the natural features in the protected area is determinant in protecting water quality.

The economic analysis was performed by estimating the marginal economic productivity of water using the production functions detailed in Box 13.2, using current market rates for the outputs of each sector. This is expressed in Table 13.3 as the incremental change in each output after a 1% increase in water use. As the analyses refer to directly usable water, the values reported reflect both the provisioning ecosystem services that provide water (water supply) and the regulating ones that keep water quality within usable levels (sediment retention).

By fitting the production function specified in Box 13.2 to the available data, we computed that for the mining sector a 1% increase in used water supply results in a 0.7% increase in the output of nickel and a 0.43% increase in the output of cobalt. Examination of the coefficients resulting from fitting the production function to data reveals that water presents *diminishing returns* for mining production, i.e., when all other inputs are held constant, an increase of water input yields a decrease in the marginal (per unit) output of nickel and cobalt.

In the agriculture and farming sector our analysis shows that a 1% increase in used water supply leads to (1) a 0.91% increase in the production of rice; (2) a 0.83% increase in the production of manioc; and (3) a 0.93% increase in the production of poultry. The coefficient estimates for the ecosystem services input water yield almost constant returns for the agriculture–farming economic sector.

In the tourism sector, lacking comprehensive data about hotel management within the Ankeniheny–Zahamena Forest Corridor, we used estimates of average daily water consumption per person from a representative hotel (Vakona Forest Lodge) and estimates of international tourism water usage from United Nations Development Programme (UNDP 2011). Assuming that the average productivity equals the marginal productivity of water,[3] we can show that the ecosystem services input water also presents constant returns for the tourism production/sector.

Table 13.3 summarizes the increase in annual productivity resulting from an additional unitary input of the ecosystem service (one cubic meter of clean, usable water per day) of water, assuming that the input is fully used.

These results can be read also in terms of opportunity costs. In this case if, for example, an additional unit of water is not used to produce cobalt, by either being allocated to an alternative production or simply becoming no longer available, the correspondent economic loss can be quantified as US$7541 over a year.

13.4 DISCUSSION

It is immediately clear that while current levels of water demand are essentially met in both the site adjacent to the forested protected corridor and the site outside the protected area, the former clearly has the potential to sustain much greater water demand. In addition, water quality, assessed as reduced sediment load, was estimated to be significantly better in the protected area than in a non-conservation area. These results, directly obtainable by biophysical analysis, clearly highlight the direct role of forested areas in providing benefits that affect society water storage and conservation and prevention of sediment contamination. The economic analysis of water, in turn, highlighted water use efficiency to be greater in the region's agricultural and tourism sectors, though the marginal value of water as a

---

[3] For instance, assuming that the per day per tourist average water consumption is totally used for showering (for instance), we can credibly infer that the same amount of water will be used by an additional tourist for taking an additional shower.

Table 13.3 *Economic productivity of water.*

Economic sector	Marginal productivity	Economic value of marginal productivity (2012 USD)	Marginal return type
Mining			
Cobalt	0.43 t year^{-1}	7,541	Diminishing
Nickel	0.70 t year^{-1}	11,906	
Agriculture			Constant
Rice	0.91 t year^{-1}	469	
Tourism			Constant
Luxury segment	0.48 tourist days year^{-1}	50	

Box 13.2 Economic production functions employed

Mining sector

$$\log(\text{quantity of mineral produced})_{i,t} = \alpha_{i,t} + \beta_1 \log(\text{machinery})_{I,t} + \beta_2 \log(\text{energy})_{I,t} \beta_3 \log(\text{work})_{I,t} + \beta_4 \log(\text{land})_{I,t} + \beta_5 \log(\text{water})_{I,t} + \beta_6 \log(\text{land})_{I,t} + \beta_7 \log(\text{primary\_material\_inputs})_{I,t} + u_{y,I,t}$$

for both cobalt and nickel in period *t*.

Agricultural sector

$$\log(\text{production})_{i,t} = \alpha_{i,t} + \beta_2 \log(\text{labor})_{I,t} + \beta_3 \log(\text{land})_{I,t} + \beta_4 \log(\text{water})_{I,t} + \beta_5 \log(\text{infrastructure})_{I,t} + u_{y,I,t}$$

for rice, manioc, and livestock in period *t*; and the explanatory variables are the logarithms of the selected production inputs, including water. We model production in the agricultural sector as a set of integrated productive activities, where a commonality of inputs of production is used (see Varian 1992).

Tourism sector

$$Q = A\ W^{\alpha} Z^{\beta}$$

where *Q* represents the number of total arrivals at the selected destination; *A* represents a technological parameter; *W* is the input water, and *Z* represents all other variables. Water was assumed to be the only input affecting output, keeping other inputs fixed. This assumption, implying hotel capacity and number of employees to not vary as much as water use (e.g. for personal hygiene) with the number of arrivals, can be defended as a first approximation within a short period such as a single tourist season.

Assuming that α equals 1 and β equals zero, we write the technological relationship as the linear relationship

$$Q = AW$$

where *A* also measures both average and marginal productivity, which are equal. This reflects the assumption that an additional tourist will use (for personal hygiene) about the same quantity of water as the other tourists.

---

production input (an estimate of the economic value per unit of output of the production sector) was greater in the mining sector.

Among the chief limitations of the biophysical study described are:

(1) Feedbacks between services and potential trade-offs, such as that between water quantity and quality in situations of resource limitation, have not been explored. It must be highlighted, however, that no such trade-offs are explored

in any of the currently available methodologies for ecosystem services assessment.

(2) Due to limitations in data availability, all water models operate annually. While water use has important annual fluctuations, the patterns of flow do not change significantly from season to season, and comparative results of forest corridor versus non-protected areas remain representative of the added values provided by the natural features of the Ankeniheny–Zahamena Forest Corridor even if seasonal dynamics is not addressed in the physical models.

From a biophysical perspective, an important next step in this analysis could entail conducting modeling studies to pinpoint more precise thresholds in the extent and quality of the natural environment that supports critical services. For example, simulation of land use conversion at different degrees can be used to determine the precise amount of forest loss that causes unrecoverable, non-linear changes in water supply for each class of users. The same methods demonstrated in this study are also suitable for integrated simulation of 'bundled' ecosystem services, so that the consequences of policy aimed at optimizing the output for one can be assessed in terms of their consequences on others. Such consequences can be fed back to the economic analysis to complete the quantification of the associated societal costs in an integrated ecosystem services perspective. One can hypothesize that this function will have thresholds that can be later studied as a function of other variables. An economic analysis can be performed on the results to establish optimal balance between forest use and conservation.

Lastly, performing a pilot analysis, as demonstrated here, in as many pilot areas as practical can help define a protocol to reduce the inaccuracies inherent in any national accounting based on partial assessments. A future study may be specifically directed to extracting this protocol from physical accounts and parametrizing a 'best case' transfer matrix that can help adjust economic estimates to the national level based on percentage coverage of biophysical and socio-economic characteristics of interest.

13.5 CONCLUSIONS

The assessment presented in this chapter has shown how specific landscape characteristics can determine sensitivity of the ecosystem services supply. In the case analyzed here we see a four-fold change in quality and a ten-fold change in sustainability of supply in protected versus non-protected areas. Such a matrix is likely to be country-specific or even region-specific within a country, and needs to be defined on a case-by-case basis. Yet this information can greatly help contextualize economic value and reduce uncertainties. The protocol can be completed with an assessment of the

Box 13.3 Key messages

- Results produced by an integrated biophysical and economic analysis of ecosystem services can help inform prioritization and design of management alternatives better than considering these dimensions in isolation.
- In the absence of quantitative models it is useful to imagine a policy framework that considers different dimensions of value in an integrated way.
- We identify four dimensions of ecosystem services value, each with a different relevance for policy assessments: input productivity, economic value, sustainability of supply, and quality of supply.
- Methodologies for the biophysical assessment of ecosystem services that place a primary emphasis on beneficiaries provide a more realistic outlook on ecosystem services value.
- Replication of additional pilot studies can help develop methods aimed at reducing inaccuracies and facilitate scale-up from local to national and transnational studies

relative uncertainty (as well as errors) and analyzed under both current conditions and projected global or local change scenarios.

Our results suggest that conservation can provide important benefits in terms of water supply and sediment regulation, upon which a variety of economic sectors are highly dependent. These results are relevant in general, and particularly relevant within the development of regional and national integrated water resources management planning and policy in Madagascar. Such policies, by definition, must address the individual and competing needs of different sectors (i.e. agricultural, industrial, energy, tourism, mining, and others). Policies must also address the industrial efficiency and profitability of a given operation; its impact on water flows and sediments; and equity in the distribution of resources.

The four key dimensions of ecosystem services addressed here (input productivity, economic value, sustainability of supply, and quality of supply) have complementary relevance for policy assessments. While these four dimensions are interrelated in complex ways, in the absence of quantitative models it is natural to imagine a policy framework that considers all of them in a multiple-criteria analysis. Such an analysis could be used to rank the opportunity value of each prospective policy instrument in regards to the policy context to which it would apply. This analysis can inform the design of policy interventions by allowing comparisons among different sites, highlight important trade-offs between competing alternatives, and ultimately guide decision-making based on locally determined priorities. Further research should focus on the determination of monetary values of services for incorporation into national accounting, emphasizing tight integration with the biophysical analysis. Lastly, the replication of additional pilot studies can help develop methods aimed at reducing the inaccuracies inherent in any national accounting based on partial assessments, and facilitate scale translation from local to national and transnational studies.

ACKNOWLEDGMENTS

We express gratitude to the WAVES Madagascar Steering Committee, which provided guidance at the onset of the research, and particularly to Leon Rajaobelina (CI) and Jean-Gabriel Randriamarison (Ministry of Economy and Industry). We are also indebted to Alison Clausen (World Bank Madagascar), Ramy Razafindralambo, and many governmental officials for their time in providing advice, guidance, and data, including Ramdrianbolanamitra Samuel (INSTAT), Naivonirina Ramananjaona (ANDEA), Rene Randriambohanginjatovo (National Park, Mantadia-Andasibe), and Ravahinimbola Yolanda Rachel (Tourism Regional Office Alaotra-Mangoro). Our thanks extend to important stakeholders from the Ankeniheny–Zahamena Forest Corridor for their time and guidance. The development of ARIES was originally funded by the US National Science Foundation. The first author also receives support from the ASSETS project funded by ESPA/NERC (grant no. NE-J002267-1). UNEP-WCMC and Conservation International also provided support for the development of specific components of ARIES. The work of Kenneth J. Bagstad, Gary W. Johnson, and Brian Voigt was instrumental in developing the methods used for the biophysical modeling. Brian Voigt also collaborated to drafting Figures 13.1 and 13.2.

We are grateful to Julia Martin-Ortega for her input to earlier versions of this chapter.

References

Bagstad, K., Villa, F., Johnson, G. W., & Voigt, B. (2011). ARIES: ARtificial Intelligence for Ecosystem Services: a guide to models and data. ARIES report series 122.

Bagstad, K. J., Johnson, G. W., Voigt, B. & Villa, F. (2013). Spatial dynamics of ecosystem service flows: a comprehensive approach to quantifying actual services. *Ecosystem Services* **4**, 117–125.

Bagstad, K. J., Villa, F., Batker, D., *et al.* (2014). From theoretical to actual ecosystem services: mapping beneficiaries and spatial flows in ecosystem service assessments. *Ecology and Society* **19**(2), art. 64.

Boyd, J. & Banzhaf, S. (2007). What are ecosystem services? The need for standardized environmental accounting units. *Ecological Economics* **63**(2–3), 616–626.

Brouwer, C., Goffeau, A., & Heilbloem, M. (1985). *Irrigation Water Management Training Manual.* FAO, Rome.

Costanza, R., dArge, R., deGroot, R., *et al.* (1997). The value of the world's ecosystem services and natural capital. *Nature* **387**(6630), 253–260.

Daily, G. C., Polasky, S., Goldstein, J., *et al.* (2009). Ecosystem services in decision making: time to deliver. *Frontiers in Ecology and the Environment* **7**(1), 21–28.

Eigenbrod, F., Armsworth, P. R., Anderson, B. J., *et al.* (2010). The impact of proxy-based methods on mapping the distribution of ecosystem services. *Journal of Applied Ecology* **47**(2), 377–385.

Fisher, B., Turner, K., Zylstra, M., *et al.* (2008). Ecosystem services and economic theory: integration for policy-relevant research. *Ecological Applications* **18**(8), 2050–2067.

Fisher, B., Turner, R. K., Burgess, N. D., *et al.* (2011). Measuring, modeling and mapping ecosystem services in the Eastern Arc Mountains of Tanzania. *Progress in Physical Geography* **35**(5), 595–611.

Hannah, L., Dave, R., Lowry, P. P., *et al.* (2008). Climate change adaptation for conservation in Madagascar. *Biology Letters* **4**(5), 590–594.

Johnson, G. W., Bagstad, K. J., Snapp, R. & Villa, F. (2010). Service Path Attribute Networks (SPANs): spatially quantifying the flow of ecosystem services from landscapes to people. *Lecture Notes in Computer Science* **6016**, 238–253.

Johnson, G. W., Bagstad, K., Snapp, R. & Villa, F. (2012). Service Path Attribution Networks (SPANs): a network flow approach to ecosystem service assessment. *International Journal of Agricultural and Environmental Information Systems* **3**(2), 54–71.

Kareiva, P. M., Tallis, H., Ricketts, T., Daily, G. C. & Polasky, S. (2011). *Natural Capital : Theory and Practice of Mapping Ecosystem Services.* Oxford University Press, New York.

Kull, C. A. (2000). Deforestation, erosion, and fire: degradation myths in the environmental history of Madagascar. *Environment and History* **6** 423–450.

Martinez-Harms, M. J. & Balvanera, P. (2012). Methods for mapping ecosystem service supply: a review. *International Journal of Biodiversity Science, Ecosystem Services and Management* **8**(1–2), 17–25.

Portela, R., Nunes, P. A. L. D., Onofri, L., Villa, F., Shepard, A., & Lange, G. M. (2012). *Assessing and valuing ecosytem services in the Ankeniheny–Zahamena Corridor, Madagascar: A Demonstration Case Study for the Wealth Accounting and the Valuation of Ecosystem Services (WAVES) Global Partnership.* World Bank, New York.

Rakotoarison, H. F. (2003). *Evaluation Economique des Bassins Versants Dans Les Regions de Fierenana et D'Andekaleka. Mémoire de fin d'Etudes.* Antananrivo. Université d'Antananarivo, Madagascar.

Tallis, H. & Polasky, S. (2009). Mapping and valuing ecosystem services as an approach for conservation and natural-resource management. *Annals of the New York Academy of Sciences* **1162**, 265–283.

Troy, A. & Wilson, M. A. (2006). Mapping ecosystem services: practical challenges and opportunities in linking GIS and value transfer. *Ecological Economics* **60**(2), 435–449.

UNDP (2011). Situation des autres usages de l'eau dans le Grand Sud Malgache. Rapport Provisoire.

Varian, H. R. (1992). *Microeconomic Analysis.* W. W. Norton, New York.

Vigerstol, K. L. & Aukema, J. E. (2011). A comparison of tools for modeling freshwater ecosystem services. *Journal of Environmental Management* **92**(10), 2403–2409.

Villa, F. (2007). A semantic framework and software design to enable the transparent integration, reorganization and discovery of natural systems knowledge. *Journal of Intelligent Information Systems* **29**(1), 79–96.

Villa, F. (2009). Semantically-driven meta-modelling: automating model construction in an environmental decision support system for the assessment of ecosystem services flow. *Information Technology in Environmental Engineering.* In I. N. Athanasiadis, P. A. Mitkas, A. E. Rizzoli, & J. Marx Gomez. Springer, New York.

Villa, F., Bagstad, K., Voigt, B., *et al.* (2014a). A methodology for robust and adaptable ecosystem services assessment. *PLoS One* **9**(3).

Villa, F., Voigt, B. & Erickson, J. (2014b). New perspectives in ecosystem services science as instruments to understand environmental securities. *Philosophical Transactions of the Royal Society of London Series B: Biological Sciences* **369**(1639).

Waage, S., Kester, C. & Armstrong, K. (2013). *Global Public Trends in Ecosystem Services, 2009–2012.* BSR, San Francisco, CA.

Wallace, K. J. (2007). Classification of ecosystem services: problems and solutions. *Biological Conservation* **139**(3–4), 235–246.

14 Rapid land use change impacts on coastal ecosystem services

A South Korean case study

Hojeong Kang, Heejun Chang, and Min Gon Chung

14.1 INTRODUCTION

Coastal zones occupy a relatively small portion of the global land surface, but play a key role in various aspects of the biophysical settings of ecosystems and human activities. Coastal areas form the interface of land, freshwater, and sea, where terrestrial and marine ecosystems and socio-economic processes are linked. As such, coastal areas are one of the most ecologically and economically productive areas on Earth, providing multiple ecosystem services, including water quality amelioration, accumulation, and conversion of carbon and nutrients, protection against floods and tidal inundation, and the provision of wildlife habitat. They also provide fisheries, agricultural production, living spaces, water resources for industry, and recreational activities. More than half of the world's population lives within 200 kilometres of a coast (Hinrichsen 1998), many of them in coastal cities. Overall, coastal zones produce more than 60% of the economic value of the Earth (Martínez et al. 2007).

Both population growth and climate change increase the importance of coastal ecosystem services and yet, paradoxically, many coastal areas have been significantly degraded, primarily because of ongoing land development driven by population growth (Hong et al. 2010). For example, expanding urban development has caused land subsidence in the Pearl River delta in China (Wang et al. 2012). Additionally, globalization of trade in natural commodities, such as shrimp and oysters, has promoted the over-exploitation of these resources, negatively affecting the regulating ecosystem services, and species diversity more generally (Vermaat et al. 2012). Together with the predicted rise in sea levels caused by global climate change, coastal ecosystems are now facing many challenges associated with changes in land cover and land intensification that result in nitrogen enrichment, exposure to toxins, and alteration of hydrological regimes. Thus, maintaining the functions and integrity of coastal ecosystems through land use and development planning in a changing climate has become a primary policy issue in coastal ecosystem-based management (Barbier et al. 2008).

Across Asia generally, and in East and South Asia in particular, mega-cities (with populations estimated at over one billion) such as Shanghai and Ho Chi Minh City are located in coastal areas. Usually located on or near the mouth of large rivers such as the Yangtze, the Han, or the Mekong, coastal zones provide fertile land for agriculture and support diverse fisheries. For many generations, coastal residents relied heavily on water-dependent ecosystem services, including rice production and fishing. Coastal wetlands and mangrove forests also provide buffers against natural disasters such as tsunamis, and therefore have both direct and indirect use values (Sanford 2009).

South Korea, in particular, serves as an interesting example of the dynamic feedbacks between coastal ecosystem services and human activities. Coastal ecosystems in South Korea play a significant role in local economies through fish production and eco-tourism; over 90% of fishery rights are located in coastal areas and 84% of industrial complexes are located in coastal cities and counties (Hong et al. 2010). As such, coastal environments in South Korea are closely linked to local human communities, forming strong socio-ecological systems that have an important influence on human wellbeing (Koh et al. 2009). In the last few decades, land use patterns have dramatically changed in South Korea. A high rate of economic growth (average annual growth rate of 8.5% between 1961 and 1991) and dramatic increases in urban populations (81.3% of the total population in 2005 compared to 20.8% in 1960) have forced both local and central governments to develop or reclaim coastal areas to secure more resources and development space (Korea Statistics 2011). Tension over whether coastal ecosystems should be protected or developed has heightened, but priority continues to be given to development-oriented projects (Hong et al. 2010). The government has introduced large-scale drainage and land fill of coastal wetlands for agricultural, housing, and industrial purposes; 43% of all coastal wetlands have been developed since 1918, and the health of coastal ecosystems has deteriorated (Koh et al. 2009). However, an ecosystem services-based approach has rarely been used to estimate the costs and benefit of various land developments and in-fill. Furthermore, land use change to maximize a single ecosystem service often decreases the provision of other ecosystem services, hence understanding the trade-offs among different ecosystem services is of great importance to support decision-making.

Previously most decision-making on coastal resource management utilized Neoclassical economic models (Weinstein & Reed 2005). In these models, non-market values of natural resources are significantly underestimated. Attempts to redress these failures have primarily focused on the provisioning ecosystem services (Liquete *et al.* 2013), and yet fall short of understanding the complex dynamics and potential synergies and trade-offs among multiple coastal ecosystem services under different land use and management scenarios (McNally *et al.* 2011; Hicks *et al.* 2013; Lester *et al.* 2013). In this chapter we explore how a more holistic ecosystem services-based approach, as defined in Chapter 2 of this book, can be operationalized to support decision-making by overcoming some of the limitations of previous approaches. Using a rapid spatial analysis to identify ecosystem services hotspots, we (1) estimate ecosystem services provision in coastal areas in South Korea; (2) visualize the distribution of ecosystems services at a county scale; and (3) determine the relationships among

components of ecosystem services to identify potential synergies and trade-offs in multiple ecosystem services and the correlation with regional gross domestic product. We map and quantify four ecosystem services (recreation, aesthetic quality, terrestrial biodiversity, and carbon storage) and two ecosystem dis-services (marine habitat risk and coastal vulnerability). We explicitly chose to focus on these regulating and cultural ecosystem services because they have been largely neglected in previous analyses, but we also recognize the importance of provisioning services (see Box 14.1).

14.2 LAND USE CHANGES AND DEVELOPMENT PRESSURES IN COASTAL AREAS OF SOUTH KOREA

Land development pressure is one of the major driving forces of coastal ecosystem services degradation in South Korea. The

---

Box 14.1 Provisioning services from aquatic coastal environments

Although we focused narrowly on coastal ecosystem services in this study by excluding direct market values of coastal ecosystems, the provisioning services from fishery and aquaculture are definitely of great importance. In particular, products from fishery and aquaculture comprise a substantial source of nutrition (e.g. protein and energy) in many Asian countries. Except for Japan, all countries considered exhibit continuous growth in fishery capture in the last 50 years (Figure 14.1a).

In addition, production from aquaculture has increased more rapidly in most Asian countries regardless of their stages of economic development (Figure 14.1b).

Such a massive increase in aquaculture is promoted not only by increasing demand from local people, but also by the global demand for fish products. Although aquaculture can generate economic revenues, it is known to affect coastal environment adversely by deteriorating water quality and decreasing biodiversity (Outeiro & Villasante 2013). Additionally, conversion of mangrove forests to aquaculture can reduce the value of other regulating or cultural coastal ecosystem services (Barbier 2012). Such aspects of trade-offs among different coastal ecosystem services warrant further studies (Schmitt *et al.* 2013).

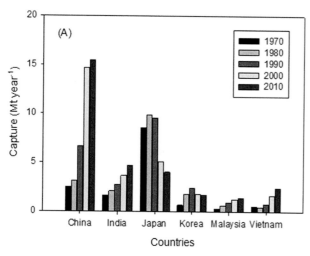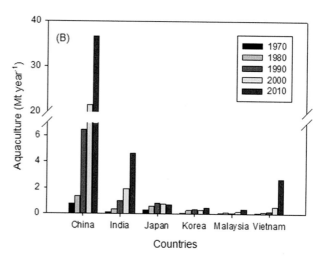

Figure 14.1 Changes in fishery production by capture (a) and aquaculture (b) in key Asian countries from 1970 to 2010. Adapted from Food and Agriculture Organization (2013)

Table 14.1 *Percentages of land cover patterns in coastal areas in Korea, 1980, 1990, and 2000.*

	Urban built-up areas	Agricultural land	Forest	Grassland	Wetland	Open field	Open water
1980	11.8	29.3	40.9	5.0	3.4	2.5	7.1
1990	15.6	26.9	40.1	5.8	2.1	4.1	5.4
2000	17.9	25.5	42.2	4.6	1.3	3.7	4.7

Source: Korea Statistics (2011)

western and southern coastal regions of South Korea, which extend from marine areas to terrestrial county boundaries, are characterized by relatively flat terrains. They have been intensely developed in the past 40 years and development pressure continues to be high. Fifty-nine cities and counties (cities and counties range from 2.82 to 1005.8 km^2 in area) lie within the coastal zone and 21% of South Korea's total population live in this area (Korea Statistics 2011). Table 14.1 shows land cover patterns in coastal areas in Korea in 1980, 1990, and 2000. Urban built-up areas increased from 11.8% in 1980 to 17.9% in 2000. In contrast, wetland areas decreased from 3.4% to 1.3% during the same period

Increases in dry land and concurrent decreases in open water are mainly due to current development projects introduced by the South Korean government. Development of tidal flats on the western coast of Korea does not require complicated civil engineering works as tidal flats occur inside bays, with little impact from waves and tidal flow. Rapid economic growth and the introduction of the Public Waters Reclamation Act of 1999 have resulted in large-scale development of wetlands to increase crop production and secure land for human habitation. This law legitimized the development of coastal wetlands until 2000, by which time 810.5 km^2 had been turned over to agricultural, industrial, or urban use (Table 14.1). Such changes have resulted in (1) reduction in wildlife habitat and biodiversity; (2) reduction in sedimentation and flood protection; (3) declining water quality; and (4) decreased primary production and fishery productivity.

To cope with environmental degradation and increasing social pressure for wetlands conservation, the Korean government introduced the Wetland Conservation Law and Coastal Management Law in 1999, which initiated systematic research and policy development for the conservation of coastal wetlands. Under these laws, a national survey of coastal wetlands is conducted every five years, and a Basic Plan for Wetland Conservation has been proposed. The result has been a decline in the rate of coastal habitat loss (Ministry of Land, Transport and Maritime 2012). However, accurate assessment of impacts on coastal ecosystems or damage to ecosystem services is still required to underpin management and policy development.

14.3 ASSESSMENT AND MAPPING OF ECOSYSTEM SERVICES

To assess and visualize coastal ecosystem services, we integrated ecosystem services mapping methods with spatial statistical analysis. Six principal ecosystem services were classified and mapped using the Integrated Valuation of Environmental Services and Tradeoffs (InVEST) model (Kareiva *et al.* 2011).[1] Among the sub-models, we selected habitat risk assessment (coastal), vulnerability (coastal), recreation, aesthetic quality, biodiversity (terrestrial ecosystems facing the sea), and carbon storage (terrestrial ecosystems facing the sea) (Table 14.2). These services were chosen as they represent a spectrum of regulating (e.g. carbon sequestration and storage) and cultural (e.g. recreation and aesthetic) ecosystem services and require non-monetary valuation that traditional Neoclassical economics do not include. The habitat risk assessment and coastal vulnerability models represent the risk and vulnerability of target ecosystems to physical (e.g. soil erosion and inundation) and biological (e.g. biodiversity loss, changes in population size) impacts, and hence can be used as a proxy for the regulating services of ecosystems (Tallis *et al.* 2013), which are looked at here as ecosystems dis-services.

The spatial database required by the Integrated Valuation of Environmental Services and Tradeoffs model was constructed using ecological, social, and geographical data supplied by several government agencies, including the Ministry of Land, Transport and Maritime Affairs and the Ministry of Environment. Ecological data include coastal habitat information, sea floor characteristics, and information on protected areas. Social data include locations of industrial complexes and tourist attractions, beaches, recreational fishing sites, and population by county. The geographical data used for basic ecosystem services mapping include a digital elevation model, the county boundaries, and coastline data. Source materials in paper copy or online image format were converted to vector or raster formats in Geographical Information Systems through digitization and geo-referencing.

[1] Using version 2.2.3.

Table 14.2 *Details about the characterization of the six ecosystem services considered in this study*

Services	Method	Unit (resolution)	Purposes	Data used for analysis
Biodiversity (coastal)	Habitat Risk Assessment	Unitless Per unit cell *(500 m × 500 m)*	To assess the risk to coastal habitats exposed to each stressor and estimate the resultant consequence of that exposure to habitat risk at a pixel level. The weighted averages of the exposure and consequence values are combined to estimate the overall score of habitat risk	Gridded seascape Habitat raster data Stressor raster data Table of habitat–stressor ratings
Coastal vulnerability	Vulnerability Index	Unitless Per unit cell *(250 m × 250 m)*	To produce a qualitative estimate of coastline exposure to storm	Digital Elevation Model Wave Watch III model data Average sea depth Natural habitat (polygon) Table of natural habitat Shoreline type (polyline)
Recreation	Overlap analysis	Number of recreation sites Per unit cell *(500 m × 500 m)*	To determine the locations of recreational activities	Gridded seascape Recreation layers Table of recreation layers
Aesthetic quality	Aesthetic views analysis	Number of scenic view sites per unit cell *(500 m × 500 m)*	To identify locations of high scenic value	Points of scenic amenity Digital Elevation Model Population raster
Biodiversity (terrestrial)	Habitat quality and rarity scores	Unitless Per unit cell *(100 m × 100 m)*	To assess the conditions of terrestrial habitats adjacent to the sea and the risk posed by human activities. The habitat quality score is based on the location and intensity of human land uses, and the habitat rarity score is based on current habitat types' relative rarity from a reference period	Land use and land cover map Threat raster data Threat data set
Carbon storage (terrestrial)	Sum of four carbon pools	Ton carbon per hectare *(100 m × 100 m)*	To estimate the carbon storage in terrestrial habitat	Land use and land cover map Table of carbon pools

Source: Tallis *et al.* (2013).

The value of six ecosystem services in each county was estimated from the results of spatial analysis using zonal statistics tools.[2] For this analysis, habitat risk assessment (coastal), coastal vulnerability, biodiversity (terrestrial), and terrestrial carbon storage are expressed as average numeric values in each county. Briefly, habitat risk assessment was conducted with surveyed data on habitat quality, and the index of presence of threat and rarity. For coastal vulnerability, information about relief, natural habitat, sea-level change, wind exposure, average depth of adjacent sea, and distances between coastlines and the edge of the continental margins were used as input variables. Recreation is represented as the number of recreational sites in each county, while aesthetic quality is calculated as the number of scenic points in a county divided by the area of that county. The sum of these numeric values in each county for each service was then ranked on a scale from 1 to 5 to represent degrees of provision of specific ecosystem services in each county. We then used these normalized values to compare different ecosystem services directly.

14.3.1 Identification of ecosystem services hotspots

Because specific ecosystem services are provided at specific places in the landscape, it is important to map each service provided and compare it to other services. The overlapping areas where the value of multiple ecosystem service provision is high can be considered as ecosystem service hotspots, providing an indication to policy makers of areas that merit further conservation. To identify ecosystem service hotspots, ecosystem

---

[2] using ArcGIS

service values were normalized by classifying them on a five-point scale (1 being the lowest and 5 being the highest) before undertaking overlay analysis. Since the six ecosystem services have different units, it is necessary to convert the absolute numbers into standardized numbers before they are overlaid on a single map. We then conducted a sensitivity analysis to determine the effects of subjective uncertainty in the weighted value of each service.

'Hotspots' are defined here as areas where the sum of six normalized ecosystems services exhibits the highest classification score, i.e. 5. Looking at the map in Figure 14.2 it is apparent that the southern coastal counties are generally associated with more service hotspots than the western counties, although some hotspots are found in the westernmost part of the west coastal counties. Such hotspots coincide with protected natural areas and national parks, maintaining high terrestrial and marine habitat quality. Some southern counties have intensive aquaculture facilities that have negative impacts on aesthetic quality (Ministry of Land, Transport and Maritime 2012), resulting in medium to low provision of ecosystem services. The map also identifies those areas with a lower density of ecosystems service provision which are scattered around the coastal counties and are potentially suitable for sustainable urban and industrial development.

Sensitivity analysis, where the relative weighting of each ecosystem service was changed, confirmed that the location of areas of high ecosystem service provision was not statistically significantly altered.

14.3.2 Synergies and trade-offs among ecosystems services

A simple map correlation and map overlay can provide insights on possible synergies and trade-offs among multiple ecosystem services to inform policy decisions (Hoyer and Chang 2014). Table 14.3 illustrates the results of map correlation between two different ecosystem services. It is noteworthy that positive correlations between recreation or aesthetic value and coastal vulnerability actually reflect trade-offs between conservation and development of coastal zones. For example, removal of sea walls increases scenic value (e.g. increases 'recreation' or 'aesthetic quality'), but also increases 'coastal vulnerability' to storms, resulting in a positive correlation.

Overall, results indicate that well-preserved, more natural ecosystems exhibit positive results for services such as recreation, aesthetic value, and carbon storage, but at the same time such areas are highly vulnerable to negative impacts such as climate change or new land developments.

However, several ecosystem services exhibit negative correlations with each other. For example, aesthetic or recreation values are closely related to areas with minimal interference from human activities. In contrast, the vulnerability of coastal areas to natural hazards such as a rise in sea level or typhoons is often higher without human intervention. For instance, the construction of artificial structures such as sea walls substantially decreases aesthetic value while reducing coastal vulnerability to natural hazards. This is well-illustrated in Figure 14.3, where there is significant positive correlation between aesthetic/recreation value and coastal vulnerability.

To maximize the delivery of ecosystem services in such areas, new technologies or management schemes, such as low-impact developments, should be considered to increase the physical stability of coastal areas without reducing other ecosystem services such as aesthetic or recreation values. Regional land use planners could use the findings of our research to identify the location and type of future land development to sustain the provision of multiple ecosystem services.

14.3.3 Ecosystem services and gross regional domestic product

The relationship between specific ecosystem services and the gross domestic product is interesting (see Table 14.3). The regulating ecosystem service (e.g. carbon storage) is significantly negatively associated with gross domestic product, while the two cultural ecosystem services (aesthetic and recreational ecosystem services) are marginally negatively related to gross domestic product. On the other hand, cultural ecosystem services and carbon storage are all positively related to each other. This suggests that increasing the number of eco-tourism sites through tree planting or other forms of conservation or restoration may increase the value of other regulating services. These land conservation or restoration activities can thus promote the regional economy in preference to traditional land development that typically leads to the degradation of some coastal ecosystem services (Tallis *et al.* 2008).

14.4 CONCLUSIONS

As the first ecosystem service mapping project in South Korea, our spatial analysis identified potential synergies and trade-offs among multiple ecosystem services in coastal areas. Our results showed that risks to recreation and terrestrial habitat values have significant negative impacts on the size of the local economy (gross regional domestic production) at the county scale. Hotspots of ecosystem services are mostly found in areas where human intervention and development are limited and hence landscape aesthetics remain high.

The spatially explicit ecosystem services-based approach to the rapid assessment of ecosystem services and analysis of trade-offs has both advantages and limitations. The approach used here provides a relatively rapid assessment of multiple ecosystem

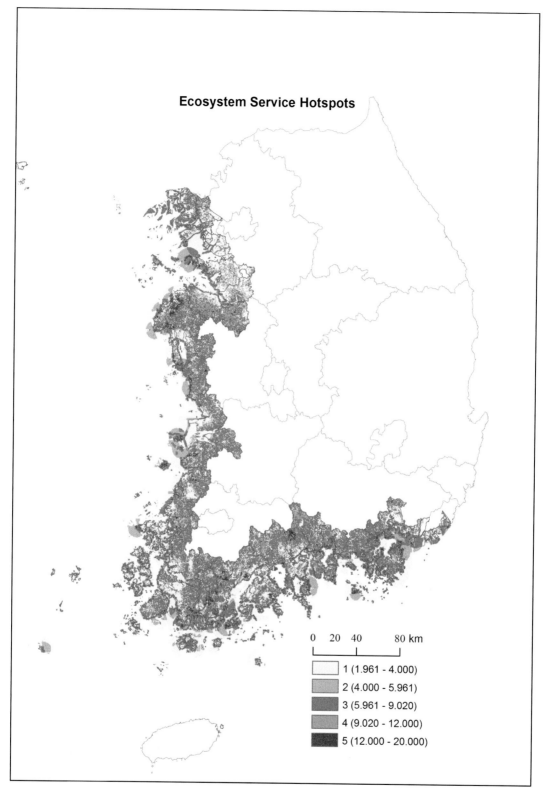

Figure 14.2 Hotspots of ecosystem services based on six components of ecosystem services in coastal areas of Korea. Source: Chung (2013). A black and white version of this figure will appear in some formats. For the colour version, please refer to the plate section.

Table 14.3 *Correlation coefficients (r) between ecosystem services and gross regional domestic production (GRDP).*

Variables	GRDP	Coastalvulnerability	Risk of coastal habitat	Recreation	Aesthetic quality	Carbon storage
Coastal vulnerability	−0.346***					
Risk of coastal habitat	−0.539***	0.456***				
Recreation	−0.226*	0.202	0.332***			
Aesthetic quality	−0.090	0.440***	0.141	0.361***		
Carbon storage	−0.332***	0.176	0.346***	0.480***	0.017	
Risk of terrestrial habitat	−0.640***	0.141	0.500***	0.447***	0.145	0.520***

Modified from Chung (2013).
* significant at the 0.1 level; ** significant at the 0.05 level; *** significant at the 0.01 level.

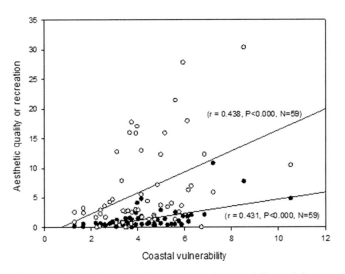

Figure 14.3 The relationship between coastal vulnerability and the calculated aesthetic value (open circle) or recreation (closed circle). The scales on the *x* and *y* axes represent the unit less normalized vulnerability index and total numbers of scenic views or recreation sites by each county.

services using widely available Geographic Information System and national statistical data, which can allow policy makers to simultaneously consider ecological information such as the health of coastal ecosystems and important ecosystem services such as recreation, aesthetic quality, and amenity that influence local economies both directly and indirectly. This can be used to help develop ecosystem-based spatial planning and management policy that considers both ecosystems and society (Katsanevakis *et al*. 2011), and can help in the identification of protected areas in which social benefits can be maximized (i.e. ecosystem services hotspots).

Among the limitations of the approach presented here is the choice of the metrics to quantify provision of ecosystem services. Monetary valuation is often criticized because the complexity of human wellbeing cannot be captured in monetary terms, which neglect intangible components. Similarly, it could be argued that

Box 14.2 Key messages

- Coastal regions provide multiple ecosystems services that are essential to humans.
- Specific ecosystem services are provided at specific areas in coastal regions.
- Spatially explicit ecosystem services analysis identifies potential synergies and trade-offs among multiple ecosystem services.
- Ecosystem service hotspots can be targeted for conservation, based on ecosystem-based spatial and land use planning.
- Rapid assessment approaches such as the one presented here are valuable for supporting decision-making, but outcomes still need to be verified with local-scale data.

the metric used here to measure cultural ecosystem services, i.e. the number of recreational sites and scenic points, is also too simplistic. A better alternative indicator could be one based on the number of visitors or perceptions through user surveys (Klain *et al*. 2012). Another limitation is the use of publicly available geographical and national statistical data to quantify specific ecosystem services at a local scale. Such data and the resultant ecosystem services estimated may need to be verified by local stakeholders and experts through workshops or surveys. Without transparent communications between stakeholders and scientists, the implementation of the ecosystem service-based approach presented in the current case study may not be practical (core element 3 of ecosystem services-based approaches as defined in this book). Additionally, although providing interesting first insights, the simple correlations used in our analysis does not show any causal relationships between different ecosystem services. In particular, the negative correlation between habitat risk and gross regional domestic product does not show any causal mechanisms of different habitat quality under different levels of gross regional domestic production. As such, caution is needed when interpreting the relationships between these services. Finally, ecosystem services are constantly changing over time and space, so a time-series analysis of changes in multiple ecosystem services is required with relevant long-term data on ecosystem status (Hughes *et al.* 2013; Hoyer & Chang 2014). Further studies are needed to elucidate the connection between increases in population and changes in ecosystem services in coastal ecosystems.

ACKNOWLEDGMENTS

A partial support for Chang was provided by NSF-BCS Grant # 1026629: Spatial Analysis of Ecosystem Services Shifts caused by Climate Change and Land Conversion in the Metropolitan Fringe.

References

Barbier, E. B. (2012). A spatial model of coastal ecosystem services. *Ecological Economics* **78**, 70–79.

Barbier, E. B., Koch, E. W., Silliman, B. R., *et al.* (2008). Coastal ecosystem-based management with nonlinear ecological functions and values. *Science* **319**(5861), 321–323.

Chung, M. G. (2013). Mapping ecosystem services and statistical analysis for ecosystem-based management of coastal areas. MS thesis, Yonsei University.

Food and Agriculture Organization (2013). *FAO Statistical Yearbook*. Available at: www.fao.org/statistics/en (last accessed 22 October 2014).

Hicks, C. C., Graham, N. A. J., & Cinner, J. E. (2013). Synergies and tradeoffs in how managers, scientists, and fisheries value coral reef ecosystem services. *Global Environmental Change* **23**, 1444–1453.

Hinrichsen, D. (1998). *Coastal Waters of the World: Trends, Threats, and Strategies*. Island Press, Washington, DC.

Hong, S. K., Koh, C. H., Harris, R. R., Kim, J. E., Lee, J. S., & Ihm, B. S. (2010). Land use in Korean tidal wetlands: impacts and management strategies. *Environmental Management* **45**, 1014–1026.

Hoyer, W. & Chang, H. (2014). Assessment of freshwater ecosystem services in the Tualatin and Yamhill basins under climate change and urbanization. *Applied Geography* **53**, 402–416.

Hughes, B. B., Eby, R., Van Dyke, E., *et al.* (2013). Recovery of a top predator mediates negative eutrophic effects on seagrass. *Proceedings of the National Academy of Sciences of the United States of America* **110**, 15313–15318.

Kareiva, P., Tallis, H., Ricketts, T.H., Daily, G.C., & Polasky, S. (eds) (2011). *Natural Capital: Theory and Practice of Mapping Ecosystem Services*. Oxford University Press, New York.

Katsanevakis, S., Stelzenmuller, V., South, A., *et al.* (2011). Ecosystem-based marine spatial management: review of concepts, policies, tools, and critical issues. *Ocean and Coastal Management* **54**, 807–820.

Klain, S. C. & Chan, K. M. A. (2012). Navigating coastal values: participatory mapping of ecosystem services for spatial planning. *Ecological Economics* **82**, 104–113.

Koh, C. H., Ryu, J. S., & Kwon, B. O. (2009). A comparative study on policy and management of the Wadden sea and Korean getbol. Ministry of Land, Transport and Maritime Affairs.

Korea Statistics (2011). *Korea Statistical Yearbook*. Korea Statistics, Gwachun, Korea.

Lester, S. E., Costello, C., Halpern, B. S., Gaines, S. D., White, C., & Barth, J. A. (2013). Evaluating tradeoffs among ecosystem services to inform marine spatial planning. *Marine Policy* **38**, 80–89.

Liquete, C., Piroddi, C., Drakou, E. G., *et al.* (2013). Current status and future prospects for the assessment of marine and coastal ecosystem services: a systematic review. *Plos One* **8**, e67737.

Martínez, M. L., Intralawan, A., Vázquez, G., *et al.* (2007). The coasts of our world: ecological, economic, and social importance. *Ecological Economics* **63**, 254–272.

McNally, C. G., Uchida, E., & Gold, A. J. (2011). The effect of a protected area on the tradeoffs between short-run and long-run benefits from mangrove ecosystems. *Proceedings of the National Academy of Sciences of the United States of America* **108**, 13945–13950

Ministry of Land, Transport and Maritime Affairs (2012). Coastal management information system. Ministry of Land, Transportation, and Maritime Affairs and Fisheries.

Outeiro, L. and Villasante, S. (2013). Linking salmon aquaculture synergies and trade-offs on ecosystem services to human wellbeing constituents. *Ambio* **42**, 1022–1036.

Sanford, M. P. (2009). Valuating mangrove ecosystems as coastal protection in post-Tsunami South Asia. *Natural Areas Journal* **29**, 91–95.

Schmitt, L. H. M. & Brugere, C. (2013). Capturing ecosystem services, stakeholders' preferences and trade-offs in coastal aquaculture decisions: a Bayesian belief network application. *PLoS ONE* **8**(10), e75956.

Tallis, H., Kareiva, P., Marvier, M., & Chang, A. (2008). An ecosystem services framework to support both practical conservation and economic development. *Proceedings of the National Academy of Sciences* **105**(28), 9457–9464.

Tallis, H. T., Ricketts, T., Guerry, A. D., *et al.* (2013). *InVEST 3.0.0 User's Guide*. The Natural Capital Project, Stanford, CA.

Vermaat, J. E., Estradivari, E., & Becking, L. E. (2012). Present and future environmental impacts on the coastal zone of Berau (east Kalimantan, Indonesia), a deductive scenario analysis. *Regional Environmental Change* **12**, 437–444.

Wang, H., Wright, T. J., Yu, Y. P., *et al.* (2012). InSAR reveals coastal subsidence in the Pearl River Delta, China. *Geophysical Journal International* **191**, 1119–1129.

Weinstein, M. & Reed, D. J., (2005). Sustainable coastal development: the dual mandate and a recommendation for 'commerce managed areas'. *Restoration Ecology* **13**(1), 174–182.

Part IV
Broadening the perspective

15 Ecosystem services-based approaches to water management

What opportunities and challenges for business?

Joël Houdet, Andrew Johnstone, and Charles Germaneau

15.1 INTRODUCTION

Water sourcing, use, and discharge management have become critical issues to numerous business activities throughout the world. For instance, securing water supply is critical to farming and food production (e.g. World Water Assessment Programme 2014), while increasing water use and pollution are starting to generate severe operational, investment, and reputational risks for the agribusiness, mining, industrial, and financial sectors (e.g. Gulati 2014; von Bormann 2014; Bizikova et al. 2014).

Increasing research and funding is being dedicated to the link between ecosystem services and business. For example, in the UK the 2011 Natural Environment White Paper contained a commitment to establish a business-led Ecosystem Markets Task Force to review the opportunities for UK business from expanding green goods, services, products, investment vehicles, and markets which value and protect nature's services.[1] Business and ecosystem services are the central focus of several professional networks including the Business for Social Responsibility's Ecosystem Services Working Group[2] and the World Business Council for Sustainable Development's Ecosystems Focus Area[3] (for a more extensive list, see Hanson et al. 2008; Waage & Kester 2014). As well as an increase in recruitment demanding business expertise in ecosystem services, a body of literature is developing to aid businesses to integrate ecosystem services into existing performance systems (Houdet et al. 2012).

Though most firms depend and/or impact, to varying extents, on the various provisioning, regulating, and cultural ecosystem services derived from aquatic ecosystems, applying an ecosystem services-based approach to their uses of water presents both opportunities and challenges. The aim in this chapter is three-fold. First, we intend to explain the interdependency links (dependencies and impacts) between water ecosystem services and various business activities. In doing so, we discuss the main water ecosystem services risks for business activities, notably in terms of water scarcity, ageing water infrastructures, changing policy environments and regulatory frameworks, and reputation.

Then, we underline the opportunities offered by using an ecosystem services-based approach as defined in Chapter 2 of this book to water management in the private sector, chiefly through cost savings, secured operations, and positive stakeholder partnerships and feedbacks.

Finally, we discuss the challenges of mainstreaming ecosystem services-based approaches to all form of business' water use and pollution, highlighting some key organisational, technological, and institutional issues.

Throughout the chapter, we make use of various business case studies, highlighting both best practices and shortcomings.

15.2 INCREASING USES AND DEGRADATION OF WATER ECOSYSTEM SERVICES: UNDERSTANDING THE RISKS FOR BUSINESS

15.2.1 Businesses depend and impact on water ecosystem services

Most businesses, both small and large, either depend or impact on more than one water ecosystem service. For instance, a pharmaceutical company may use genetic resources from water organisms. A tour operator may use a specific coastal landscape and a hotel which relies on local water resources. An agribusiness may depend on specific water quantities – within a given range of quality or content properties – through its supply chains.

Uses of water-related ecosystem services are thus diverse and may involve trade-offs, which may result in water ecosystem disservices (Zhang et al. 2010) when water use by a stakeholder compromises that of another. For instance, a farmer involved in intensive crop production or a coal mine may use, pollute, and waste large volumes of water. This can have detrimental effects on downstream users (e.g. breweries; Kissinger 2013) in terms of both water availability and quality.

[1] www.valuing-nature.net/taskforce   [2] www.bsr.org   [3] www.wbcsd

Indeed, surface water and groundwater around the world are threatened by pollution from the cumulative impacts of agricultural, industrial, extractive, and urban water uses. For instance, microbial pathogens are often the most pressing water quality issue in many developing countries (Grossman *et al.* 2013). Eutrophication (from excessive nutrient pollution) is also pervasive, and at least 169 coastal areas around the world are considered hypoxic (Grossman *et al.* 2013). In that context, business sources of water pollution may be direct (e.g. point-source pollution of a food factory) or indirect (e.g. through pharmaceutical and personal care product use and disposal by customers) and are increasing to such an extent that persistent toxic chemical pollutants are now found in 90% of the world's water bodies (Grossman *et al.* 2013).

In other words, firms who use water ecosystem services are also potentially negatively impacting their future availability (in terms of quantity and quality) for both themselves and other users.

15.2.2 Impaired water ecosystems and risks to businesses

With respect to degraded or impaired water ecosystems and ecosystem services, several key risks for business can be emphasised (Institute of Directors Southern Africa 2012):

- *Water availability/biophysical and ecological risks*: water resources and ecosystem services are increasingly scarce and degraded in many regions of the world. Global water withdrawals have tripled over the last 50 years to meet agricultural, industrial, and domestic demands (Grossman *et al.* 2013). Most projections of water demand and water withdrawals through 2050 indicate a large global net increase, but with significant regional variations (United Nations Environment Programme 2012). In other words, in the future, water availability and quality will most likely be further restricted as a result of changing climatic conditions and increased/competing demand for water resources due to population growth, urbanisation, and economic growth. This risk is critical to consider when looking for new investment or expansion opportunities (e.g. McCarthy 2011; Institute of Directors Southern Africa 2012; Kissinger 2013). It may result in constraints on business growth as well as operational and supply-chain disruptions. In addition, climate change may exacerbate flood and drought disasters. Between the 1980s and the 2000s, the number of extreme events increased 230% and 38%, respectively, causing economic losses of billions of dollars (Grossman *et al.* 2013). In addition, rises in sea levels will have major impacts on businesses operating in coastal environments, including insurance companies and their clients (e.g. potentially no

more insurance cover for physical assets located in areas prone to flooding or wave damage; Stephenson *et al.* 2007; Nel *et al.* 2011; Otto-Mentz *et al.* 2011).

- *Water infrastructure/energy-related risks*: in many countries, water infrastructure is ageing and in very poor condition (e.g. in South Africa; CSIR 2010). Maintenance backlogs, water leaks, and necessary investments may potentially affect water prices for business. Coupled with energy-related water risks (e.g. energy costs of water pumping or wastewater treatment), infrastructure risks are a critical aspect of both current and future business operations, especially in farming and mining. The role of water ecosystem services may be critical in that context: malfunctioning water ecosystems can further increase operational costs (e.g. increased water pumping costs due to increased reliance on artificial water infrastructures) and generate further investment needs (e.g. wastewater treatment plant necessary to bring water quality within the required range) if the water ecosystem services derived from them are not available, unreliable, or of low quality.

- *Policy environment/regulatory risks/legal licence to operate*: the water sector is often governed by complex sets of legislation, regulations, and policies, which tend not to be fully implemented in many countries. In South Africa, for instance, to ensure more sustainable water use and the protection of water resources, policy reviews are being progressively undertaken as regards to water rights and water use permitting processes (including tariff structures and pricing). This generates uncertainty for businesses and prevents them from making informed decisions about their future water needs and the associated capital and operating expenses. Increasing concerns about water quality may lead to greater regulatory restrictions on stormwater runoff from construction sites, wastewater from industrial sites, and acid mine drainage from mining sites (see case study 1 in Box 15.1), so as to ensure that their operations do not pollute waterways, aquifers, and other ecological infrastructures. Companies may face considerable additional costs to implement pollution prevention measures as well as discharge monitoring and treatment, among other compliance requirements.

- *Reputational risks/social licence to operate*: reputational risks become more apparent as stakeholders become aware of their basic human right of access to clean water (Grossman *et al.* 2013). More than two billion people currently live in water-stressed areas (mostly in Asia), and that figure is expected to rise substantially (e.g. four-fold growth in Africa) due to population growth, increased water use, and climate change (United Nations Environment Programme 2012). Competing uses of water ecosystem services can put pressure on business if their operations impact on the livelihoods of their stakeholders (e.g. local communities, farmers). Businesses risk losing their social licence to operate, and hence not being able to sustain their right of access to existing water resources or secure additional ones. For instance, in some water-scarce countries and regions (e.g. small islands), the tourism sector can be the major water user. Water may be used for golf courses, irrigated gardens, swimming pools, and guest rooms, among other purposes. This can put pressure on already scarce local water supplies, compete with other local sectors, threaten the subsistence needs of local populations, and result in stark inequity between the water use of tourists and neighbouring communities – thus creating reputational risks and potential increases in costs (e.g. transaction costs linked to stakeholder engagement, negotiations, public relations).

15.3 USING ECOSYSTEM SERVICES-BASED APPROACHES: EMERGING OPPORTUNITIES FOR BUSINESSES

Any risk can be turned into an opportunity by a business or entrepreneur (Institute of Directors Southern Africa 2012). Water scarcity, poor water quality, degraded water infrastructure, and stricter regulations are already generating demand and new markets for:

- water-efficient products;
- water quality and pollution monitoring/control devices and systems;
- increasingly effective wastewater treatment solutions, including for acid mine drainage and nanoparticles/chemicals found in pharmaceutical and personal care products;

Box 15.1 Case study 1. Acid mine water and the associated loss of water ecosystem services in South Africa: what will be the impacts on business?

South Africa is well endowed with vast mineral resources, and the wealth created through mining, particularly gold mining, has funded the development of the country. However, as the gold mining industry enters its twilight years we are now beginning to grasp the environmental damage this industry has caused and will continue to cause in the decades to come (McCarthy 2011). Acid mine drainage has been reported from a number of areas within South Africa, including the Witwatersrand Gold Fields, Mpumalanga and KwaZulu-Natal Coal Fields, and the O'Kiep Copper District.

Risks identified include:

- contamination of shallow groundwater resources required for agricultural use and human consumption and of surface streams, with devastating ecological impacts (e.g. impacts on the Blesbokspruit Ramsar Site – MacFarlane and Muller 2011);
- rising mine water levels have the potential to flow towards and pollute adjacent groundwater resources;
- geotechnical impacts, such as the flooding of underground infrastructure in areas where water rises close to urban areas;
- increased seismic activity which could have a moderate localised effect on property and infrastructure.

The Western, Central, and Eastern Basins are currently identified as priority areas requiring immediate action. This is due to the lack of adequate measures to manage and control the problems related to acid mine drainage, the urgency of implementing intervention measures before problems become more critical, and their proximity to densely populated areas (Ramontja et al. 2010). Other regions are also being closely monitored, especially the Mpumalanga Coal Fields where mining has severely impacted the freshwater sources in the upper reaches of the Vaal and Olifants River Systems.

Accordingly, this crisis is generating several developments with significant potential impacts on mining companies and other business water users (e.g. farmers, agribusiness, insurance companies, banks), including but limited to:

- commissioning of studies on the apportionment of financial liability due to acid mine drainage;
- new litigation processes against specific mines, especially state entities (i.e. owners easier to target than those of abandoned and derelict mines);
- refusals and/or delays in the expansion of mining operations or opening of new mines (e.g. unsuccessful water licence applications) in response to pressures from other water ecosystem services users;
- tighter regulatory frameworks for existing and new mining operations (e.g. acid mine drainage treatment capital investment on-site);
- discussions and potential policy reviews as regards to the pricing of potable water supplied by municipalities.

Innovative solutions, based on ecosystem services-based approaches, are required now to change behaviour and restore water ecosystems.

- water consulting services to find innovative solutions for water sourcing and permitting, as well as for water cost management and reduction.

However, thinking in terms of water ecosystem services requires firms to go beyond purely technological and/or artificial abiotic solutions. Over the past decade, many organisations worldwide have attempted to help companies better understand the importance of biodiversity and ecosystems by making them aware of their dependencies on healthy ecosystems and specific ecosystem services (Hanson *et al.* 2008; Houdet 2008). Awareness-raising tools have been developed to help businesses assess their dependencies and impacts on priority ecosystem services so as to help the firm develop new strategies, gain competitive advantage, or reduce/avoid costs. Business perceptions, attitudes, behaviours, and strategies regarding ecosystem services are hence progressively changing (Houdet *et al.* 2012). Firms can no longer exclusively consider biodiversity and ecosystem services as external constraints on their activities (impact mitigation approach). They are becoming increasingly aware that they are managing – or that they need to manage – the biodiversity and geodiversity which underpin the ecosystem services that are directly or indirectly influencing their activities, and hence their bottom line. This calls for embedding the management of dependencies on water ecosystem services into their organisational objectives, strategies, plans, budgets, and routines.

An increasing number of firms are managing the spatial and temporal dimensions of desired water ecosystem services (sources/origins, diffusion modes – trajectories, distance) because they need to secure the ensuing benefits (quantity, quality, delivery timing, low costs). At the same time, some firms are starting to develop strategies and mechanisms aiming to avoid any form of negative impacts (ecosystem dis-services) linked to changes in ecosystem functions and processes affecting their activities: e.g. damages to assets due to floods or droughts, disruptions to production processes via the decrease in the quantity and/or quality in water ecosystem services benefits effectively secured (e.g. see case study 2 on Vittel's strategy to secure clean natural mineral water in the long term, Box 15.2). In other words, thinking in terms of interdependencies with water ecosystem services allows firms to better assess their internal (e.g. critical successes factors) and external (e.g. opportunities, constraints/pressures, market positioning) strategic diagnosis. This helps them to identify their *ecological water infrastructure* – i.e. all their key interactions with water ecosystem services and the associated ecosystems structures and processes (core element 2 proposed in Chapter 2 of this book) – so as to precisely target key 'relationships' which need to be closely managed or developed towards maintaining or improving their competitive advantage (adapted from Houdet *et al.* 2012). Besides, using an ecosystem services-based approach to understanding business interactions with water ecosystem services leads to a transversal assessment and management of environmental and socio-economic issues affecting stakeholders throughout water ecosystems (coherently with core element 3 on trans-disciplinarity), notably the interconnections between land use, waste, emissions, and climate change issues.

For instance, food and beverage companies have specific water quality, volume, and delivery timing requirements so as to be financially viable. This may lead them to develop an ecosystem services-based landscape and multi-stakeholder approach to water sourcing, use, and quality management. For instance, Vittel (Perrot-Maître 2006; Déprés *et al.* 2008; case study 2, Box 15.2) and SAB Miller (Kissinger 2013; case study 3, Box 15.3) have taken steps:

- to improve the water quality in agricultural supply chains and/or in processing plants and other facilities; and
- to engage with local farmers, communities, scientists, and NGOs to address local and watershed-level water challenges and generate reputational benefits, in addition to operational savings and competitive advantages (core element 3).

The Ingula pumped storage scheme in the Little Drakensberg Escarpment developed by Eskom (main electricity producer in South Africa) to produce hydroelectricity is another good example of using an ecosystem services-based approach. To mitigate the environmental impacts of the scheme on a sensitive wetland site and maximise reputational benefits, Eskom conducted an assessment of the socio-economic values of ecosystem services, and discussed with stakeholders (non-governmental organisations, farming communities) potential actions for improving the management of threatened wetlands and associated grasslands.[4] Beyond strict environmental management, monitoring, and compliance measures throughout the scheme, a new conservation area has been created and is being effectively restored and managed (e.g. erosion control, invasive species eradication).

15.4 BEYOND INDIVIDUAL COMPANY APPROACHES: OVERCOMING CHALLENGES FOR MAINSTREAMING ECOSYSTEM SERVICES-BASED APPROACHES TO WATER MANAGEMENT

Pro-active actions towards using water ecosystem services-based approaches tend to be limited to a small number of companies. This is due to a combination of factors, including:

---

[4] www.eskom.co.za/live/content.php?Category_ID=361 as at 30 August, 2013.

Box 15.2 Case study 2: paying for water ecosystem services – the case of Vittel (Déprés *et al.* 2008; Perrot-Maître 2006)

Vittel natural mineral water originates from Grande Source, located in the town of Vittel in the foothills of the Vosges Mountain in France. Water comes from a 6000 ha aquifer 80 m below ground. Maintaining water quality is essential to the entire water bottling business. Selling 'natural mineral water' is the activity where the legislation is the most constraining and the reputational risk is especially high. This implies water must come from a well-protected specific underground source, the composition of the water must be stable, and the water must be bottled at the source.

In the early 1980s, the owners of the Vittel brand realised that the intensification of agriculture in the Vittel catchment posed a risk to the nitrate and pesticide level in Grande Source and consequently to the Vittel brand. The traditional hay-based cattle-ranching system had been replaced by a maize-based system while free-range cattle grazing was limited and stocking rates were increased. This coupled with heavy leaching of fertilisers from the maize fields in the winter and poor management of animal waste led to increased nitrate concentrations.

Because French legislation prohibits any treatment for 'natural mineral water', the family considered five alternatives to ensure water quality over the next 50 years: (1) doing nothing; (2) relocate to a new catchment where risks are lower; (3) purchase all lands in the spring catchment; (4) require farmers to change their practices through legal action; (5) provide incentives to farmers to voluntarily change their practices. While the first three options were prohibitively costly, the third alternative was not feasible as French legislation prevents the sale of agricultural land for non-agricultural purposes. For the fourth option, even if nitrate rates had reached a level above the allowed level for mineral water, legal action was found to be questionable as the existing legislation related to natural mineral water production was unclear or incomplete, notably with respect to the protection of water quality from other activities.

Accordingly, only one alternative was left: to convince the farmers to change their farming practices, and develop a system of incentives attractive enough for them to want to do so. Extensive hydro-geological modelling was conducted in the perimeter and showed that ensuring a nitrate rate of 4.5 mg l^{-1} in Grande Source required maintaining nitrate levels at the root zone (up to 1.5 m below the surface) at 10 mg l^{-1}. The area was modelled at sub-catchment, farm, and plot level to test the technical and economic feasibility of the proposed alternatives. A four-step methodology was then developed:

(1) Understand the farming systems and why farmers do what they do.
(2) Analyse the conditions under which farmers would consider changing farming behaviour.
(3) Identify, test, and validate in farmers' fields the management practices necessary to reduce the nitrate threat.
(4) Provide financial and technical support to farmers willing to enter the programme (i.e. payments for securing water ecosystem services).

In the end, the changes in farming practices were quite drastic because Vittel and its partners addressed the land, labour, and capital shortages farmers faced in implementing the required changes.

- large companies with corporate image issues (e.g. Eskom's Ingula scheme) and strong stakeholder pressures (e.g. Eskom's Ingula scheme, SABMiller);
- production assets which cannot be moved away because of huge capital investment and long life-span of assets to be financially profitable (e.g. Vittel and SABMiller);
- new projects under intensive public and NGO scrutiny (i.e. social licence to operate needs to be secured) and increasing/stricter environmental regulations (Eskom's Ingula scheme).

While such initiatives generate many environmental and social benefits and may be taken up by other stakeholders (e.g. suppliers, clients, neighbouring communities, and stakeholders including local authorities and government agencies), they tend to deal with ad hoc or geographically restricted problems and, therefore, fail to address the cumulative impacts of water

ecosystem services users throughout watersheds. The mainstreaming of ecosystem services-based approaches to corporate water management thus faces many challenges. Following are several key issues to take into account or address for the effective mainstreaming of the water ecosystem services approach throughout the private sector.

15.4.1 The challenge of water ecosystem services accounting and mapping

Precisely quantifying and mapping the different uses and impacts of water ecosystem services is challenging. This involves defining ecosystem boundaries, including spatial and temporal relationships across different scales between economic agents as regards to dependencies, and impacts on ecosystem services. Several ecosystems may exist within a larger one and their

---

Box 15.3 Case study 3: SABMiller's landscape and multi-stakeholder approach to water ecosystem services management

SABMiller is one of the world's largest brewers, with a strong interest in water security. The company faced operational, reputational, and regulatory risks linked to water quantity and quality concerns in many countries. This was argued to be due to a combination of factors including climate change, water scarcity, competition for water resources, unsustainable land use upstream, as well as social dimensions of water use and interactions with business.

Accordingly, SABMiller developed the Water Futures Partnership in 2009, with support from the World Wildlife Fund and Gesellschaft für Internationale Zusammenarbeit (Kissinger 2013). The initiative looked 'beyond the breweries' to the landscape and communities the company operates in so as to identify shared responsibilities to craft shared solutions. Landscape approaches were thus created to craft integrated management solutions across all key users in catchment areas of Peru, Tanzania, South Africa, Ukraine, Colombia, Honduras, India, and the USA. This involved:

- assessing water risks throughout SABMiller's value chain and identifying how to mitigate them;
- proving the business case for private sector engagement in promoting the sustainable management of water resources;
- sharing the lessons learnt throughout the business' global operations with other stakeholders to promote better water stewardship.

In other words, SABMiller's landscape approaches involved quantifying, analysing, and managing its water ecological infrastructure – i.e. its water ecosystem structures, processes, services – and collaborating with its key associated water stakeholders. They have resulted in significant reduction of operational risk, and have generated reputational benefits for the company, primarily due to the resulting significant benefits for other water users.

---

boundaries may expand and contract over time in response to various drivers of change, including anthropogenic influences. What's more, researchers and practitioners find it difficult to trace ecosystem services from their source(s), which may be discrete, ambient, or variable, to their ultimate user(s) (point, diffuse, or spotty). For instance, how far downstream can the benefits of wetland water purification by a specific wetland be captured by users?

Yet, as argued by Ruhl *et al.* (2007), identifying service provision timing, delivery channels, distance delivery, and delivery timing is a critical challenge to effective ecosystem services management. In the case of Vittel (case study 2, Box 15.2), the quantification and mapping of key water ecosystem services sources, users, and polluters was undertaken upstream of the water bottling plant: it was a prerequisite to developing an effective strategy and action plan to reverse negative trends in water quality due to conventional farming practices (Deprés *et al.* 2008; Perrot-Maître 2006). In line with core element 3, cooperation between businesses, scientists, water management agencies, and other stakeholders will be instrumental towards the development and sharing of improved knowledge on water ecosystem services sources/localisation, uses/users, and trends (quantity, quality).

15.4.2 Financial aspects are critical to driving organisational changes

The critical importance of tangible monetary flows (expenses, revenues, investments, liabilities, and contingent liabilities) in influencing business decision-making cannot be overemphasised

(core element 4). To drive the required organisational and corporate behaviour changes, one needs to directly influence critical success factors such as specific product pricing requirements (Houdet *et al.* 2012). Firms will always assess the return on investment and project alternatives, making sure (at best) that the maintenance of specific water ecosystem services and the associated ecological infrastructure would not affect the viability of their venture (e.g. minimising costs of legal compliance, voluntary actions targeting legitimate stakeholders, and impact mitigation measures).

In other words, for businesses to systematically and rigorously integrate water ecosystem services stewardship into their strategies and operations, their degradation/loss needs to imply immediate and tangible costs while changes in practices required for their conservation, restoration, or sustainable uses need to become financially viable (accounting for opportunity costs – i.e. forgone alternative revenue streams). In that context, the economic valuation of water ecosystem services (core element 4) may be useful to identify and rank priority ecosystem services for business and their stakeholders, but are not sufficient to influence corporate behaviours in favour of an ecosystem services-based approach (e.g. Billé *et al.* 2012). Regulated markets making use of clear rights and responsibilities for water users and polluters as well as of clear water ecosystem services protection and/or restoration targets will without any doubt be more effective. For instance, clear legal natural mineral water production and quality requirements were key drivers for Vittel's payments for water ecosystem services to upstream farmers (case study 2, Box 15.2).

15.4.3 Mainstreaming monetary incentives and disincentives

Monetary incentives (e.g. direct payments, premiums, state subsidies) and disincentives (e.g. environmental taxes, charges, and penalties) can provide tangible reasons for corporate behaviour changes towards mainstreaming water ecosystem services stewardship in the private sector (core element 4), provided they are significant enough when compared to alternative undesirable behaviours and the associated costs and income streams. For instance, by requiring that project developers offset their residual impacts on wetlands, wetland mitigation banking schemes worldwide have added a significant cost to wetland destruction (Madsen et al. 2011). In the case of the Payment for Ecosystem Services project in Naivasha, Njenga and Muigai (2013) explain that it is the yield increase due to changes in farming practices rather than the annual payment for environmental services that represents the main incentives for small-scale farmers to join in the programme. Besides, it appears that many potential Payment for Ecosystem Services buyers (large businesses) downstream failed to see the benefits of joining in, hence highlighting the institutional challenges of implementing large-scale Payment for Ecosystem Services schemes.

Yet, combining market-based and regulation-based strategies for mitigating water ecosystem services loss and remunerating water ecosystem services supply opens the door to new forms of arbitrage with respect to land use and development, as well as core business processes and practices. This approach would see water ecosystem services provision and/or stewardship becoming an integral part of the business plan of the firm, first as a strategic core variable among others for decision-making and management (beyond impact mitigation) and, perhaps more importantly, as a source of new assets and liabilities (ecosystem services trading rights and/or contractual agreements), new skills or competencies, as well as technological and organisational innovations. The development of such incentives and disincentives for water ecosystem services may hence lead to major changes in business routines, practices, intra- and inter-organisational norms and organisation of the workplace.

15.4.4 Collective action, watershed-based policies and regulations, industry-specific standards

However, for mainstreaming such initiatives and getting support from firms and stakeholders, numerous uncertainties would need to be resolved at different scales. An efficient sharing of ecosystem services advantages (Perrings et al. 2009; Pascual et al. 2010) would need clarity of the level of excludability and rivalry of such ecosystem services by beneficiaries and providers, to make sure there would be sufficient demand or willingness to pay for such services by the beneficiaries. It would be important

to delineate and enforce clear regimes of rights surrounding land use and ecosystem services and invest in social capital to foster collective action and cohesion between the providers and beneficiaries of ecosystem services. This calls for clear watershed-based policies and regulations, with adequate provisions for cross-boundary issues (e.g. rivers crossing different countries or provinces, and hence involving several water policies and management agencies).

In addition, sector-specific (voluntary or state-driven) standards should be developed for all water ecosystem services business users (e.g. farmers, energy producers, food and beverage producers). These would be broad in scope, from water efficiency measures (e.g. for irrigation) to water ecological infrastructure maintenance and restoration best practices (e.g. invasive species, soil, fire, wetland, runoff, and wastewater management).

15.4.5 Improving corporate water performance disclosure and accountability

The disclosure of a company's water-related information – such as water consumption in production processes or the management of wastewater discharges and pollution events – in annual sustainability reports can become a critical tool for improving corporate water management and accountability, notably for adopting and mainstreaming an ecosystem services-based approach. It provides opportunity for stakeholder scrutiny and engagement, water management performance comparison with other companies, and the context to identify strategies, targets, and the associated action plans so as to demonstrate progress to relevant stakeholders. From that perspective, the Carbon Disclosure Project's water programme is becoming, in our view, the pre-eminent platform and tool for independent corporate water management disclosure assessment.[5] The process of responding to the Carbon Disclosure Project water disclosure questionnaire aims 'to help businesses and institutional investors to better understand the risks and opportunities associated with water scarcity and other water-related issues whilst promoting water stewardship and delivering insight that enables companies to take intelligent action to manage this critical resource'.[6]

Yet, only a minority of companies participate in the Carbon Disclosure Project's water disclosure programme, while comprehensive water footprint disclosure has yet to emerge: i.e. the Carbon Disclosure Project water disclosure assessment criteria do not contain all aspects covered by the Corporate Water Footprint Standard;[7] notwithstanding the current lack of standards

---

[5] The Global Reporting Initiative reporting framework and guidelines only include a limited number of water-relevant key performance indicators.
[6] www.cdproject.net/water as at 30 August 2013.
[7] www.waterfootprint.org/?page=files/CorporateWaterFootprints as at 7 January 2014.

Box 15.4 Key messages

There is an urgent need to:

- promote and fund the precise mapping of water ecological infrastructure (including ecosystem structures, processes, and services) and their users;
- share the information with all businesses and their stakeholders, providing them with the skills and tools to make informed decisions;
- promote sector-specific water ecosystem services stewardship guidelines and standards;
- develop watershed-based policies and frameworks for engaging with business, especially in emerging and developing countries;
- study and promote the development and mainstreaming of cost-effective incentive and disincentive mechanisms to promote pro-water ecosystem services organisational changes in all types of industries, notably by providing support for the empowerment of all stakeholders, especially those belonging to disadvantaged communities;
- develop science-based Water Footprint assessment guidelines and measurement tools for business which are adapted to national contexts and international trade realities, and take into account water ecological infrastructure and services; and legislate their mandatory disclosure in formats (e.g. annual sustainability reports) easily accessible to all stakeholders.

and tools that account for the links and trends between business activities, their water footprints, and the associated/impacted water ecosystem services. What is needed by stakeholders, from institutional investors to local communities, is a precise picture of firms' green, blue, and grey water footprints (Hoekstra 2008; Hoekstra *et al.* 2011) and the associated water ecological infrastructures and services which they depend on and/or impact at different decision-making levels:

- at the overall group/company level, to showcase global trends and performance;
- at the product and service level, to be able to make informed assessments and comparisons;
- at the level of new projects, typically those subject to socio-environmental impact assessment and the associated authorisation processes (e.g. water use licences); and
- at the local sub-catchment level where site-specific information is required.

Indeed, improving current water disclosure standard/tools and practices would be more than instrumental to significantly increase transparency and stakeholder empowerment (link with core element 3) towards the equitable governance and benefit-sharing of water ecosystem services.

References

Billé, R., Laurans, Y., Mermet, L., Pirard, R., & Rankovic, A., (2012). Valuation without action? On the use of economic valuations of ecosystem services. Policy Briefs N°07/2012. IDDRI.

Bizikova, L., Dimple, R., Venema, H., *et al.* (2014). *The Water-Energy-Food Nexus and Agricultural Investment: A Sustainable Development Guidebook.* IISD, Canada.

CSIR (2010). A CSIR perspective on water in South Africa. CSIR Report No. CSIR/NRE/PW/IR/2011/0012/A.

Déprés, C., Grolleau, G., & Mzoughi, N. (2008). Contracting for environmental property rights: the case of Vittel. *Economica*, **75**, 412–434.

Grossman, D., Erikson, J. & Patel, N. (2013). *GEO-5 for Business. Impacts of a Changing Environment on the Corporate Sector.* UNEP, SustainAbility, Green Light Group, Nairobi.

Gulati, M. (2014). *Understanding the Food Energy Water Nexus: Through the Energy and Water Lens.* WWF-SA, South Africa.

Hanson, C., Ranganathan, J., Iceland, C., & Finisdore, J. (2008). *The Corporate Ecosystem Services Review. Guidelines for Identifying Business risks and Opportunities Arising from Ecosystem Change.* WRI, WBCSD, and Meridian Institute, Geneva.

Hoekstra, A. Y. (2008). *Water Neutral: Reducing and Offsetting the Impacts of Water Footprints.* UNECSO, Delft.

Hoekstra, A. Y., Chapagain, A. K., Aldaya, M. M., & Mekonnen, M. M. (2011). *The Water Footprint Assessment Manual. Setting the Global Standard.* Water Footprint Network and Earthscan, London.

Houdet, J. (2008). *Integrating Biodiversity into Business Strategies: The Biodiversity Accountability Framework.* FRB – Orée, Paris.

Houdet, J., Trommetter, M., & Weber, J. (2012). Understanding changes in business strategies regarding biodiversity and ecosystem services. *Ecological Economics*, **73**, 37–46.

Institute of Directors Southern Africa (2012). Water as a risk to business. Sustainable Development Forum, Position Paper 6.

Kissinger, G. (2013). *Reducing Risk: Landscape Approaches to Sustainable Sourcing. SAB Miller Case Study.* EcoAgriculture Partners, Washington, DC.

Macfarlane, D. M. & Muller, P. J. (2011) Blesbokspruit Ramsar Management Plan: draft. Report prepared for the Department of Environmental Affairs.

Madsen, B., Carroll, N., Kandy, D., & Bennett, G. (2011). State of biodiversity markets report: offset and compensation programs worldwide. Available at: www.thegef.org/gef/sites/thegef.org/files/publication/Bio-Markets-2011.pdf (accessed December 2013).

McCarthy, T. (2011). The impact of acid mine drainage in South Africa. *South African Journal of Science*, **107**(5/6), art 712.

Nel, J., le Maitre, D. C., Forsyth, G., Theron, A., & Archibald, S. (2011). *Understanding the Implication of Global Change for the Insurance Industry: The Eden Case Study.* CSIR, Stellenbosch.

Njenga, N. & Muigai, P. (2013). Case studies on remuneration of positive externalities (RPE)/Payments for environmental services (PES): engaging local business in PES – Lessons from Lake Naivasha, Kenya. Prepared for the Multi-stakeholder Dialogue 12–13 September 2013, FAO. Available at: www.fao.org/fileadmin/user_upload/pes-project/docs/FAO_RPE-PES_WWF-Kenya.pdf (last accessed 7 January 2014).

Otto-Mentz, V., de Meyer, H., Lee, E., *et al.* (2011). Insurance in a changing landscape. Report by the research partnership between the Santam Group, the WWF, the University of Cape Town and the Council for Scientific and Industrial Research.

Pascual, U., Muradian, R., Rodriguez, L. C., *et al.* (2010). Exploring the links between equity and efficiency in payments for environmental services: a conceptual approach. *Ecological Economics* **69** (6), 1237–1244.

Perrings, C., Brock, W. A., Chopra, K., *et al.* (2009). The economics of biodiversity and ecosystem services. In: S. Naeem, D. Bunker, A. Hector, *et al.* (eds), *Biodiversity, Ecosystem Functioning and Ecosystem Services.* Oxford University Press, Oxford.

Perrot-Maître, D. (2006). *The Vittel Payments for Ecosystem Services: A 'Perfect' PES Case?* IIED, London.

Ramontja, T., Coetzee, H., Hobbs, P. J., *et al.* (2010). Mine water management in the Witwatersand Gold Fields with special emphasis on acid mine drainage. Report to the Inter-Ministerial Committee on Acid Mine Drainage.

Ruhl, J. B., Kraft, S. E., & Lant, C. L. (2007). *The Law and Policy of Ecosystem Services.* Island Press, Washington, DC.

Stephenson, J., Elstein, S., Huntley, C., *et al.* (2007). Climate change: financial risks to federal and private insurers in coming decades are potentially significant. Report to the Committee on Homeland Security and Governmental Affairs, US Senate. GAO-07-285.

United Nations Environment Programme (2012). *Global Environment Outlook 5 (GEO-5): Environment for the Future we Want.* UNEP, Nairobi.

von Bormann, T. (2014). *Understanding the Food Energy Water Nexus: Through the Water and Food Lens.* WWF-SA, South Africa.

Waage, S. & Kester, C. (2014). Making the invisible visible: analytical tools for assessing business impacts & dependencies upon ecosystem services. BSR's Environmental Services Working Group.

World Water Assessment Programme (2014). *The United Nations World Water Development Report 2014: Water and Energy.* UNESCO, Paris.

Zhang, C.-B., Wang, J., Liu, W.-L., *et al.* (2010). Effects of plant diversity on microbial biomass and community metabolic profiles in a full-scale constructed wetland. *Ecological Engineering* **36**, 62–68.

16 Key factors for successful application of ecosystem services-based approaches to water resources management

The role of stakeholder participation

Jos Brils, Al Appleton, Nicolaas van Everdingen, and Dylan Bright

16.1 INTRODUCTION

Ecosystem services-based approaches for water resources management may help to select and engage with stakeholders whose interests are affected by any changes in their water environment (Blackstock *et al.*, this book). In addition, Brauman *et al.* (2014) state that the notion of ecosystem services provides a framework for organizing stakeholders, identifying those whose actions affect the provision of river basin ecosystem services and those whose wellbeing will be impacted by changes in the provision of ecosystem services, and by delineating the mechanisms by which land use changes affect stakeholders. Furthermore, ecosystem services may also be a common language to facilitate communication with, and thus participation of stakeholders. It seems to offer an easy language to communicate their positions and interests and to identify common interests (Brils *et al.* 2014). Some results from regional case studies are already available that confirm this expectation (e.g. Granek *et al.* 2009; van der Meulen & Brils 2011; van der Meulen *et al.* 2013). But what are key factors for the successful application of ecosystem services-based approaches to water resources management? This chapter aims to identify some of these key factors with a focus on stakeholder participation.

We start by analysing the application of the ecosystem services-based approaches in the Catskill watershed management in the USA as this is probably one of the best practical examples known of successful application. The key success factors are extracted from this case study and their applicability to Northern Europe – the 'River Tamar' case in the UK and the 'Farmers around Amsterdam as Water Managers' case in the Netherlands – is discussed. This chapter serves as an illustration of the stakeholder participation aspects of core element 3 (transdisciplinarity) of this volume.

16.2 CATSKILL WATERSHED MANAGEMENT (USA)

16.2.1 Issues at stake

The Catskill watershed protection programme was a response by the city of New York to one of its most prized infrastructure assets, the pure, unfiltered water of the Catskill Mountains (Figure 16.1) that flows to New York City through a long series of aqueducts, representing the engineering vision of generations of water managers. This system gave New York City pure drinking water of such high quality that it has often been referred to as 'the champagne of drinking waters'.

New York City had this enormous asset because it had chosen to gather its drinking water from distant rural sources that it assumed would be perpetually free of the kinds of pollution that have forced most other urban areas to filter their source water. However, in the 1980s that reality began to change, as altered economic conditions in North American farming forced the Catskill farming community – largely an area of dairy farms – to adopt the highly polluting practices of industrial agriculture in an attempt to remain economically viable. As this trend became apparent, numerous water regulators began to foresee a time when New York City would be forced to build filtration works for its Catskill source waters, at a cost of many billions of dollars (Appleton 2002).

16.2.2 Application of an ecosystem services-based approach

The New York City water system leaders recognized that the high cost of filtration works meant that, if they could develop a pollution prevention strategy based on environmental stewardship, the value added provided by that strategy would be more

Figure 16.1 New York City's water supply system. A black and white version of this figure will appear in some formats. For the colour version, please refer to the plate section. Source and for more information: http://www.nyc.gov/html/dep/html/drinking_water/wsmaps_wide.shtml (last accessed 9 December 2014).

than sufficient to pay for the costs of watershed farmers participating in such a programme and become water quality stewards, i.e. enablers of the ecosystem service 'water purification'.

In short, pollution prevention was recognized as a cost-effective strategy that would preserve the economic value of the ecosystem services the Catskill watershed was providing, and generate a surplus of environmental wealth that would not only pay the out-of-pocket costs of the programme, but would enable local landowners to profit from their environmental stewardship. It was estimated – and later proved in practice – that billions of US dollars could be saved in this way (Chichilnisky and Heal 1998). Since the needs of environmental stewardship were distinctly hands-on management of site-specific situations – e.g. stream corridor protection, soil erosion, isolation of water bodies from livestock manure, and the byproducts of animal husbandry – such a strategy dovetailed perfectly with both the needs of the water system and the economic and social culture of the affected landowners, making for a perfect potential partnership.

In terms of distinctions normally used, the enablers were New York City's Department of Environmental Protection leaders, who were determined not to allow traditional water quality thinking, institutional inertia, or political opposition to steer them into what they saw as a course that was both economically sterile and counter to the ideal environmental dynamics of clean water. The beneficiaries were the farmers, landowners of the watershed whose basic goal was to preserve a cherished farming community and a local way of life from what they feared would be insensitive and top-down bureaucratic diktats that would not solve the problem, but would only add costs and friction that could further undermine their already fragile agricultural economics.

The Catskill watershed programme was literally ground-breaking in many senses, not least of which is that it took what had always been given as an ideal value – watershed protection – and did something no one in the American water community, either academic or governmental, believed to be possible: it made watershed protection for one of the largest water systems in the world a practical, functioning reality. Ultimately, it did so by insisting on what is common sense and obvious – that pollution prevention is better than pollution clean-up, and that the best guarantee of good water is a good environment. This was just so obvious that it had to be made to work.

In making landowner-based environmental stewardship work, New York City blazed a trail for future ecosystem services programmes, and provided what remains one of the most successful examples of ecosystem services-based approaches applied to watershed management.

The Catskill programme – even though it was not originally conceived as an ecosystem services programme – stands as a model for how to apply ecosystem services-based approaches (as defined in Chapter 2 of this book) for several reasons. Most importantly, it used the wealth of the environment to protect the environment, by sharing that wealth with those who could protect it. Second, using ecosystem services ideas of sharing environmental wealth gave reality to something the American water quality community had long dismissed as an impossibly idealistic dream: effective watershed protection. Today, this cynical dismissal has been replaced in the USA by an embracing of watershed protection that makes it, in American terms, an 'apple pie and motherhood idea', and not some 'tree hugger's dream'. The Catskill watershed programme may be a model of how the ecosystem services-based approaches, that were intuitively drawn on to make the Catskill programme work, could become a future model of how to build a sustainable economy.

16.2.3   Possible key factors for successful application of the ecosystem services concept

The success in this case may be due to the recognition of, and use by New York City's Department of Environmental Protection leaders of, a series of important factors. The most important was a successful community partnership with the Catskill farmers. Rather than making the typical bureaucratic mistake of trying to impose a preconceived model of partnership or community activity on the farmers, New York City wisely allowed the farmers to work through how they would participate, and to devise a programme that was not only consistent with, but built on the strengths of their own farm culture. The only important qualification New York City made was to insist that the farmer partnership not only met farmer needs, but would deliver New York City the environmental results it needed to protect its water resource. Once it was clear that New York City did not intend to attempt to impose a predetermined model on farmers, the farm community readily accepted meeting New York City's goals as an appropriate *quid pro quo* for being given the lead role in a partnership of environmental stewardship. This clarity of purpose enabled both sides to avoid distraction by side issues or political point scoring, and to stay focused on designing a programme based on only two objectives: providing pure water and helping maintain the viability of Catskill farming. It was a classic entrepreneurial exercise in seeing an opportunity and pursuing it, drawing on any sources of innovation or community expertise that could realize the opportunity for mutual assistance in preserving and utilizing the high-value environmental asset – pure water – that the Catskills provided. This exercise produced a broad spectrum of individuals who, each in their own sphere, provided facilitative leadership, as best exemplified by the solution to the potential deal-breaker of the degree of voluntary engagement. The farmers insisted that the programme be voluntary; from New York City's point of view a voluntary programme would not guarantee the kind of critical mass of engagement that was needed to guarantee the preservation of water purity. The compromise eventually struck was that the programme would be voluntary for any individual farmer, but that the farm community itself would recruit the voluntary participation of a minimum of 85% of all farmers within five years. The farm community accepted and delivered this obligation in a way that deeply rooted New York City farmer environmental partnership in the Catskill community.

The Catskill programme also highlights the critical need to understand and work with local social capital. The leadership of the farm community played a critical role in bringing about the watershed programme, even though the Department of Environmental Protection can rightfully claim an important role in empowering that leadership and welcoming it, instead of following the normal governmental impulse to try and channel and control it. But the Catskill programme would not have existed were it not for the farm leaders who saw working with New York City as an opportunity rather than a threat, and who realized that if they could persuade New York City to work with them on their terms, they had to include meeting New York City's substantive need for water quality. In this way, they were wise enough to realize that they could do something dramatic to help preserve and improve Catskill farming and the farming lifestyle they loved.

In summary, four key factors for successful application of ecosystem services-based approaches in participatory water resources management can be distilled from the Catskill case experience:

(1)   a clear and urgent need by an important stakeholder for action;
(2)   an entrepreneurial approach;
(3)   defining and sticking to clear targets;
(4)   facilitative leadership.

16.3   RIVER TAMAR (UK)

16.3.1   Issues at stake

The Westcountry Rivers Trust was founded in 1994 because the local private fisheries and river associations wanted to do something about declining salmon numbers in Westcountry rivers. Westcountry Rivers Trust quickly realized that salmon declines were a symptom of a wider problem, and that modern intensive agriculture was, to some degree, responsible, leading to the conclusion that a solution would require working across the entire landscape, with all land managers and those benefiting from, and impacted by, land use. Meanwhile, government had failed to deliver either an integrated or a spatially planned view of the environment, contributing to the decline of many

ecosystem services (Brown et al. 2011). Additionally, declines are currently often dealt with in isolation, using end-of-pipe solutions on small, fragmented pockets of land, often without the involvement of local communities. This is commonly referred to as 'fortress conservation' (Brockington 2002) and represents an old-fashioned view that perpetuates the modern problem of an increasing disconnect between society and the environment that nurtures it.

Westcountry Rivers Trust recognized the propensity of our centralized environmental protection organizations to enact their duties in sector-specific groups; wildlife and biodiversity treated separately from water resource protection, for example. If conservation was delivered pro-actively in the wider landscape at all, it was delivered by sectioning-off protected areas using legislation and then managing those areas for the provision of one ecosystem service only. The rest of the land, representing the vast majority of the landscape, was managed in order to achieve profit, using the only market mechanisms available at the time: the provision of food and fuel. Westcountry Rivers Trust took the view that this approach, while necessary in its time, was no longer the way forward.

16.3.2   Application of an ecosystem services-based approach

Westcountry Rivers Trust developed methods that aimed to deliver sustainable management across the wider landscape. For this, economic analyses were used to incentivize landowners in the delivery of a wider set of ecosystem services; this is referred to as Community Conservation (Gezon 1997). The mode of action was to educate land managers about the direct economic self-interest of undertaking certain farming best practices. For example, if farmers test their soils for nutrients they can then target nutrient delivery, which saves money on input costs and decreases loss of nutrients to the environment.

The first of these community conservation projects took place in the River Tamar catchment (Figure 16.2) and was funded by European Structural Funds. Interestingly, the requirement to report economic outcomes from EU Structural Funds (jobs created, economic uplift in a sector), in part led to Westcountry Rivers Trust's recognition that many of the targeted farming best practices were actually cost-beneficial to the farmers directly. Additionally, an appraisal of the wider ecosystem service value of the project identified a 100:1 benefit-to-cost-ratio (Everard 2009).

Community conservation worked well in the Tamar case, but it was realized that global commodity prices could quickly overwhelm the fragile economics of this approach and – because uptake was voluntary – location-specific changes in environmental conditions could not be guaranteed. However, while it lasted this approach was very successful (Everard 2009). The

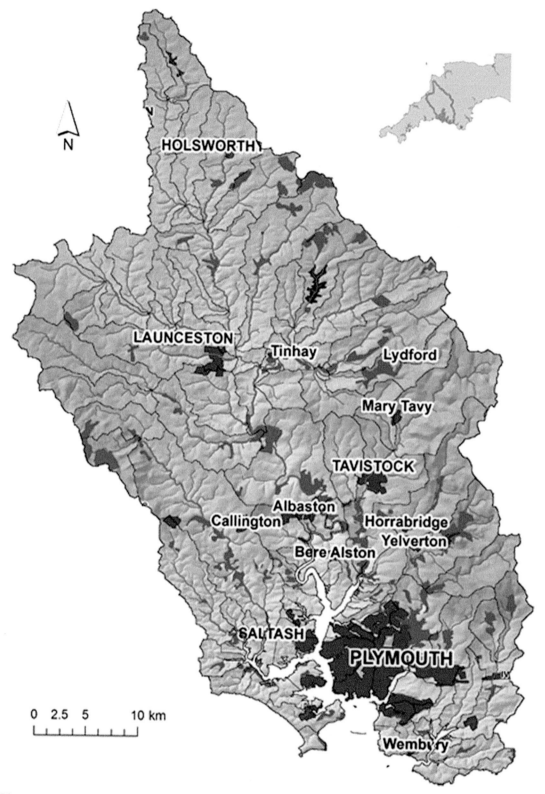

Figure 16.2 River Tamar catchment. A black and white version of this figure will appear in some formats. For the colour version, please refer to the plate section (source and more information: http://river-gateway.org.uk/catchments/tamar.html).

economic-led guidance that was developed as a result has now been adopted by the UK government through their Catchment Sensitive Farming Initiative.

The recommended farming best practice measures had significant economic benefit for the Tamar land managers, and were taken up instantly and maintained. However, the more economically marginal actions often fell by the wayside. Good land management led to economic benefits for third parties, such as the water company and people living in flood areas downstream. Additionally, improvements were seen in the ecosystem services that are very difficult to recognize economically, such as biodiversity and pollination, and cultural and landscape character value.

This recognition only served to strengthen earlier observations, i.e. that it is people rather than nature who need management, and the tools for managing people are education, economics, and regulation. It was felt that regulation was failing but, as a charity, it was not in the Westcountry Rivers Trust's ambit to solve this. However, education was well within the charitable purpose of the Trust, which had a growing understanding of the power of economics as a conservation tool when used both directly and indirectly to define costs and benefits to different groups.

Westcountry Rivers Trust became involved in 2006 in an integrated national research project called the Rural Economy and Land Use Programme.[1] This project investigated how to extend to the UK the scientific and social accomplishments of innovative catchment management programmes in the USA, Australia, and other European countries. A catchment management 'template' was derived, which compiles and assimilates scientific understanding and governance procedures, as tested in actual decision-making and management practice in case study catchments.

Thereafter, two catchments served as case studies against which the lessons from international experience were tested: the River Tamar and the River Thurne. These case studies involved partnership with, among others, the New York City Catskill watershed. At the same time, there was also cooperation with the regional water company, South West Water, to solve eutrophication problems in one of their reservoirs. This was done by working with the landowners who managed the catchment upstream. South West Water used to be a public authority but was privatized in the 1990s. Within the privatization process it was recognized that water companies are both an essential utility provider and a *de facto* monopoly. Accordingly, the Water Services Regulator in England defined the objective, i.e. to ensure that water consumers had a secure water supply at a reasonable cost per unit.

Westcountry Rivers Trust – with support from Rural Economy and Land Use Programme partners and South West Water – extended its understanding of the wider economic implications of improved land management to a landscape scale. With South West Water in particular, Westcountry Rivers Trust was able to develop a detailed understanding of the cost of delivering improved raw water quality pre-abstraction, through working with landowners, compared with the cost of filtering and treating the water post-abstraction. The water company was able to demonstrate a positive cost–benefit ratio, which then enabled them to secure funds through the household water bill, with the approval of the Water Services Regulator.

Westcountry Rivers Trust also learnt from their partners in the Rural Economy and Land Use Programme projects that success depends on an iterative process of project governance, and that the groups within that process, in successful examples, possessed certain attributes (Smith *et al.* 2011). For example, it seemed to be consistent across successful groups that regulation information and advice was delivered by separate, independent interests. Integrated spatial planning was essential to ensure that ecosystem services were being derived from the most suitable areas, as were local ownership of plans and an iterative planning cycle. Experience has further honed Westcountry Rivers Trust's knowledge: there is great value in understanding how each of the local stakeholders can interact with a plan, either as potential providers or beneficiaries of services, or as brokers linking the two groups.[2]

Thus was born the UK's first water company-funded Payment for Ecosystem Services project, called 'Upstream Thinking'. The water company funded a third-sector, ethical, not-for-profit broker (the Westcountry Rivers Trust) to deliver farming best practice advice and to deliver grants incentivizing best practices. The grants provide funding for essential infrastructure (covered yards, clean and dirty water separation, slurry pits) and landowners are tied in to good practices for the long term, using a deed of covenant linked to a land management contract. The project, which is ahead of its time in the UK, is still in progress. As such, the funding and governance structures are only just aligning with the delivery model, and as work continues, delivery of the project itself is still being refined.

16.4 FARMERS AROUND AMSTERDAM AS WATER MANAGERS (THE NETHERLANDS)

16.4.1 Issues at stake

Regional water resources management in the Netherlands is mostly the concern of water boards. Water boards are in charge of selecting and executing the measures needed to achieve

---

[1] For further details see: www.relu.ac.uk.

[2] Much of this derived understanding has been recorded in project documentation that is available via www.theriverstrust.org/seminars/archive/water.

Water Framework Directive objectives. The countryside around the city of Amsterdam can be characterized mostly as polders, i.e. agricultural areas that lie below or just at sea level. Water quantity and quality in these polders is managed by the water board Amstel, Gooi en Vecht. Its challenge in achieving Water Framework Directive objectives relates mainly to poor water quality at agricultural sites. At these sites the groundwater quality standard for nitrates is exceeded. Solving this problem needs more than reducing the amount of manure used to fertilise farmland. Besides groundwater, the surface water quality (chemically and ecologically) in this area is rather poor. To solve all these problems Amstel, Gooi en Vecht realized that they had to involve and cooperate with individual, local farmers. Hence, to reach the top-down Water Framework Directive goals, Amstel, Gooi en Vecht realized that they had to work bottom-up. On their part, farmers feel the continuous pressure of the 'big city' (Amsterdam), and know they have to earn 'their licence to produce' every day. Thus the request of the water board for farmers' cooperation was answered by a wide array of very useful ideas. Since 2009, several farmers around the city of Amsterdam have helped Amstel, Gooi en Vecht to reach Water Framework Directive goals in the project 'Farmers as Water Managers' (Pelsma & van Everdingen 2012).

16.4.2   Application of an ecosystem services-based approach

Most of the local farms have dairy cows. For draining of their wet, lowland polders, farmers use a large number of ditches, reducing the effective farmland area considerably. This makes it hard to run a profitable farm without compensation for any decline in production area. Thus, to date, 24 farmers have received financial compensation from the water board for taking voluntary measures such as: digging out their ditches to create nature-friendly (i.e. less steep) banks (Figure 16.3), installing fish-friendly diver tubes to stimulate fish migration, and digging deeper ditches than usual to give fish a chance to hibernate. This money comes from citizens who, via 'water board tax', pay the water boards for regional water quantity and quality management.

The citizens of Amsterdam make intensive use of their surrounding polders for relaxation. Although it often seems that citizens take for granted that farmers will deliver them the green landscape with cows and wetland birds for free, the ecosystem service of 'landscape enjoyment' is very much appreciated. Thus, the water board feels strengthened to act on behalf of these citizens as beneficiaries of the ecosystem services enabled by farmers on their farmland. Water board tax is also used to pay for introducing more natural elements in and around the water

Figure 16.3 Nature friendly banks created voluntarily by farmers at their farmland. Photograph by N. van Everdingen). A black and white version of this figure will appear in some formats. For the colour version, please refer to the plate section.

system, in a way that does not immediately result in a better water quality, but which provides more consumer-oriented or consumer-friendly water-related landscapes.

Agricultural nature associations exist in the Netherlands, which are associations among farmers having agricultural nature-management as their key objective. This involves taking measures to conserve or improve the quality of nature and landscapes. Such measures may include: creation of amphibian pools, leaving the shores of ditches unmowed, not mowing grassland during the nesting period, and not using, or at least not excessively using, fertilizers. Thus these associations can be seen as associations of environmental stewards or associations of countryside stewards (Oerlemans *et al.* 2007). Agricultural nature-management is also a financial instrument that allows the government to compensate farmers for taking nature-enabling measures on their farmland (RLI 2013). Many of these associations also involve citizen participation. The associations have a long tradition of protecting grassland birds, thus enabling the ecosystem service of 'enjoyment of biodiversity'. Cooperating with those associations provides an opportunity to cost-effectively introduce a wider focus on other ecosystem services.

In the 'farmers as water managers' case, 48 individual 'kitchen table talks' were held among the water board representatives and farmers. This resulted in 24 individual contracts, and through these the farmers provided 20 km of nature-friendly banks. This was much more than the water board had anticipated, i.e. more than their original target of >500 m^2 of ditches per participating partner to be deepened, including the creation of nature-friendly banks. Besides that, the farmers were very enthusiastic about this new form of cooperation because nature-friendly banks are more stable, and also enrich their farmland (Pelsma & van Everdingen 2012). The banks act as a buffer to nutrients – thus improving local water quality – and function as a kind of 'high pressure cooker' for local biodiversity. They bring back the traditional agricultural landscape and, with that, International Union for the Conservation of Nature red-listed bird species, such as the skylark and godwit.

An added benefit of involving farmers in nature-management is that use can be made of the ample local knowledge of farmers in management of their land. This includes much more than knowledge on how to increase agricultural productivity. Farmers in this case proved to have many valuable ideas for benefiting nature, and thus enhancing the ecosystem services provided by their farmland. Of course, financial compensation was a clear incentive to make this all work. But in the end, money was not the key trigger; rather, the insight that cooperation can make the work much easier – for the water board to achieve its Water Framework Directive goals as well as for the farmer to run a profitable farm – and provide a lot of fun for both parties.

16.5 APPLICABILITY OF THE CATSKILL KEY FACTORS IN TWO EUROPEAN CASES

16.5.1 A clear and urgent need by an important stakeholder for action?

As in the Catskill case there were also clear, but varying, water resources management issues to solve in each of the two European cases: restoring the balance in the use of ecosystem services in the British case, and complying with Water Framework Directive objectives in the Dutch case. There were also important stakeholders who felt committed to act: Westcountry Rivers Trust and the water board Amstel, Gooi en Vecht. In contrast to the American case, neither of these two stakeholders faced the huge costs (billions of dollars) the Department of Environmental Protection was facing if they did not act. Thus, the urgent need for action was considerably lower in the European cases. The prime reason to act in the UK and the Netherlands cases was probably more the personal commitment of Westcountry Rivers Trust and Amstel, Gooi en Vecht to actually, and jointly with other stakeholders, solve the water resource management issues at stake, although latterly Westcountry Rivers Trust was able to engage with the water company in a significant programme of investment.

16.5.2 An entrepreneurial approach?

Similarly to the American case, the European cases brought the local communities together. They left their comfort zones and demonstrated willingness to 'learn together to manage together' (Ridder *et al.* 2005). It paid off to invest time in getting to understand how local farmers run their businesses, as well as educating them about how a better environment provides many more benefits (ecosystem services) to society than food production alone. However, similarly to the American case, the beneficiaries had to provide economic incentives to at least achieve a zero balance for the European farmers, i.e. to compensate for income foregone. Thus the farmers were enabled to implement more environmentally friendly management practices. And it proved possible for farmers and water managers to jointly explore the multiple benefits from ecosystem services if farmers behaved more like environmental stewards. The limited use of top-down control mechanisms (regulation) to make their programmes work and the voluntary basis for involvement were features of all three cases.

16.5.3 Defining and sticking to clear targets?

The US case focused on only two simple and clear objectives: New York City desired pure water provision, and the upstream farmers wanted help to maintain the viability of Catskill farming. Westcountry Rivers Trust and the water board Amstel, Gooi en Vecht also defined clear targets: Westcountry Rivers Trust wanted to reverse the decline in ecosystem services provision by the Tamar watershed, and Amstel, Gooi en Vecht wanted to achieve Water Framework Directive objectives. Each of the three case studies took – with success – what can now be understood as an ecosystem services-based approach to achieve these goals. And in each case they realized that in order to reach their goals they had to cooperate with farmers. All managed to 'gain' their cooperation by investing time in 'mutual education', and thus also learnt about the concerns and key issues of local farmers. Certainly, a key issue for each of the farming communities involved in the three cases was to (continue to) run a profitable farm. The provision of economic incentives to compensate for any loss in income resulting from more environmentally friendly farming was without doubt a key element of success in each case.

16.5.4   Facilitative leadership?

A facilitative leader guides and assists stakeholders in designing the participatory water resources management process, and thus helps to improve decision-making, as opposed to leadership that is primarily focused on influencing the content of the decision (van Maasakkers *et al.* 2014). The UK and Netherlands cases were 'gently' facilitated by Westcountry Rivers Trust and Amstel, Gooi en Vecht, respectively. The same gentle facilitation is evident in the US case (provided by the Department of Environmental Protection). It was the committed individuals working for these organizations who 'made it work', i.e. who demonstrated clear facilitative leadership. More generally it can be said that local authorities (e.g. water managers) play a crucial role: they can facilitate the application of ecosystem services in support of participatory water resources management enormously if they act less like a 'controller/regulator' and more as out-of-the-box-thinkers, persuaders and motivators, entrepreneurs who exploit unrecognized opportunities, and as boundary-spanners, communicators, or mediators.

16.5.5   Other factors?

Additionally, there is scope to use policy and regulation to develop markets for the delivery of more broad-based actions. Setting up and recognizing ecosystem service-based planning structures and ensuring incentive payments will create a market for integrated delivery. Furthermore, new regulations, for example to offset the impact of development, can be linked to this integrated spatial planning, creating a fully integrated, but locally hypothecated delivery fund and an integrated participatory mechanism. Local hypothecated payments, rather than general taxation, create a better link between the beneficiary and the environment, which contributes to the restoration of older social norms, when people were more directly linked with, and aware of, the health of the environment.

Another factor evident in all three cases is the use of proper communication to enable stakeholder participation. 'To make it

work' the facilitative leaders had to talk the 'farmers' language', not European Commission policy language such as 'I have to achieve the Water Framework Directive goals', or scientific language such as 'I want to have a better balance in ecosystem services provision'. Farmers' language is, for example, 'I have a problem with water quality; with your help/insights I can overcome that problem. Help us to improve infrastructure for fish, and structure for flora and fauna, and we will help you (continue) to run a profitable farm.'

Although each of the three cases operated in line with an ecosystem services-based approach, the term 'ecosystem service' was hardly or not at all used in communication with farmers. The experiences in these cases thus show that for successful application of this approach, it is not necessary to use these terms in the discussion or to explain the concept to farmers. Ecosystem services intrinsically offer an easy language to communicate stakeholder positions and interests, and to discover common interests in water resources management. It provides a common language in that anyone can understand the question 'what is nature providing you?' or 'what benefits do you get from nature?' By receiving answers to these questions, programme leaders have already started involving respondents in an ecosystem services-based approach. This was clearly demonstrated in each of the three case studies addressed in this chapter, and has also been demonstrated in other case studies (for example, van der Meulen & Brils 2011).

16.6   CONCLUSIONS

Each of the key factors extracted from the US Catskill experiences applied remarkably well to the UK and the Dutch case. This is no surprise if these key factors are seen as simply common sense.

It is not the use of the right terminology (definitions, semantics), but the application of common sense that is the prerequisite for successful applications of ecosystem services-based approaches in support of participatory water resources management. Therefore, leaders should:

*   Spend ample time in framing and thereafter communicating the need for water resources management to those whose interests are affected by that management. Take the time to understand from stakeholders how they are affected.
*   Take an entrepreneurial approach:
    *   leave comfort zones, take an adventurous road;
    *   learn together to manage together;
    *   regard the environment not as a cost but as a profit centre; and
    *   consider other than only command-and-control solutions.
*   Spend ample time in defining SMART[3] targets that can be explained and thus understood by all stakeholders involved

---

[3] Specific, measurable, attainable, realistic, and timely.

```
Box 16.1  Key messages

It is not the use of the right terminology, but the application of
common sense, that is the pre-requisite for successful applica-
tions of ecosystem services-based approaches in support of
participatory water resources management. Therefore:

•   Spend ample time in framing and thereafter communi-
    cating the need for action.

•   Take an entrepreneurial approach.

•   Spend ample time in defining SMART formulated targets.

•   Provide facilitative leadership.

•   Try to speak the language of the stakeholders.
```

(realizing that different stakeholders have different targets).
Make sure to stick to these targets. It should also be made
clear what the consequences will be, and for whom, if
targets are not met.

• Provide facilitative leadership. Here, authorities can play a
  key role by acting less like a 'controller or regulator' and
  more as an 'enabler, persuader, motivator, or mediator'.

Above all, be aware of misunderstandings around the use of
economics; the absolute need for ecosystem services-based spatial
planning; and try to speak the language of the stakeholders.

References

Appleton, A. (2002). How New York City used an ecosystem services
strategy carried out through an urban–rural partnership to preserve the
pristine quality of its drinking water and save billions of dollars – and –
What lessons it teaches about using ecosystem services. Paper for Forest
Trends presented at the Katoomba Conference, Tokyo, November 2002.

Brauman, K., van der Meulen, S., & Brils, J. (2014). Ecosystem services in
river basin management. In: J. Brils, W. Brack, D. Müller, P. Negrel,
J. Vermaat (eds), *Risk-Informed Management of European River Basins*.
Springer, Berlin.

Brils, J., Barcelo, D., Blum, W., *et al.* (2014). Synthesis and recommenda-
tions towards risk-informed river basin management. In: J. Brils, W. Brack,
D. Müller, P. Negrel, J. Vermaat (eds), *Risk-Informed Management of
European River Basins*. Springer, Berlin.

Brockington, D. (2002). *Fortress Conservation: The Preservation of
the Mkomazi Game Reserve, Tanzania*. International African Institute,
Oxford.

Brown, C., Walpole, M., Simpson, L., & Tierney, M. (2011). Introduction to
the UK National Ecosystem Assessment. In: *The UK National Ecosystem
Assessment Technical Report*. UK National Ecosystem Assessment,
UNEP-WCMC, Cambridge.

Chichilnisky, C. & Heal, G. (1998). Economic returns from the biosphere.
Nature **391**, 629–630.

Everard, M. (2009). *Ecosystem Services Case Studies*. Environment Agency
Publications, Bristol.

Gezon, L. (1997). Institutional structure and the effectiveness of integrated
conservation and development projects: case study from Madagascar.
Human Organization **56**(4), 462–470.

Granek, E. F., Polasky, S., Kappel, C. V., *et al.* (2009). Ecosystem services as
a common language for coastal ecosystem-based management. *Conservation Biology* **24**(1), 207–216.

Oerlemans, N., Guldemond, J. A., & Visser, A. (2007). *Role of Farmland
Conservation Associations in Improving the Ecological Efficacy of a
National Countryside Stewardship Scheme: Ecological Efficacy of Habitat
Management Schemes*. Statutory Research Tasks Unit for Nature and the
Environment, Wageningen.

Pelsma, T. & van Everdingen, N. (2012). *Boeren als waterbeheerders*.
Waternet, Amsterdam, (in Dutch). Available at: http://watermaatwerk.nl/
Site_Watermaatwerk/Publiciteit_files/Eindrapport_Boeren_als_Waterbe-
heerders.pdf (last accessed 21 October 2014).

Ridder, D., Mostert, E., & Wolters, H. A. (eds) (2005). *Learning Together to
Manage Together : Improving Participation in Water Management*. University of Osnabrück, Osnabrück.

RLI (2013). *Preserved for Ever: Towards Robust Nature Management*. The
Hague: the Netherlands (in Dutch).

Smith, L., Bright, D., & Inman, A. (2011). A model for piloting new
approaches to catchment management in England and Wales. Defra.

van der Meulen, S. & Brils, J. (2011). Do ecosystem services provide a
common language to facilitate participation in water management? Report
of special session at Resilience 2011 conference, Tempe, Arizona, March
2011.

van der Meulen, S., Brils, J., Borowski-Maaser, I., & Sauer, U. (2013).
Payment for Ecosystem Services (PES) in support of river restoration.
Water Governance **4**, 40–44.

van Maasakkers, M, Duijn, M., & Kastens, B. (2014). Participatory
approaches and the role of facilitative leadership. In J. Brils, W. Brack,
D. Müller, P. Negrel, J. Vermaat (eds) *Risk-Informed Management of
European River Basins*. Springer, Berlin.

17 Cultural ecosystem services, water, and aquatic environments

Andrew Church, Rob Fish, Neil Ravenscroft, and Lee Stapleton

17.1 INTRODUCTION

Cultural ecosystem services have proved a highly challenging area for undertaking ecosystem assessments and developing ecosystem services-based approaches that can be incorporated within decision-making. Cultural ecosystem services are clearly core to understanding how ecosystems relate to human wellbeing since they focus on the cultural and social processes by which humans and the non-human interact. This involves activities, such as recreation, and spaces, such as parks and gardens, that are at the centre of everyday life. The problematic dimension of cultural ecosystem services is that it brings the longstanding philosophical and social theory debates over the meaning of culture into an approach to understanding the natural environment that has largely emerged out of natural science and economics. Ecosystem services-based approaches encourage classification and measurement which is often highly problematic when considering cultural entities and practices that resist simple definitions. Furthermore, the cultural aspects of ecosystems are not just confined to cultural ecosystem services. For example, Holmlund and Hammer (1999) discuss food production as a cultural service, whereas the Millennium Ecosystem Assessment (2005) would define food as a provisioning service. As Fish (2011, pp.674–675) notes, it 'is probably more accurate to think of "culture" less as a separate "box" within the services typology [...]. In what sense, for instance, is the provision of "food" not also a cultural ecosystem service?'

The same arguments can be applied to water in that how water is treated and 'mis-treated' in any society will to a significant degree be influenced by cultural attitudes. There is a large literature on the cultural significance of water which highlights how different societies currently and historically have understood the significance of water and attached distinct political, cultural, and social meanings to water (Toussaint 2006). Strang (2006) argues that it is also important to recognise the cross-cultural 'flows' relating to water and points out that in Western society and for some Australian Aboriginal peoples immersion in water can be a highly significant spiritual act.

Clearly the multifaceted cultural dimensions of water raise significant challenges for attempts to incorporate cultural ecosystem services into an understanding of water ecosystems. Nevertheless, there is now a substantial body of writing on cultural ecosystem services that is starting to provide some conceptual and empirical clarity. A proportion of this literature has been concerned with water and aquatic environments, which has begun to indicate how cultural ecosystem services can be incorporated into the analysis of water-related ecosystem services. This chapter focuses mainly on summarising and discussing the literature on cultural ecosystem services in general and studies that focus on water found in peer-reviewed journals and academic books, but reference is also made to some policy-instigated ecosystem assessments. The body of research discussed in this chapter reveals a number of challenges relating to understanding water and cultural ecosystem services. These include the definition and scope of cultural ecosystem services and also the conceptual and empirical approaches used to study cultural ecosystem services. The chapter also seeks to consider the implications of these challenges for the incorporation of cultural ecosystem services into an ecosystem services-based approach as defined in Chapter 2 of this book, which can underpin decision-making.

17.2 DEFINITIONS OF CULTURAL ECOSYSTEM SERVICES

In the Millennium Ecosystem Assessment (2005) cultural ecosystem services were defined as 'the non-material benefits people obtain from ecosystems through spiritual enrichment, cognitive development, reflection, recreation, and aesthetic experience' (p.29). This definition presents cultural ecosystem services as involving a series of these five processes that generate wellbeing benefits. The Millennium Ecosystem Assessment listed ten *benefits* provided through these five *processes* and these were: cultural diversity, spiritual and religious values, knowledge systems, educational values, inspiration, aesthetic values, social relations, sense of place, cultural heritage values, recreation and eco-tourism. The inclusion of recreation in both the list of benefits and processes confirms the widely recognised view that the Millennium Ecosystem Assessment approach equates services and benefits (e.g. Nahlik *et al.* 2012). However, what is less well recognised is an important innovation in the Millennium Ecosystem Assessment in terms of the conceptual consideration of cultural ecosystem services, which is that benefits are mediated by processes.

Where other definitions of cultural ecosystem services are put forward in the literature, these tend to correspond to the Millennium Ecosystem Assessment definition, either explicitly or implicitly. Alternative definitions are limited. Chan *et al.* (2012, p. 9) modified the Millennium Ecosystem Assessment definition so that cultural ecosystem services are defined as 'ecosystems' contributions to the non-material benefits (e.g., capabilities and experiences) that arise from human–ecosystem relationships' (p.9). This definition is adopted by Plieninger *et al.* (2013) and Klain and Chan (2012).

Beyond these alternative definitions of cultural ecosystem services, sometimes the term *cultural ecosystem service* is not used but a series of benefits are described under different terminologies that might be recognised as cultural ecosystem services according to the Millennium Ecosystem Assessment definition. Examples include *amenity functions* (Pinto-Correia & Carvalho-Ribeiro 2012), *benefits* (Fisher *et al.* 2008, 2009), *information functions* (de Groot 2006; de Groot *et al.* 2002) *life fulfilling services* (Chee 2004), and *socio-cultural fulfilment* (Wallace 2007). Fisher *et al.* (2008, 2009) adopt a rather different approach, seeking clarity by distinguishing between cultural ecosystem services and benefits rather than equating the two: 'we define ecosystem services to be about ecological phenomena (e.g. not cultural services which we see as very valuable benefits derived from ecosystems and services)' (Fisher *et al.* 2009, p.644).

The variations in definitions of cultural ecosystem services raise a problem for ecosystem services-based approaches as a whole which are designed in part to encourage certain stakeholders to consider a range of environmental issues in decision-making. As Potschin and Haines-Young (2013) argue, if proponents of ecosystem services approaches lack some form of agreement over definitions then this limits the advocacy role of the approach. A recent attempt to refine the definition of cultural ecosystem services occurred as part of the UK National Ecosystem Assessment (UK NEA 2011) follow on research. The UK National Ecosystem Assessment was completed in 2011 and further research was commissioned to address identified knowledge gaps (see Schaafsma *et al.* in this book). This included a study that sought to develop a definition of cultural ecosystem services in conjunction with certain national policy bodies responsible for conservation and the development of the ecosystem services policy approaches in the UK (Church *et al.* 2014). The definition used

in this study disaggregated cultural ecosystem services into environmental spaces (e.g. lakes and beaches) where people interact with the natural environment and a series of cultural practices, such as exercising and playing, that shape these interactions and spaces. More generally these cultural practices and environmental spaces are understood to reflect and influence the cultural values people individually and collectively hold concerning the natural environment (Fish & Church 2015). This definition sought to identify measurable entities such as places and practices that can be utilised in decision-making while also acknowledging the complex value-based dimensions to cultural ecosystem services. Such definitional problems are always likely to persist, partly because cultural ecosystem services links ecosystems services-based approaches to complex, longstanding, and contested debates over culture–nature (Descola 2013) and the interactions between humans and non-humans (Hinchcliffe 2008) that have recently been addressed through relational ontologies drawing on concepts such as hybridity (Whatmore 2002) and the gift of nature (Church *et al.* 2013). These current writings suggest how humans relate to nature will always be subject to philosophical and scientific discourse and contestation, which means that cultural ecosystem services will be similarly debated in terms of how they are defined.

17.3 SCOPE OF CULTURAL ECOSYSTEM SERVICES

The variations that can be found in discussions of the definitions of cultural ecosystem services are also reflected in the literature which considers the scope of cultural ecosystem services using one or more categories of service. The most common category of cultural ecosystem services is recreation/tourism, featuring in the majority of the literature (Hernandez-Mocillo *et al.* 2013). There are only limited examples of writing which do not directly consider this and instead focus on other categories: Grêt-Regamey *et al.* (2008) – scenic beauty; Grêt-Regamey *et al.* (2013) – landscape aesthetics; Raudsepp-Hearne *et al.* (2010a 2010b) – so-called proxies;[1] Sandhu *et al.* (2008) – aesthetics; and van Berkel and Verburg (2014) – aesthetic beauty, cultural heritage, spirituality, and inspiration.

A number of the studies of cultural ecosystem services and tourism/recreation focus on specific water-based activities such as diving (Worm *et al.* 2006; Ruiz-Frau *et al.* 2013) and whale watching (Beaumont *et al.* 2007; Cisneros-Montemayor & Sumaila 2010). Whereas other ecosystem services studies of water and tourism/recreation examine particular freshwater bodies, usually lakes and rivers (see Raymond *et al.* 2009 for

---

[1] '[N]on-food ecosystem services, including forest cover and percentage of land under protected-area status (proxies for many cultural and regulating services)' (p.579).

the Murray-Darling river basin in Australia) and sub-areas of marine environments (see Worm *et al.* 2006 for the Caribbean).

More generally, the other processes put forward by the Millennium Ecosystem Assessment, which lead to cultural ecosystem services/benefits, feature to varying extents in the literature. In papers in refereed journals aesthetics is second to recreation/tourism as the most mentioned cultural ecosystem service (Hernandez-Mocillo *et al.* 2013). This is reflected in some studies of cultural services in freshwater habitats such as the Gariep River Basin in Lesotho and South Africa (Bohensky *et al.* 2006). Cultural ecosystem services are framed in terms of recreation/tourism in each case, as well as aesthetics in Peterson *et al.* (2003). The tendency to focus on recreation/tourism and aesthetics, with less attention paid to other service categories, is concerning because of its partiality. Fish (2011, p.674) notes that 'the danger is that an [ecosystems approach] ends up addressing a rather underwhelming and predictable set of activities, such as types and patterns of recreation and (undertheorized) to aesthetic value'.

Spiritual enrichment and cognitive development, incorporating education, both feature to a similar, more moderate extent. It is difficult to find examples which mention the process of reflection included in the Millennium Ecosystem Assessment definition although, relevant to this process, the benefit of inspiration is variously covered (Chee 2004; Raymond *et al.* 2009; Maynard *et al.* 2010; Schaich *et al.* 2010; Klain & Chan 2012; Norton *et al.* 2012; Piwowarczyk *et al.* 2013; Plieninger *et al.* 2013; van Berkel & Verburg 2014; see also, Maynard *et al.* in this book). Other benefits related to reflection and inspiration are covered, such as calm (Norton *et al.* 2012), peace (Klain & Chan 2012), serenity (Chee 2004), and tranquilising effects (Piwowarczyk *et al.* 2013).

There is also a wide diversity in terms of the number of cultural ecosystem services categories covered, ranging from one in several studies (usually recreation/tourism) to 30 in Klain and Chan (2012), split between economic activities (e.g. eco-tourism), tangible non-monetary benefits (e.g. biodiversity/wildlife), and intangible non-monetary benefits (e.g. community/identity).

17.4 APPROACHES TO STUDYING CULTURAL ECOSYSTEM SERVICES

The peer-reviewed literature also indicates the variety of theoretical approaches adopted to understanding cultural ecosystem services. Theoretical approaches have investigated the relationship among different ecosystem services, including cultural ecosystem services (Bennett *et al.* 2009), and advocated greater attention to how these relationships affect cultural ecosystem services in future ecosystem assessments (Carpenter *et al.* 2009). Daniel *et al.* (2012) advocate greater attention to the spatial dimension of cultural ecosystem services, framed in terms

of the Millennium Ecosystem Assessment understanding that cultural ecosystem services equate to non-material benefits, thus mapping places where particular benefits are realised rather than places where any, not necessarily specified, cultural benefits are realised. Indeed, in many regional or national ecosystem assessments a place-based approach is adopted, which considers how places and landscapes are central to cultural ecosystem services and the related human wellbeing benefits (Church *et al.* 2011). De Groot *et al.* (2002), however, note that a challenge of ecosystem services-based approaches to decision-making is whether places and landscapes with cultural meaning are considered as services in themselves or proxies for services and/or benefits. For de Groot *et al.* (2002) landscapes are proxy indicators of services.

Whatever the distinction made between proxies and services, studies have identified a number of the theoretical and empirical advantages of a place-based approach to studying ecosystem services compared to habitat-, systems-, or process-based approaches (Potschin & Haines-Young 2013). The merits of linking cultural ecosystem services research with that concerning cultural landscapes have also been explored (Schaich *et al.* 2010). In terms of the latter, it is argued that research into cultural landscapes studies encounters the same problem as research using cultural ecosystem services in terms of the variations in terminology and definitions, but the analyses of cultural landscapes has had substantially more academic attention applied to it, thus offering a bank of acquired knowledge that could be exploited for understanding cultural ecosystem services (Schaich *et al.* 2010). Other theoretical discussions have emphasised that more account must be taken of the types of values implied by the different benefits realised from cultural ecosystem services and individual versus group values (Chan *et al.* 2012) have been explored. Theoretical critiques have also been put forward in terms of the monetary methods used to value cultural ecosystem services (Chee 2004) and the prevailing conceptualisation of cultural ecosystem services (Fish 2011).

These differences in the scope of cultural ecosystem services covered and the theoretical approaches adopted are to be expected and to some degree encouraged since cultures are locally and socially specific. Nevertheless, these differences make comparisons between studies of cultural ecosystem services difficult for decision-makers.

17.5 APPROACHES TO STUDYING WATER-RELATED CULTURAL ECOSYSTEM SERVICES

The studies of water and cultural ecosystem services to some degree reflect the general literature on cultural ecosystem services in terms of the scope of activities studied, often with a focus on recreation and tourism, the emphasis on examining particular places associated with cultural ecosystem services, and the variety of approaches adopted. In addition, the existing studies of cultural ecosystem services, water, and aquatic environments do reveal a number of significant issues facing organisations and key actors seeking to incorporate ecosystems services-based approaches into water-related decision-making.

The range of approaches adopted in relation to water-related cultural ecosystem services is revealed in the studies that review existing literature or utilise secondary data in an attempt to draw globally relevant conclusions. Such studies include a theoretical discussion of the cultural ecosystem services associated with fish populations (Holmlund & Hammer 1999) to an analysis based on secondary data that argues for assessing cultural ecosystem services arising from marine-based recreational activities globally in terms of levels of participation, employment, and expenditure (Cisneros-Montemayor & Sumaila 2010). A global Geographic Information System-based approach to cultural ecosystem services by Ghermandi and Nunes (2013) developed a meta-analytical framework of the values of coastal recreation based on 253 existing valuation studies that used primary data to assess the recreational benefits of coastal ecosystems. The findings were used to argue that such an approach could be used to incorporate socio-economic data into decision-making that sort to rank coastal areas in terms of their importance for conservation. The study by Holmlund and Hammer (1999) is also a reminder of one of the limitations of a place-based approach in that some cultural ecosystem services are linked to species that are 'mobile' which is a challenge for decision-makers. For example, the wellbeing benefits gained from some forms of recreational angling and sea fishing while linked to certain aquatic spaces is also reliant on mobile fish stocks.

Many of the empirical studies of water and cultural ecosystem services that use primary data do, like cultural ecosystem services research more generally, adopt a spatially explicit approach and focus on specific places associated with cultural ecosystem services. In the cases where primary research data are used to examine cultural ecosystem services in aquatic environments it is possible to distinguish in terms of the approach adopted between those that are based on expert knowledge (usually researchers and decision-makers) and those that collect primary data from the public. An example of the former approach used participatory research with experts (Maynard et al. in this book) for a case study in South East Queensland, Australia that included coastal areas, river basins, and wetlands. Here, an expert panel was convened to assess the magnitude of total ecosystem functions (one cultural function is given: landscape opportunity) for identified ecosystem reporting categories, producing a map of the case study area delimited in terms of high to low total ecosystem functions. According to the nomenclature used by these authors, services

(11 categories in the case of cultural ecosystem services, including aesthetics and recreational opportunities) emanate from biophysical functions and further work is underway to produce a map at the service level. Another participatory expert-focused approach was adopted for a case study in the Murray-Darling basin region of Australia (Raymond et al. 2009; Crossman et al. in this book). Fifty-six local and regional decision-makers were each asked to assign 40 green dots (natural capital and ecosystem service values) and ten red dots (threats) to a map of the area. Recreation/tourism was the most valued of all the ecosystem services assessed and aquatic places were seen as particularly valuable to delivering cultural ecosystem services.

Other expert-based studies use policy documentation as a data source. For example, Piwowarczyk et al. (2013) use content analysis of municipality documents for towns and cities situated on the Polish coast of the North Sea to determine how ecosystem services including cultural ecosystem services 'are perceived in the practice of urban planning and long term management'. Similarly, Worm et al. (2006) use a range of existing documentation to analyse quantitatively the effect on ecosystem services of 138 protected ocean areas in the Caribbean and, as with many marine studies of cultural ecosystem services, tourism is the main form of cultural ecosystem services considered. Expert-based studies have a value for organisations seeking to include the assessment of ecosystem services into decision-making in that they allow a relatively rapid assessment of some of the key issues that may arise when seeking to take account of cultural ecosystem services. Such approaches, however, are limited in terms of identifying the benefits of cultural ecosystem services to human wellbeing, which require data drawn from the public concerning people's practices and interactions with the natural environment and how this relates to wellbeing.

Public-focused approaches to studying cultural ecosystem services are varied in terms of the extent to which people were involved, ranging from small to large sample studies. Larger sample studies of cultural ecosystem services have gathered data using a variety of methods. Ruiz-Frau et al. (2013) gathered data using an online survey to attempt to devise spatially varied economic values of cultural ecosystem services in coastal environments and to explore how values were affected by levels of marine biodiversity. Martín-López et al. (2009) used face-to-face questionnaires with 525 respondents to study cultural ecosystem services in the Donana national park on the south-western coast of Spain, and the findings suggested that assessing the economic value of cultural ecosystem services in aquatic environments requires advanced methodologies that take account of spatial and temporal heterogeneity in valuation. As part of the research that followed on from the UK National Ecosystem Assessment, Kentner et al. (2014) undertook a large-scale study of the value of marine protected areas through an online survey of 1683 divers and sea anglers, as well as 11 deliberative monetary valuation workshops with 130 participants who were also divers and anglers. These methods revealed that the economic benefits of marine protected areas to recreational users who are benefiting from cultural ecosystem services could be considerable and should be more closely considered in protected area designation decisions since these benefits can outweigh the economic costs emphasised by fisheries stakeholders.

Smaller sample studies of cultural ecosystem services and water spaces are often designed to undertake some form of non-monetary valuations that explore the harder-to-measure cultural values associated with aquatic environments. These studies have highlighted how marine environments have particular cultural value to coastal communities. Focusing on the seascape of Northern Vancouver Island, Canada, Klain and Chan (2012) carried out 30 semi-structured interviews with people whose jobs are linked to the sea. Each drew on a map to show where and why areas were important both for their income as well as for non-monetary reasons; degree of importance was quantified by distributing tokens on the completed maps and 'people allocated the highest non-monetary values to places notable for wildlife, outdoor recreation, then cultural heritage'. Kentner *et al.* (2014), in the follow-on research to the UK National Ecosystem Assessment, explored similar issues in relation to the marine environment through a case study of Hastings on the south coast of England, which examined the cultural values, defined as shared principles and virtue, as well as the communal values that are specifically shared by the local community. This study also examined how deliberative research methods can reveal complex communal values arguing that such research methods are essential if decision-makers wish to understand fully the range of values likely to be expressed when consulting on environmental decisions relating to ecosystem services. The follow-on research to the UK National Ecosystem Assessment also involved an analysis of the large-scale national survey in England – the Monitor of Engagement with the Natural Environment (MENE). This survey collects data on people's use and enjoyment of the natural environment in England and involves over 45000 interviews per year with members of the public. Analysis of this data set revealed that 35% of the respondents reported that beaches were the most wellbeing-enhancing outdoor environments and blue spaces generally (beaches, rivers, seaside/coast) were identified by 65% of respondents as spaces that enhanced wellbeing (Church *et al.* 2014). Similarly, a study by MacKerron and Mourato (2013), which used a mobile phone-based wellbeing tracking technology identified marine and coastal margins as the outdoor environments where people felt 'happiest'. Clearly, in the English context at least, cultural ecosystem services linked to blue spaces are important to human wellbeing.

Finally, the study of ecosystem services for the incorporation of an ecosystems services-based approach into decision-making (core element 4, Chapter 2 of this book) has often involved the use of scenarios (Ash *et al.* 2010)– see Box 17.1.

Box 17.1 Cultural ecosystem services, water, and scenarios

The incorporation of an ecosystems services-based approach into decision-making has often involved the use of scenarios (Ash *et al.* 2010). Such scenarios devise 'plausible and often simplified descriptions of how the future may unfold based on a coherent and internally consistent set of assumptions about key driving forces, their relationships, and their implications for ecosystems' (Henrichs *et al.* 2010, p.152).

The existing literature on scenarios for ecosystem services which consider cultural ecosystem services can also be divided into those that are expert-focused (researchers and decision-makers) approaches and those that are more public-focused. Expert-driven scenario exercises that consider cultural ecosystem services in aquatic environments or specific water bodies are available for case studies covering the Gariep River Basin in Lesotho and South Africa (Bohensky *et al.* 2006) and the Northern Highlands Lake District in Wisconsin, USA (Peterson *et al.* 2003). In the context of Hawai'i, Daily *et al.* (2009) introduce an application of quantitative modelling software that can involve a range of stakeholders 'for quantifying [monetary and non-monetary] ecosystem service values across land- and seascapes' (p.22) across a range of different future scenarios. Cultural ecosystem services are approached in terms of recreation, tradition, and community. In this study it is not fully clear as to the extent to which stakeholders were involved in the design and validation of scenarios or whether this conception of stakeholders extends beyond researchers and decision-makers to less empowered people. Nevertheless, it shows the potential of scenarios for involving different stakeholders through scenario development in the operationalisation of ecosystems services-based approaches to decision-making. Overall, these studies of scenarios highlight the importance of eliciting non-expert perspectives on the nature of cultural ecosystem services (Ash *et al.* 2010).

17.6 VARIABILITY IN USES AND PREFERENCES FOR CULTURAL ECOSYSTEM SERVICES BY SOCIO-ECONOMIC AND DEMOGRAPHIC GROUPS

Despite the importance of involving stakeholders in ecosystem assessments as described in core element 3 of Chapter 2, there has been only limited attention in the literature on cultural ecosystem services and aquatic environments to variability in uses and preferences for cultural ecosystem services by people from different socio-economic and demographic groups. Maass *et al.* (2005), in the context of the Pacific coastal forests, compare expert perceptions and stakeholder perceptions of benefits of different ecosystem services for the following groups of stakeholders: farmers; tourist industries; landless locals; and external users. In terms of one of the cultural ecosystem services they consider, scenic beauty, the greatest divergence between expert and stakeholder perceptions was in terms of the landless locals group who tended to be very aware of the benefits of this service, yet experts determined that this group only slightly benefited in terms of cultural ecosystem services. Martín-López *et al.* (2009), in a study of tourism as a cultural ecosystem service in the coastal Donana national park in Spain, found that older and better educated people were more likely to visit the area and thus receive wellbeing benefits. In a larger piece of research across eight case study sites in Spain which included coastal areas, Martín-López *et al.* (2009) noted differences in the perceived importance of different cultural ecosystem services depending upon whether respondents lived in urban or rural areas, 'nature tourism, aesthetic values, environmental education, and the existence value of biodiversity were mostly perceived by urban inhabitants [...] recreational hunting and local ecological knowledge obtained higher value scores from inhabitants of rural areas'.

In relation to marine environments, an examination of sense of place (often portrayed as a cultural ecosystem service) in the Great Barrier Reef region of Australia based on a survey of 372 residents found that determinants of the values people assigned to the natural environments included place and length of residence in the region, country of birth, and involvement in community activities (Larson *et al.* 2013). In addition, residents of coastal areas tended to assign greater value to environmental wellbeing compared to those living in other areas. Larson *et al.* (2013) note the need for in-depth qualitative research in the planning process to complement their quantitative study.

Recent research has sought to use qualitative methods and arts and humanities-based approaches to data collection to enable the views in relation to cultural ecosystem services of certain social groups to be more clearly represented in decision-making linked to an ecosystem services-based approach. As part of the UK National Ecosystem Assessment follow-on research, qualitative workshops involving participatory mapping were used in marginalised rural communities in north Devon, England which revealed how wellbeing benefits linked to cultural ecosystem services were restricted even for rural residents by limited access to the countryside, especially to riversides (Church *et al.* 2014). The same study also used arts-based approaches with a local community arts group to highlight what young primary school-age children value in their local environment (Church *et al.* 2014).

17.7 CONCLUSIONS

Despite these challenges to addressing cultural ecosystem services in the analysis of aquatic ecosystems, recent research suggests that water-related cultural ecosystem services are an important area for further study. As this chapter has revealed, the diversity of definitions of cultural ecosystem services and approaches used to analyse cultural ecosystem services raise a number of challenges for both researchers and decision-makers. Place-based approaches to identifying and analysing cultural ecosystem services relating to water are common and offer considerable potential, although mobile cultural ecosystem services linked to species of cultural significance also need to be considered.

There is clearly a need to move beyond the limited focus of many studies on tourism and recreation and consider a wider range of practices and benefits associated with the spiritual, creative, and educational dimensions of cultural ecosystem services. Indeed, a focus simply on tourism and recreation may lead to cultural ecosystem services being considered as purely social and cultural phenomenon, thus losing sight of the core focus of ecosystem services-based approaches, which is to appreciate how ecosystem services are related to biophysical processes.

Box 17.2 Key messages

- The problems of defining cultural ecosystem services are likely to persist because definitions are influenced by longstanding and contested debates over culture–nature relations.
- There is a tendency in the approaches to studies of water-related cultural ecosystem services to focus on recreation/tourism and aesthetics, with less attention paid to other service categories such as knowledge systems, sense of place, and spiritual and religious values.
- Differences in definitions and approaches to cultural ecosystem services make it difficult to compare the findings of studies of cultural ecosystem services in aquatic environments.
- Empirical studies of water and cultural ecosystem services often adopt a spatially explicit approach and focus on specific places associated with cultural ecosystem services.
- Both quantitative and qualitative approaches and data are often needed to analyse cultural values associated with cultural ecosystem services in aquatic environments.
- More research is needed to understand the uses and preferences for cultural ecosystem services of people from different socio-economic and demographic groups.

Recent research also suggests that ensuring the views of different social groups and stakeholders are addressed by decision-making linked to ecosystem services will require quantitative methods of data collection and analysis to be used alongside qualitative and deliberative techniques and approaches drawn from the arts and humanities.

The challenges relating to cultural ecosystem services are being addressed but, for now at least, cultural ecosystem services will remain a core but elusive element of ecosystem services-based approaches to water and aquatic environments.

ACKNOWLEDGEMENTS

The authors gratefully acknowledge that research for this chapter was funded by the UK Arts and Humanities Research Council as part of the Towards Hydrocitizenship project (grant number AH/L008165/1).

References

Ash, N., Blanco, H., Brown, C., *et al.* (2010). *Ecosystems and Human Well-Being: A Manual for Assessment Practitioners.* Island Press, Washington DC.

Beaumont, N. J., Austen, M. C., Atkins, J. P., *et al.* (2007). Identification, definition and quantification of goods and services provided by marine biodiversity: implications for the ecosystem approach. *Marine Pollution Bulletin* **54**(3), 253–265.

Bennett, E. M., Peterson, G. D., & Gordon, L. J. (2009). Understanding relationships among multiple ecosystem services. *Ecology Letters* **12**(12), 1394–1404.

Bohensky, E. I., Reyers, B., & Van Jaarsveld, A. S. (2006). Future ecosystem services in a Southern African river basin: a scenario planning approach to uncertainty. *Conservation Biology* **20**(4), 1051–1061.

Carpenter, S. R., Mooney, H. A., Agard, J., *et al.* (2009). Science for managing ecosystem services: beyond the Millennium Ecosystem Assessment. *Proceedings of the National Academy of Sciences* **106**(5), 1305–1312.

Chan, K. M.A, Satterfield, T., & Goldstein, J. (2012). Rethinking ecosystem services to better address and navigate cultural values. *Ecological Economics* **74**, 8–18.

Chee, Y. E. (2004). An ecological perspective on the valuation of ecosystem services. *Biological Conservation* **120**, 549–565.

Church, A., Burgess, J., & Ravenscoft, N. (2011). Cultural services. In: *The UK National Ecosystem Assessment.* Defra, London. Available from: http://uknea.unep-wcmc.org/LinkClick.aspx?fileticket=QLgsfedO70I%3D&tabid=82 (last accessed 21 October 2014).

Church, A., Ravenscroft, N. & Gilchrist, P. (2013). Property ownership, resource use and the 'gift of nature'. *Environment and Planning D: Society and Space* **31**(3), 451–466

Church, A., Fish, R., Haines-Young, R., Mourato, S. & Tratalos. J. (2014). *Cultural Ecosystem Services and Indicators, UK National Ecosystem Assessment Follow On, Work Package 6.* Defra, London. Available from: http://uknea.unep.wcmc.org/NEWFollowonPhase/Whatdoesthefollowon-phaseinclude/WorkPackage4/tabid/147/Default.aspx (last accessed 21 October 2014).

Cisneros-Montemayor, A. M. & Sumaila, U. R. (2010). A global estimate of benefits from ecosystem-based marine recreation: potential impacts and implications for management. *Journal of Bioeconomics* **12**(3), 245–268.

Daily, G. C., Polasky, S., Goldstein, J., *et al.* (2009). Ecosystem services in decision making: time to deliver. *Frontiers in Ecology and the Environment* **7**(1), 21–28.

Daniel, T. C., Muharb, A., Arnbergerb, A., *et al.* (2012). Contributions of cultural services to the ecosystem services agenda. *Proceedings of the National Academy of Sciences* **109**, 8812–8819.

de Groot, R. (2006). Function-analysis and valuation as a tool to assess land use conflicts in planning for sustainable, multi-functional landscapes. *Landscape and Urban Planning* **75**(3–4), 175–186.

de Groot, R. S., Wilson, M. A., & Boumans, R. M. J. (2002) A typology for the classification, description and valuation of ecosystem functions, goods and services. *Ecological Economics* **41**(3), 393–408.

Descola, P. (2013). *Beyond Nature and Culture.* University of Chicago Press, London.

Fish, R. D. (2011). Environmental decision making and an ecosystems approach: some challenges from the perspective of social science. *Progress in Physical Geography* **35**(5), 671–680.

Fish, R. & Church, A. (2015). Cultural ecosystem services: stretching out the concept. Forthcoming *Environmental Scientist.*

Fisher, B., Turner, K., Zylstra, M., *et al.* (2008) Ecosystem services and economic theory: integration for policy-relevant research. *Ecological Applications* **18**(8), 2050–2067.

Fisher, B., Turner, R. K., & Morling, P. (2009). Defining and classifying ecosystem services for decision making. *Ecological Economics* **68**(3), 643–665.

Ghermandi, A. & Nunes, P. (2013). A global map of coastal recreation values: results from a spatially explicit meta-analysis. *Ecological Economics* **86**, 1–15

Grêt-Regamey, A., Bebi, P., Bishop, I. D., & Schmid, W. A. (2008). Linking GIS-based models to value ecosystem services in an Alpine region. *Journal of Environmental Management* **89**, 197–208.

Grêt-Regamey, A., Celio, E., Klein, T. M., & Hayek, U. W. (2013). Understanding ecosystem services trade-offs with interactive procedural modeling for sustainable urban planning. *Landscape and Urban Planning* **109**, 107–116.

Henrichs, T., Zurek, M., Eickhout, B., *et al.* (2010). Scenario development and analysis for forward-looking ecosystem assessment. In: *Ecosystems and Human Well-Being: A Manual for Assessment Practitioners.* Island Press, Washington, DC.

Hernandez-Morcillo, M., Plieninger, T., & Bieling, C. (2013) An empirical review of cultural ecosystem service indicators. *Ecological Indicators* **29**, 434–444.

Hinchcliffe, S. (2008). Reconstituting nature conservation: towards a careful political ecology. *Geoforum* **39**(1), 88–97.

Holmlund, C. M. & Hammer, M. (1999). Ecosystem services generated by fish populations. *Ecological Economics* **29**(2), 253–268.

Kentner, J., Reed, M., Irvine, K., *et al.* (2014). *Shared, Plural and Cultural Values of Ecosystems: UK National Ecosystem Assessment Follow On, Work Package 5.* Defra, London.

Klain, S. C. & Chan, K. M. A. (2012). Navigating coastal values: participatory mapping of ecosystem services for spatial planning. *Ecological Economics* **82**, 104–113.

Larson, S., De Freitas, D. M., & Hicks, C. (2013). Sense of place as a determinant of people's attitudes towards the environment: implications for natural resources management and planning in the Great Barrier Reef, Australia. *Journal of Environmental Management* **117**, 226–234.

Maass, J. M., Balvanera, P., Castillo, A., *et al.* (2005). Ecosystem services of tropical dry forests: insights from long term ecological and social research on the Pacific Coast of Mexico. *Ecology and Society* **10**(1), 17.

MacKerron, G. & Mourato, S. (2013). Happiness is greater in natural environments. *Global Environmental Change* **23**(5), 992–1000.

Martín-López, B., Gómez-Baggethun, E., Lomas, P. L., & Montes, C. (2009). Effects of spatial and temporal scales on cultural services valuation. *Journal of Environmental Management* **90**(2), 1050–1059.

Maynard, S., James, D., & Davidson, A. (2010). The development of an ecosystem services framework for South East Queensland. *Environmental Management.* **45**(5), 881–895.

Millennium Ecosystem Assessment (2005) *Current State & Trends Assessment.* Island Press, Washington, DC.

Nahlik, A. M., Kentula, M. E., Fennessy, M. S., & Landers, D. H. (2012). Where is the consensus? A proposed foundation for moving ecosystem service concepts into practice. *Ecological Economics* **77**, 27–35.

Norton, L. R., Inwood, H., Crowe, A., & Baker, A. (2012). Trialling a method to quantify the 'cultural services' of the English landscape using Countryside Survey data. *Land Use Policy* **29**(2), 229–455.

Peterson, G. D., Beard Jr, T. D., Beisner, B. E., *et al.* (2003). Assessing future ecosystem services: a case study of the Northern Highlands Lake District, Wisconsin. *Conservation Ecology* **7**(3), 1.

Pinto-Correia, T. & Carvalho-Ribeiro, S. (2012). The Index of Function Suitability (IFS): a new tool for assessing the capacity of landscapes to provide amenity functions. *Land Use Policy* **29**(1), 23–34.

Piwowarczyk, J., Kronenberg, J., & Dereniowska, M. A. (2013). Marine ecosystem services in urban areas: do the strategic documents of Polish coastal municipalities reflect their importance? *Landscape and Urban Planning* **109**(1), 85–93.

Plieninger, T., Dijks, S., Oteros-Rozas, E., & Bieling, C. (2013). Assessing, mapping, and quantifying cultural ecosystem services at community level. *Land Use Policy* **33**, 118–129.

Potschin, M. & Haines-Young, R. (2013). Landscapes, sustainability and the place-based analysis of ecosystem services. *Landscape Ecology* **28**(6), 1053–1065.

Raudsepp-Hearne, C., Peterson, G. D., & Bennett, E. M. (2010a). Ecosystem service bundles for analyzing tradeoffs in diverse landscapes. *Proceedings of the National Academy of Sciences* **107**, 5242–5247.

Raudsepp-Hearne, C., Peterson, G. D., Tengö, M., *et al.* (2010b). Untangling the environmentalist's paradox: why is human wellbeing increasing as ecosystem services degrade? *BioScience* **60**(8), 576–589.

Raymond, C. M., Bryan, B. A., Hatton MacDonald, D., *et al.* (2009). Mapping community values for natural capital and ecosystem services. *Ecological Economics* **68**(5), 1301–1315.

Ruiz-Frau, A., Hinz, H., Edwards-Jones, G., & Kaiser, M. J. (2013). Spatially explicit economic assessment of cultural ecosystem services: non-extractive recreational uses of the coastal environment related to marine biodiversity. *Marine Policy* **38**, 90–98.

Sandhu, H. S., Wratten, S. D., Cullen, R., & Case, B. (2008). The future of farming: the value of ecosystem services in conventional and organic arable land. An experimental approach. *Ecological Economics* **64**(4), 835–848.

Schaich, H., Bieling, C., & Plieninger, T. (2010). Linking ecosystem services with cultural landscape research. *GAIA* **19**(4), 269–277.

Strang, V. (2006). Aqua culture: the flow of cultural meanings in water. In: M. Leybourne & A. Gaynor (eds), *Water, Histories, Cultures and Ecologies*. University of Western Australia Press, Crawley.

Toussaint, S. (2006). Introducing water: a symposium and this volume. In: M. Leybourne & A. Gaynor (eds), *Water, Histories, Cultures and Ecologies*. University of Western Australia Press, Crawley.

UK National Ecosystem Assessment (2011). *The UK National Ecosystem Assessment: Synthesis of the Key Findings*, UNEP-WCMC, Cambridge.

van Berkel, D. B. & Verburg, P. H. (2014). Spatial quantification and valuation of cultural ecosystem services in an agricultural landscape. *Ecological Indicators* **37**, 163–174.

Wallace, K. J. (2007). Classification of ecosystem services: problems and solutions. *Biological Conservation* **139**(3–4), 235–246.

Whatmore, S. (2002). *Hybrid Geographies: Natures, Cultures, Spaces*. Sage, London.

Worm, B., Barbier, E. B., Beaumont, N., *et al.* (2006). Impacts of biodiversity loss on ocean ecosystem services. *Science* **314**(5800), 787–790.

18 The psychological dimension of water ecosystem services

Victor Corral-Verdugo, Martha Frías-Armenta, César Tapia-Fonllem, and Blanca Fraijo-Sing

18.1 INTRODUCTION

Discussions of ecosystem services and ecosystem services-based approaches have so far focused on their ecological, social, and economic aspects (Fisher *et al.* 2011). These are aspects of our common guiding core elements 2 and 3 as presented in Chapter 2 of this book. Core element 4 has also been dominated by economic and monetary quantification of benefits, including the environmental and socio-economic paybacks that result from the protection of water sources (Syme *et al.* 2008). The focus on these three aspects neglects to include psychological factors involved in water ecosystem services and their enjoyment by people. Also neglected is the study of the relationship between water ecosystem services and water conservation (i.e. individuals' actions aimed at avoiding water waste and contamination). Understanding the *psychological components* of the benefits of water ecosystem services is important because: (1) individuals value aspects of the environment that provide them with positive psychological consequences (i.e., satisfaction of needs, pleasure, wellbeing, etc.);[1] and (2) if water ecosystem services are clearly perceived, people tend to conserve water in order to guarantee the continued provision of those services. These two components are psychological because they include mental (perceptions, values, emotions, psychological) benefits as well as behavioural (water conservation) aspects.

Although a number of *psychologically positive consequences* of the relationship between humans and water have been studied, ecosystem services-based approaches rarely incorporate the understanding of those consequences within their explanatory models and empirical studies. Therefore, little is known about people's perceived psychological benefits when (directly or indirectly) they come into contact with a water ecosystem. These psychologically positive consequences are fundamental

to the appreciation of both water ecosystem services and water conservation efforts because pro-environmental behaviours are more likely to be maintained if those behaviours produce desirable outcomes in the form of positive mental states and psychological experiences (De Young 2000). Psychologically positive consequences also stimulate people's appreciation of water ecosystem services since some of these services include the experience of psychological restoration, happiness, and psychological wellbeing (which are aspects of core element 1 as defined in Chapter 2). Great emphasis is being put on informing decision and policy making, but a very crucial aspect is to promote pro-environmental behaviours at the individual level that contribute to reversing the decline of ecosystem services. Core element 4 of ecosystem services-based approaches as proposed in Chapter 2 appears so far too narrowly focused on decision-making at higher levels (i.e. management and policy levels), a situation that might lead to neglecting the importance of individual pro-environmental behaviour (water conservation included). The absence of information about psychological gains that water ecosystems provide might limit our ability to understand what drives people to conserve those ecosystems. Core element 4 of this book's definition of ecosystem services-based approaches includes assessment of social values and preferences in identifying the benefits that the public derives from ecosystem services. This chapter deals with a fundamental aspect of this core element 4: the psychological dimension involved in those benefits.

The conservation of water resources also depends on psychological predispositions (beliefs, environmental emotions, human capacities) that make people prone to caring for their integrity. These factors, which we have elsewhere called the '*psychological antecedents* of water conservation' (Corral-Verdugo *et al.* 2012a), have also been under-researched within ecosystem services-based approaches.

Consequently, this chapter explores some expected psychologically positive consequences that water ecosystem services and water ecosystem conservation might provide to individuals, and also analyses the likely psychological antecedents predisposing people to both value water ecosystem services and engage in

[1] By 'positive consequences' we mean the psychological benefits that water provides. Those benefits are individually experienced in the form of pleasure, aesthetic joy, spirituality, psychological wellbeing, intrinsic motivation, feelings of happiness, and psychological restoration, among others.

water conservation practices. A number of psychological factors are identified as components of water ecosystem services. Some of them lead people to perceive, enjoy, and appreciate the benefits of water ecosystems. Other psychological factors predispose individuals to pursue the goal of water (and other natural resource) conservation. In fact, the psychological benefits of water ecosystems also instigate water conservation practices. This set of psychological factors includes mental states, psychological inclinations, and behavioural capacities that are generated in the course of interactions between individuals and water. Since these predispositions emerge from past and present water–person interactions, they are conceived as historical-psychological factors (Vanderbeeken & Weber 2002). The psychological benefits that a water ecosystem provides include pleasure, aesthetic joy, psychological restoration, the promotion of spirituality, and psychological wellbeing, among others. In turn, those psychological benefits, and also environmental beliefs, positive and negative emotions, as well as pro-environmental skills or abilities, are psychological states, tendencies, or capacities that lead individuals to the protection of a water ecosystem service.

A general framework for the development of the psychological dimension of ecosystem services-based approaches is presented in Figure 18.1, showing that psychological factors interact with ecological, social, and economic aspects. In order to impact on people, ecosystem services must first be perceived (i.e. the individual has to be aware of them). The perceived (ecological, social, economic, as well as the psychological) benefits of water ecosystem services then influence the practice of water conservation behaviours (because the individual anticipates gains from a conserved water ecosystem), which subsequently produce

psychologically positive consequences, such as intrinsic motivation, happiness, wellbeing, and mental restoration. Additional psychological factors such as environmental beliefs and emotions also result from the exposure of people to water ecosystem services, and those factors influence water conservation along with environmental abilities.

18.2 PSYCHOLOGICAL BENEFITS OF WATER ECOSYSTEM SERVICES

Writing about the positive psychological consequences of access to water, Syme and Nancarrow (2008, p.22) state that 'a water benefit is not the quantity of water itself'. Instead, it is the subjective benefit that results from using water; for example, the pleasure that derives 'from drinking it in terms of refreshment, or the feeling of enjoyment of the amenity of a vase of flowers.' These positive psychological states are the main topics of this section.

18.2.1 Psychological wellbeing as a water ecosystem service

For some time, human wellbeing has been considered a positive benefit of a good ecosystem condition. In this vein, DeFries *et al.* (2005) conceive of human wellbeing as having a number of 'key' components: material needs satisfied, freedom and choice, health, good social relations, and personal security. Strangely, most of the positive factors identified in the classical definition of 'psychological wellbeing' (Ryff 1989) are not included in this list. In Ryff's classification, these factors include environmental mastery, purpose in life, self-acceptance, personal growth, autonomy, and positive relations (the only one considered in DeFries *et al.*'s (2005) list). The inclusion of the psychological dimension within the human wellbeing concept will help to understand the human benefits of water ecosystem services.

18.2.2 Hedonic psychological states, aesthetic joy, happiness, mental restoration, and access to water ecosystem services

Some studies explicitly suggest that people gain pleasure and aesthetic joy from observing aquatic scenes, and also from direct contact with water (Dutton 2003). According to Brady (2006), the aesthetic experience provided by contact with water (and the natural environment) promotes psychological wellbeing and mental restoration, yet she does not empirically test this assumption. Water or water features in urban areas have the potential to induce psychological restoration, as the study by Pradhan (2012) demonstrates. Water produces a calming effect in people who

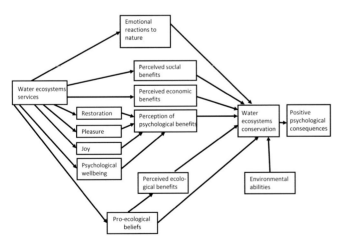

Figure 18.1 A representation of the water ecosystem services and water ecosystem conservation outcomes predicted by psychological predispositions (beliefs, emotions, abilities), perceived (economic, socio-ecological, and psychological) benefits, and the positive psychological consequences of water ecosystem conservation.

engage in behaviours such as resting, reading, conversing, eating, meditating, or contemplating when coming in contact with water.

18.2.3 Environmental beliefs and water ecosystem services

Environmental beliefs reflect the way people think about nature and their role in the natural world. In traditional societies, some of those beliefs have a religious or spiritual nature. For instance, Perez-Castro (2011) mentions the prayers of indigenous people in the Mexican region of *La Huasteca,* who ask mountains, caves, and water springs for rain.

Offersmans *et al.* (2011) offer a classification of water beliefs using three types of perspectives: Hierarchist, Egalitarian, and Individualist. The *Hierarchist* believes in a need for controlling water and nature; water is mainly seen as a threat to human safety so that recommended water policy options emerging from this perspective would include building dikes and channelling. *Egalitarians* ask for ecological recovery and natural development. They think that more space for nature, animal, plants, and water are needed, and believe that human control over water has gone too far. Thus, Egalitarians tend to be preservationist and recommend the creation of natural spaces free of human activities. *Individualists* do not see water as a threat, but as an opportunity in terms of economy, images, creativity, self-development, and recreation. They are in favour of an adaptive approach, involving high levels of trust in technology and in free enterprise. These three perspectives coexist in modern societies, and depending upon the circumstance, one of them might prevail over the others.

Current classifications of environmental beliefs also include ecocentric (pro-ecological), anthropocentric (utilitarian or pro-human), and interdependence beliefs. Corral-Verdugo *et al.* (2003) report that the general environmental beliefs of the *New Environmental Paradigm* (Dunlap *et al.* 2000) (i.e. the idea that the physical environment might be protected by humans and that people need nature in order to survive) have a positive effect on people's specific ecocentric beliefs regarding water, seeing it as a valuable, exhaustible resource that should be conserved. In turn, those pro-ecological beliefs directly promote water conservation. Martimportugués *et al.* (2007) report an effect of water pro-ecological beliefs on the water conservation behaviours of Spanish teenagers, while Fraijo *et al.* (2010) found that ecocentric water beliefs are significantly associated with the water conservation skills of Mexican children and, that together, they result in water conservation behaviours.

In recent reports, Corral-Verdugo *et al.* (2008), Carrus *et al.* (2010), and Hernández *et al.* (2012) found that the general beliefs of a *New Human Interdependence Paradigm* directly predict water conservation behaviours in respondents from France, Italy, Mexico, India, and Spain. This finding closely resembles the description of indigenous Mexican people, made by Perez-Castro (2011, p.142), as holders of beliefs in 'nature and society [as] united in a constant dialogue [in which] the principle of reciprocity prevails'. Such reciprocity implies that people depend on water, so they should protect this resource.

18.2.4 Environmental emotions from exposure to water ecosystem services

The perceived status of water resources might result in emotional states, which depend on whether or not the 'liquid' itself is available or potable. People experience negative emotional responses such as frustration, anguish, bother, anxiety, worry, and anger if a water shortage is present (Ennis-McMillan 2001; Coelho *et al.* 2004). Negative emotional responses also arise if a water contamination crisis is experienced (Woolloff 2009). Conversely, positive emotions (pleasure, subjective wellbeing, joy, etc.) could result from direct contact with clean water (Dutton 2003).

Kelley (2005) argues that positive emotions bring humans in contact with environmentally beneficial resources, such as water, so that the origin of positive emotions is closely related to the satisfaction that water and other environmental services offer to human needs.

18.3 POSITIVE PSYCHOLOGICAL ASPECTS LINKED TO WATER ECOSYSTEM CONSERVATION

Everybody agrees with the fact that access to fresh water produces positive psychological consequences or benefits to people. The question is whether or not this situation also applies to water conservation behaviours, which, at first sight, seems counter-intuitive. In fact, some researchers and theorists expect *negative* psychological consequences from the practice of pro-environmental behaviour. For instance, Lindenberg and Steg (2007) and Kaplan (2000) suggest that discomfort and sacrifice are more likely to occur than the experience of rewards as consequences of engaging in sustainable actions. However, a growing body of research indicates that individuals also experience psychological benefits from acting in a pro-environmental way (e.g. Bechtel & Corral-Verdugo 2010; Brown & Kasser 2005; Corral-Verdugo *et al.* 2011b; De Young 2000). They are also inspired to conserve water by intrinsic motives, positive emotions, and environmental beliefs and abilities. In addition, psychological wellbeing, mental restoration, and happiness are psychological states that lead people to water ecosystem conservation. We discuss each of these issues next.

18.3.1 Psychological wellbeing and water conservation

As far as we know, the only study that has empirically addressed the relationship between psychological wellbeing and the conservation of water is the one conducted by Corral-Verdugo et al. (2011b). In this study, a significant association between psychological wellbeing and a measure of pro-ecological behaviour (including water conservation) was found. This finding seems to indicate that environmental (and water) conservation behaviour is higher in individuals who report salient levels of psychological wellbeing. Yet, further studies are needed to firmly establish a significant link between the protection of water resources and psychological wellbeing.

18.3.2 Positive psychological states promoting water conservation

The importance of positive psychological states as consequences of water conservation has not been widely considered in psycho-environmental studies. As far as we know, no research has been published showing that water conservation produces pleasure, although there are indirect indicators suggesting that this is the case. For instance, Pelletier et al. (1998) found that individuals are more likely to engage in conservationist actions when they derive pleasure from them. Also, De Young (2000) found that people consider that such actions are worthy because they are pleasurable.

Although the evidence linking water conservation with happiness is incipient, some research establishes such a link. Bechtel and Corral-Verdugo (2010) and Corral-Verdugo et al. (2011a) studied responses provided by American and Mexican students to both an instrument assessing pro-environmental behaviour (water conservation included) and a scale measuring happiness. Their results show that levels of happiness were higher in those students reporting more engagement in conservationist activities.

The possibility that water conservation might result in psychological restoration is also discussed. Psychological restoration involves the recovery of mental resources that are lost due to stress or attentional fatigue. Attention, positive mood states, and mental health are among those resources. There is evidence suggesting that people can be motivated to act pro-environmentally by anticipating restorative effects of those actions (Hartig et al. 2007), and a recent study suggests that the practice of some pro-environmental behaviours (water conservation included) may produce restorative consequences (Corral-Verdugo et al. 2012b).

18.3.3 Intrinsic motives and self-regulated water-consumption behaviour

Water conservation, as an instance of sustainable behaviour, is conceived as deliberate, purposeful behaviour (Corral-Verdugo et al. 2012a). This means that a conservationist position is self-determined and aimed at achieving an individual's pro-environmental goals. A self-determined individual sets her or his own goals and is internally motivated to act (i.e. does not require external pressures to engage in any behaviour). Promoting self-determined (i.e. self-regulated and intrinsically motivated) behaviour constitutes one of the best ways to enhance the protection of water ecosystems. However, the literature on this approach is very limited.

Self-regulation is self-control directed towards a valued goal (Maddux 2009), and water conservation – as a goal – requires individuals to exercise control over their water consumption. Self-regulation has been found to be significantly linked to pro-environmental behaviours (Villacorta et al. 2003). Intrinsic motivation, in turn, is driven by interest or enjoyment in a task itself, and emerges from the individual without relying on external factors (pressures or reward). Intrinsic motivation is identified when a person 'feels good' about conserving the environment (De Young 2000), or when (s)he reports a sense of self-efficacy, satisfaction, or autonomy (Ryan & Deci 2000). In other words, self-regulation instigates environmental conservation, which results in intrinsic motives that maintain those conservationist behaviours.

Kessler (2007), in describing the soil and water conservation practices of Bolivian farmers, highlights the value of intrinsic motivations as a solid basis for the promotion of those pro-environmental practices. More studies documenting the self-regulatory nature of water conservation and its intrinsic psychological benefits are needed, especially in the context of water ecosystem conservation.

18.3.4 Environmental beliefs and emotions associated to water conservation

There are few empirical studies addressing the impact that environmental beliefs have on water ecosystem services conservation (Spash 2000; Spash et al. 2009). The former reveals that people are willing to pay for re-creating a wetland in order to protect endangered species, while the latter shows that improving biodiversity is a good reason for conserving a water ecosystem.

Both negative and positive emotional states might induce water conservation behaviour. In the first case, people are motivated to conserve water by a desire to avoid unpleasant emotional consequences of water scarcity or contamination (Ennis-McMillan 2001; Woolloff 2009). In the second case, individuals anticipate the emotional benefits that an available source of freshwater might provide (Corral-Verdugo et al. 2012a) and this anticipation leads them towards the goal of water conservation.

The role that negative (shame, guilt, hypocrisy) and positive (affinity towards nature and biodiversity, happiness) emotions play in inducing such conservation is mentioned in the literature. Dickerson et al. (1992) used cognitive dissonance to induce a feeling of hypocrisy (and guilt) in participants of their study. Subjects in an experimental group were asked extreme questions ('Do you always turn off the water while soaping up or shampooing?') to remind them that they occasionally over-consume water. Then, they were asked to print their names on a flyer that read 'Save water, turn showers off while soaping or shampooing'. They found that this guilt-inducing procedure was effective in reducing water consumption.

In the case of positive emotions, Corral-Verdugo *et al.* (2009) found that affinity towards diversity predicts a person's involvement in pro-environmental behaviours, water conservation included. Affinity towards diversity implies appreciation and liking for the variety of living things and for the diversity of socio-cultural manifestations that individuals face in their daily lives. Thus, the more attracted a person feels towards 'bio' and 'socio' diversity, the more (s)he is prone to protect water and other natural resources.

Happiness has also been related to water conservation, either as an instigator or as a consequence (or both) of such pro-environmental behaviour. Studies conducted by Bechtel and Corral-Verdugo (2010), and Corral-Verdugo *et al.* (2011a) suggest that happy people are also individuals committed to act pro-environmentally, including the practice of water-conservation behaviours. The authors conclude that the experience of positive feelings might induce sustainable actions, but also that those feelings are potential consequences of being a pro-environmental individual.

Although both types of emotions (positive and negative) are effective in predisposing people towards water conservation, the advantages of positive psychological states – over negative ones – are stressed, since the former result not only in a protected environment but also in psychological wellbeing for people.

18.3.5 Pro-environmental abilities

Water conservation ability is the capacity to effectively respond to water conservation requirements. Therefore, abilities and related human capacities (knowledge, competence) are required in the effort to produce water-sustaining behaviours.

The pertinent literature shows that pro-environmental abilities are important direct determinants of water conservation, at least in residential and educational scenarios. Corral-Verdugo (2002) reports that water-saving ability (indicated by water problem-solving situations, for example the ability to fix water leaks) significantly influences water conservation, while Bustos (2004) found that freshwater conservation in personal cleaning and meal preparation results, to a large extent, from conservationist skills (for instance, being able to cook without wasting water). Fraijo *et al.* (2010) demonstrated that children with higher levels of that ability exhibited reduced consumption of water. This study also found that the more versatile the set of abilities was, the more effective the children's conservation efforts proved to be. Those authors consider that the ultimate goal of environmental education is the establishment of an environmentally able citizenry to deal with water problems. Nevertheless, and despite pro-environmental ability being one of the fundamental antecedents of water conservation, there have been no studies investigating skills for the conservation of water ecosystems and their potential for delivering ecosystem services.

18.4 CONCLUSIONS

This chapter has presented evidence supporting the idea that water ecosystem services provide positive psychological consequences to people, in addition to the economic and social benefits already mentioned by the pertinent literature. Our review also shows that conservation of water ecosystems is motivated by psychological antecedents and maintained by subjective (besides the tangible) benefits people obtain when they protect water sources. Clearly, these psychological aspects refer to the value that individuals place on water ecosystem services and also their attitudes towards water conservation. The concept of water ecosystem services is currently based on the idea that those ecosystems provide benefits to people, especially in the form of positive economic, ecological, and social consequences. A number of psychological benefits should be added to this list, including hedonic psychological states, aesthetic joy, happiness, and mental restoration, which are fundamental for people's wellbeing and functioning.

Aspects related to ecosystem services and human wellbeing have been tackled from the perspective of the cultural and social processes by which humans and the non-human interact, through the study of the so-called *cultural ecosystem services* (see Church *et al.* in this book). Those services provide benefits in the form of cultural diversity, spiritual and religious values, knowledge systems, educational values, inspiration, aesthetic values, social relations, sense of place, cultural heritage values, recreation, and eco-tourism (Millennium Ecosystem Assessment 2005), and all of these pertain to water ecosystem services. Cultural ecosystem services are related to the psychological components of water ecosystem services because both affect the wellbeing of people. Yet, the psychological components are *specific* in their impact on *individual* wellbeing, while the cultural ecosystem services have a *collective* influence. In addition, the psychological perspective studies positive emotions, mental restoration, personal wellbeing, happiness, joy, and other mental states that result from water system services, which are not, in our view, sufficiently covered in the category of cultural ecosystem services.

Why is the study of the psychological components of water ecosystem services important? The answer to this question has at least two facets: (1) people, as other living organisms, behave in a way that is motivated by the psychological consequences of their actions. If those consequences are positive (i.e. reinforcing, pleasant) they will likely engage in them again; otherwise, if they receive unpleasant consequences they will likely avoid further engagement in those actions. Thus, the study of the psychological benefits that water provides will be helpful in understanding what leads people to conserve water ecosystems. (2) the *perception* or *anticipation* of the psychological benefits of water ecosystem services seems also to function as an instigator of water conservation (i.e. if someone expects to feel good by practising a water conservation behaviour, (s)he will be motivated to act pro-environmentally). This also applies to the perception/anticipation of the economic consequences of water ecosystem services because monetary or material benefits ultimately represent psychological gains for individuals. Perceived social and ecological benefits provided by water ecosystem services are also determined by psychological factors (favourable attitudes, altruistic and ecocentric orientations) so that a person is only able to perceive those benefits if (s)he is pro-environmentally or pro-socially oriented. This means that in order to anticipate an ecological or social benefit of water ecosystem services, the individual has to be interested in the

protection of the environment. Therefore, the study and promotion of conditions that favour the development of beliefs, attitudes, emotions, and abilities favourable to conservation are necessary.

This framework, and others to be proposed, could be tested in prospective studies on the psychological correlates of water ecosystems conservation. Interventional strategies for the promotion of water ecosystems conservation systems, from a psychological perspective, should be implemented and evaluated.

In using those frameworks, the exploration of positive psychological consequences of water ecosystems conservation such as feelings of autonomy, self-efficacy, and self-regulation is crucial, as well as the study of perceptual factors associated with the benefits (ecological, social, economic, psychological) of water ecosystem services. The importance of these psychological factors linked to water ecosystems protection has been stressed in this chapter since they could motivate people to self-determine their protectionist behaviour, enhancing their pro-environmental ability, and maintaining their conservationist effort.

Box 18.1 Key messages

- The debate on ecosystem services has so far focused on decision-making, social gains and on monetary valuation of benefits.
- But people also obtain psychological benefits from water ecosystems (in addition to the economic and social benefits), which are critical to the notion of wellbeing underpinning ecosystem services-based approaches.
- People also develop environmental beliefs and environmental emotions when coming in contact with water ecosystems.
- The psychological benefits (pleasure, spirituality, happiness, mental restoration, wellbeing) make people value water ecosystem services.
- People who perceive the psychological benefits of water ecosystems tend to engage in water conservation practices.
- Environmental beliefs, emotions, and abilities also predict water ecosystems conservation.
- As part of efforts to reverse the decline of ecosystem services, the promotion of water conservation behaviours is necessary.
- An ecosystem services-based approach that acknowledges the psychological dimension of ecosystem services can help promote conservation of water ecosystems.

References

Bechtel, R. B. & Corral-Verdugo, V. (2010). Happiness and sustainable behavior. In: V. Corral-Verdugo, C. García, & M. Frías (eds), *Psychological Approaches to Sustainability*. Nova Science Publishers, New York.
Brady, E. (2006). Aesthetics in practice: valuing the natural world. *Environmental Values* 15, 277–291.
Brown, K. W. & Kasser, T. (2005). Are psychological and ecological wellbeing compatible? The role of values, mindfulness, and lifestyle. *Social Indicators Research* 74, 349–368.
Bustos, M. (2004). *Modelo de conducta proambiental para el estudio de la conservación del agua potable*. Unpublished doctoral dissertation, Facultad de Psicología, UNAM.
Carrus, G., Bonnes, M., Corral, V., Moser, G., & Sinha, J. (2010). Social-psychological and contextual predictors of sustainable water consumption. In: V. Corral, C. García, & M. Frías (eds), *Psychological Approaches to Sustainability*. Nova Science Publishers, New York.
Coelho, A., Adair, J., & Mocellin, J. (2004). Psychological responses to drought in northern Brazil. *Interamerican Journal of Psychology* 38, 95–103.
Corral-Verdugo, V. (2002). A structural model of pro-environmental competency. *Environment & Behavior* 34, 531–549.
Corral-Verdugo, V., Bechtel, R., & Fraijo, B. (2003). Environmental beliefs and water conservation: an empirical study. *Journal of Environmental Psychology* 23, 247–257.
Corral-Verdugo, V., Carrus, G., Bonnes, M., Moser, G., & Sinha, J. (2008). Environmental beliefs and endorsement of Sustainable Development principles in water conservation: towards a *New Human Interdependence Paradigm* scale. *Environment & Behavior* 40, 703–725.
Corral-Verdugo, V., Bonnes, M., Tapia, C., Fraijo, B., Frías, M., & Carrus, G. (2009). Correlates of pro-sustainability orientation: the affinity towards diversity. *Journal of Environmental Psychology* 29, 34–43.
Corral-Verdugo, V., Mireles, J., Tapia, C., & Fraijo, B. (2011a). Happiness as a correlate of sustainable behavior: a study of proecological, frugal, equitable and altruistic actions that promote subjective wellbeing. *Human Ecology Review* 18, 95–104.
Corral-Verdugo, V., Montiel, M., Sotomayor, M., Frías, M., Tapia, C., & Fraijo, B. (2011b). Psychological wellbeing as correlate of sustainable behaviors. *International Journal of Hispanic psychology* 4, 31–44.
Corral-Verdugo, V., Frías, M., Tapia, C., & Fraijo, B. (2012a). Protecting natural resources: psychological and contextual determinants of freshwater conservation. In: S. Clayton (ed.), *Handbook of Environmental and Conservation Psychology*. Oxford University Press, Oxford.

Corral-Verdugo, V., Tapia, C., García, F., Varela, C., Cuen, A., & Barrón, M. (2012b). Validation of a scale assessing psychological restoration associated with sustainable behaviours. *Psyecology* **3**, 87–100.

DeFries, S., Pagiola, S., Adamowicz, W., *et al.* (2005). *Analytical Approaches for Assessing Ecosystem Condition and Human Well-being.* Vol. 1, *Current State and Trends.* Island Press, Washington, DC.

De Young, R. (2000). Expanding and evaluating motives for environmentally responsible behavior. *Journal of Social Issues* **56**, 509–526.

Dickerson, C. Thibodeau, R., Aronson, E., & Miller, D. (1992). Using cognitive dissonance to encourage water conservation. *Journal of Applied Social Psychology* **22**, 841–854.

Dunlap, R., Van Liere, K., Mertig, A., & Jones, R. E. (2000). Measuring endorsement of the New Ecological Paradigm: a revised NEP scale. *Journal of Social Issues*, **56**, 425–442.

Dutton, D. (2003). Evolutionary aesthetics. In: J. Levinson (ed.), *The Oxford Handbook of Aesthetics.* Oxford University Press, New York and Oxford.

Ennis-McMillan, M. (2001). Suffering from water: social origins of body distress in a Mexican community. *Medical Anthropology Quarterly* **15**, 368–390.

Fisher, B., Bateman, I., & Turner, R. (2011). Valuing ecosystem services: benefits, values, space, and time. Ecosystem services economics (ESE) Working Paper Series, UNEP.

Fraijo, B., Corral-Verdugo, V., Tapia, C., & González, D. (2010). Promoting pro-environmental competency. In: V. Corral, C. García, & M. Frías (eds), *Psychological Approaches to Sustainability.* Nova Science Publishers, New York.

Hartig, T., Kaiser, F., & Strumse, E. (2007). Psychological restoration in nature as a source of motivation for ecological behavior. *Environmental Conservation* **34**, 291–299.

Hernández, B., Suárez, E., Corral, V., & Hess, S. (2012). The relationship between social and environmental interdependence as an explanation of proenvironmental behavior. *Human Ecology Review* **19**, 1–9.

Kaplan, S. (2000). Human nature and environmentally responsible behavior. *Journal of Social Issues* **56**, 491–508.

Kelley, A. E. (2005). Neurochemical networks encoding emotion and motivation. In: J. M. Fellous & M. A. Arbib (eds), *Who Needs Emotions.* Oxford University Press, Oxford.

Kessler, C. A. (2007). Motivating farmers for soil and water conservation: a promising strategy from the Bolivian mountain valleys. *Land Use Policy* **24**, 118–128.

Lindenberg, S. & Steg, L. (2007). Normative, gain and hedonic goal frames guiding environmental behavior. *Journal of Social Issues* **63**, 117–137.

Maddux, J. (2009). Self-regulation. In: S. J. López (ed.), *The Encyclopedia of Positive Psychology.* Wiley-Blackwell, Chichester.

Martinportugués, C., Canto, J., & Hombrados, M. I. (2007). Habilidades pro-ambientales en la separación y depósito de residuos sólidos urbanos. *Medio Ambiente y Comportamiento Humano* **8**, 71–92.

Millennium Ecosystem Assessment (2005). *Current State & Trends Assessment.* Island Press, Washington, DC.

Offersmans, A., Haasnoot, M., & Valkering, P. (2011). A method to explore social response for sustainable water management strategies under changing conditions. *Sustainable Development* **19**, 312–324.

Pelletier, L. C., Tuson, K. M., Green-Demers, I., & Noels, K. (1998). Why are we doing things for the environment? The motivation toward the environment scale (MTES). *Journal of Applied Social Psychology* **25**, 437–468.

Perez-Castro, A. B. (2011). Actores sociales y su relación con el agua. *Revista Líder* **19**, 133–144.

Pradhan, P. (2012). The role of water as a restorative component in small urban places. Unpublished Master's Thesis in Urban Landscape Dynamics, Swedish University of Agricultural Sciences.

Ryan, R. M. & Deci, E. (2000). Intrinsic and extrinsic motivations: classic definitions and new directions. *Contemporary Educational Psychology* **25**, 54–67.

Ryff, C. (1989). Happiness is everything, or is it? Explorations on the meaning of psychological well-being. *Journal of Personality and Social Psychology* **57**, 1069–1081.

Spash, C. (2000). Ecosystems, contingent valuation, and ethics: the case of wetland recreation. *Ecological Economics* **34**, 195–215.

Spash, C., Urama, K., Burton, R., Kenyon, W., Shanon, P., & Hill, G. (2009). Motives behind willingness to pay for improving biodiversity in a water ecosystem: economics, ethics and social psychology. *Ecological Economics* **68**, 955–964.

Syme, G. & Nancarrow, B. (2008). Justice and the allocation of benefits from water. *Social Alternatives* **27**, 21–25.

Syme, G., Porter, N., Goeft, U., & Kington, E. (2008). Integrating social well being into assessments of water policy: meeting the challenge for decision makers. *Water Policy* **10**, 323–343.

Vanderbeeken, R. & Weber, E. (2002). Dispositional explanations of behavior. *Behavior and Philosophy* **30**, 43–59.

Villacorta, M., Koestner, R., & Lekes, N. (2003). Further validation of the Motivation toward the Environment Scale. *Environment & Behavior* **35**, 486–505.

Woolloff, D. (2009). O.K. We've got a problem, so who do we tell? Inter-agency communications – a water company view. In: J. Gray & C. Thompson (eds), *Water Contamination Emergencies, Collective Responsibility.* RSC Publishing, Cambridge.

19 The interface between human rights and ecosystem services

Stephen J. Turner

19.1 INTRODUCTION

Ecosystem services and human rights are usually discussed separately. The development of theory and practice in the field of ecosystem services has generally been the domain of ecologists, economists, and policy makers; lawyers have also engaged in the dialogue and development, but to a lesser extent (Mertens *et al.* 2012, p.31). The purpose of this chapter is to play a part in the development of a deepening dialogue between ecologists, economists, and lawyers by discussing the integral links between ecosystem services and human rights. It draws from references relating to water ecosystem services and the use of human rights to protect them, but these examples provide an illustration of a wider relationship that exists between human rights and ecosystem services generally.

The chapter is divided into two main sections. The first section considers the conceptual and practical relationships between ecosystem services and human rights. It therefore considers the human rights that can be impacted through the loss of ecosystem services and also the practical relationships between certain rights and ecosystem services. The requirement of brevity does not allow an exhaustive examination of all of the human rights that are affected, but those that are selected represent some of the core issues. Therefore, it includes the rights to life, health, water, food, and property. In addition, it considers the role of procedural environmental rights in enabling citizens to become involved in decision-making processes related to the environment and also the role of those rights designed to set specific environmental standards (substantive environmental rights). The crucial role that work within the United Nations has had in developing the understanding of the links between human rights, the environment, and ecosystem services is integrated within the discussion.

The second section discusses possible ways that human rights can be further developed to operationalize ecosystem services-based approaches. While acknowledging the leadership that has taken place in certain regions of the world through the development of innovative systems for the protection of the services that ecosystems provide, it highlights the overall absence within the global legal architecture of an approach that provides clear legal obligations for decision-makers to ensure that ecosystems are protected.

Therefore, it considers contemporary research and a draft treaty, which could lay the groundwork for the development of a rights-based framework of legal duties for both state and non-state actors that would in turn create robust mechanisms for the operationalization of the protection of ecosystem services (Turner 2014). It goes on to discuss the components and institutional arrangements that would be required to enable such a framework to operate on national and international levels in a manner that would lead to 'no net loss' or 'ecological impact neutrality' (Salzman 2005, p.908; Achterman and Mauger 2010, p.306; McGillivray 2012, p.417–418).

The chapter concludes by summarizing the interface between ecosystem services-based approaches and human rights-based approaches to environmental protection in the context of the proposed core elements posited in Chapter 2, and it provides key policy messages relating to the ways that the two fields can usefully complement and support each other in the future.

19.2 THE EXISTING RELATIONSHIP BETWEEN HUMAN RIGHTS AND ECOSYSTEM SERVICES

The relationship between human rights and the environment is a strong and well-documented one. It therefore follows that the relationship between human rights and ecosystem services is also a strong one, although it is much less well analysed or documented. Human rights within the modern understanding have developed since the end of the Second World War (Anton & Shelton 2011, p.174); contemporary concerns for the environment have developed since the 1960s and 1970s (Bell *et al.* 2013, p.22) and since that time there has been much written relating to human rights and the environment (Turner 2009, p.46). As the study and analysis of ecosystem services only became a mainstream area of scholarship and practice from the late 1990s (Costanza *et al.* 1997; Daily *et al.* 1997) it is a natural consequence that research into the relationship between human rights and ecosystem services *per se* is not so well established.

There have been numerous developments within the United Nations that have acknowledged and sought to expand the relationship between human rights and the environment (Turner 2009; Anton & Shelton 2011). By way of example, in 2012 the Human Rights Council appointed an expert for a three-year period to consult, study, and make recommendations pertaining to the enforcement of a 'safe, clean and healthy and sustainable environment' (UN Doc. A/HRC/Res/19/10).

The links between human rights and the environment have led to the development of specific environmental rights. Over 100 national constitutions now contain provisions related to the protection of the environment (Boyd 2012, p.47). Additionally, there are two regional human rights treaties (with complaint mechanisms) that contain specific provisions relating to environmental protection (The African Charter of Human and Peoples' Rights 1981; and The Protocol of San Salvador 1989). The Arab Charter on Human Rights also contains an environmental provision although that system does not yet have a mechanism for hearing specific complaints.

It is from this background that this section discusses the specific links between human rights and ecosystem services. Therefore, it considers the relationship between a number of human rights and ecosystem services. It will do this by reference to the categorization of ecosystem system services detailed in the Millennium Ecosystem Assessment, i.e. provisioning services, regulating services, cultural services and supporting services (Millennium Ecosystem Assessment 2005, p.1–2), although it is recognized that those categorizations have been subject to debate and refinement within the literature in recent years (Lele 2009, p.149; Haines-Young & Potschin 2010; Ojea et al. 2012).

19.2.1 The right to life

The right to life is one of the best-established human rights and has a strong relationship with the environment and ecosystem services. A contemporary understanding of the right to life can be derived from the United Nations Human Rights Council General Comment 6 (UN Doc. HRI/GEN/1/Rev.7, 1982), which states inter alia:

> 5. . . [t]he expression 'inherent right to life' cannot properly be understood in a restrictive manner, and the protection of this right requires that States adopt positive measures. In this connection, the Committee considers that it would be desirable for States parties to take all possible measures to reduce infant mortality and to increase life expectancy, especially in adopting measures to eliminate malnutrition and epidemics.

A 'General Comment' is not binding on states but is issued by a treaty body of the United Nations to provide a contemporary understanding of the meaning of a specific human right, and as such is

often widely accepted as providing a helpful interpretation for the purposes of complaints procedures and policy development (Anton & Shelton 2011, pp.436–463). Therefore, if the term 'increase life expectancy', for example, is to be read to include obligations by states to take positive measures relating to their management of the environment, then it could be asserted that governments have an obligation to manage ecosystem services accordingly.

By way of illustration, it is possible that where deforestation exacerbates flooding in times of heavy rainfall, a human rights obligation may come into play owing to the impact on people's right to life. For example, in Honduras deforestation of roughly 50% of the tree cover led to a major loss of vegetation cover which worsened the effects of Hurricane Mitch in 1998. (The deforestation had been made worse through fire, which had damaged an additional 11 000 km^2 of forest.) The overall environmental degradation exacerbated the resultant flooding which ultimately led to the deaths of 11 000 people (Botkin & Keller 2003).

From an ecosystem services perspective, the loss of forests meant the undermining of the valuable regulating services that they provide in terms of water absorption and storage in times of heavy rainfall. However, from a human rights perspective it would be possible to regard the decision-making by the national authorities, in permitting the levels of deforestation concerned, as indirectly impacting upon the right to life of those people who lost their lives.

In terms of practical case examples in the courts, there have been instances where the failure of states to take positive steps to protect specific aspects of the environment have led to determinations by regional human rights tribunals that violations of the right to life have occurred (Anton & Shelton 2011, pp.136, 236–263, 310; Turner 2014, p.18). However, there are few examples related directly to the maintenance or protection of ecosystem services. Having said this, one example is that of Social and Economic Rights Action Center and the Center for Economic and Social Rights v. Nigeria (2001) (SERAC v. Nigeria 2001). In this case extensive degradation to the provisioning services of the surface and groundwaters of the Ogoni region in Nigeria caused by the oil extraction industry meant that, among other severe problems, people were unable to access clean drinking water. The Commission stated that the pollution levels were 'humanly unacceptable' (SERAC v. Nigeria 2001: para. 67) and confirmed that the right to life (and numerous other rights) had been violated.

National constitutions often contain rights that mirror the human rights that have developed at regional and international levels. In certain jurisdictions, such as those of India, Pakistan, and some in South America, the courts have also held that the degradation of surface and groundwaters can amount to a violation of the right to life (Fabra & Arnal 2002; Razzaque 2002). While the language that is used by the courts and tribunals may not refer directly to 'ecosystem services' as such, it can be concluded that such decisions are based on the understanding that if specific provisioning services such as water supplies are sufficiently degraded, then the right to life can potentially be violated.

It must be noted, however, that where human rights litigation is used in such cases, it is usually as a last resort when all other avenues for redress have failed. It should also be noted that in most developed countries, it is not necessary to resort to human rights law to ensure that clean water supplies are provided, as states are obliged to do this under other legal provisions.

19.2.2 The right to health

In a similar way to that in which the meaning of the right to life has been broadened by the United Nations, General Comment 14 (UN Doc. E/C. 12/2000/4) provides helpful guidance on the interpretation that should be given to the right to attain the highest attainable standard of health. It states, *inter alia*, that the right:

> 4....embraces a wide range of socio-economic factors that promote conditions in which people can lead a healthy life, and extends to the underlying determinants of health, such as food and nutrition, housing, access to safe and potable water and adequate sanitation, safe and healthy working conditions, and a healthy environment.

There are examples where complaints relating to the right to health have been brought, in which the cause was degradation to water supplies or other aspects of the environment (Anton & Shelton 2011, pp.263, 438–439; Turner 2009, pp.18, 20). Indeed, in SERAC *v.* Nigeria 2001 (see above) the Commission confirmed that a violation of the right to health had occurred.

Therefore, the significance for ecosystem services-based approaches lies in the obligation that this right potentially creates in terms of a state's decision-making and policy making related to various different aspects of the environment. For example, non-point source pollution affecting water sources can have a negative effect upon the associated provisioning services. As such, policy making relating to the eutrophication of watersheds through the intensification of agriculture could potentially activate obligations, within the meaning of the right to health, depending on the severity of the impacts concerned. However, in practice, as was noted in relation to the right to life in Section 19.2.1, at this time although the link between ecosystem services and human rights can be made conceptually, relevant substantive human rights tend only to become activated as a last resort in cases of extreme pollution at a local level.

19.2.3 The right to water

Over the last decade the right to water has developed significantly. Although there is no specific treaty relating to it, in 2010 the United Nations General Assembly formally recognized

that access to clean water and sanitation is a human right (UNGA Res. 64/292, 2010). In 2011 the Human Rights Council also made a resolution requiring states to finance the sustainable delivery of water and sanitation services (UNGA Res. 18/1, 2011).

Additionally, General Comment 15 of the International Covenant on Economic Social and Cultural Rights (UN Doc. E/C. 12/2002/11, 2003) states, *inter alia*, that:

> 6 ... [w]ater is required for a range of different purposes. For instance, water is necessary to produce food (right to adequate food) and ensure environmental hygiene (right to health). Water is essential for securing livelihoods (right to gain a living by work) and enjoying certain cultural practices (right to take part in cultural life). Nevertheless, priority in the allocation of water must be given to the right to water for personal and domestic uses. Priority should also be given to the water resources required to prevent starvation and disease, as well as water required to meet the core obligations of each of the covenant rights.

Therefore, seen through this lens, the right to water not only has a direct relationship with provisioning services but also an indirect relationship with supporting and cultural services (the latter being discussed in depth by Church *et al.* in Chapter 17). It can therefore be said that where there are instances in which policy makers have taken economic and financial initiatives to protect ecosystems, which result in clean water supplies (ten Brink *et al.* 2013, p.9), they are taking positive action that satisfies a state's responsibilities relating to the right to water. Conversely, of course, decision-making that can have negative impacts upon freshwater supplies can violate the right to water of those individuals and communities that suffer as a result (Gathii & Hirokawa 2012, p.21). While the conceptual understanding of the right to water and its relationship with ecosystem services is growing, the development of practical mechanisms to enforce the obligations that it creates represents a growing challenge owing to the multiple demands on water usage.

19.2.4 The right to food

The concept of the right to food has also been further developed by the United Nations. The 1966 International Covenant on Economic Social and Cultural Rights states that everyone has the right to 'an adequate standard of living for himself and his family, including adequate food' (Art. 11(1)).

General Comment 12 (UN Doc. E/C.12/1999/5) states, *inter alia*, that the right is:

> 6. ... not to be interpreted in a narrow restrictive sense which equates it with a minimum package of calories,

proteins or other specific nutrients. The right to adequate food will have to be realized progressively. However, States have a core obligation to take the necessary action to mitigate and alleviate hunger.

Therefore, it can be noted that government decision-making relating to the regulating, provisioning, or supporting ecosystem services that have an effect on the provision of food can be regarded as impacting this right.

As with the right to water, it is a challenge to develop mechanisms that ensure that this obligation is realized through policy and practice.

19.2.5   The right to property

The interface between the environment and the right to property can lead to differing, and sometimes contentious conclusions relating to the approaches that should be taken for the protection of ecosystem services.

Property rights can be exercised in ways which are harmful to ecosystem services owing to the fact that under the laws of many countries, landowners have a high degree of autonomy in the way that they use their land (Davis 2010, p.342). Examples of this can be seen in the use of agricultural techniques by landowners, which sometimes lead to the pollution of surface and groundwaters. Such instances have on occasion led to the development of Payment for Ecosystem Services schemes in which landowners are paid to protect the ecosystem services that are affected, by managing their land in specific ways (Salzman 2005, pp.878ff).

The concept of paying landowners to protect the ecosystem services which originate from their land is controversial and raises a number of issues, which include: whether the wider community should have to pay landowners for the protection of vital ecosystem services (Lugo 2008, p.243); the overall costs of protecting ecosystem services in this way on a global scale (Costanza et al. 1997, p.256); and whether such a system could lead to further 'land grabbing' or 'green grabbing' especially in developing countries (McAfee 2012, p.124). However, others would argue that providing payments to those with property title over lands for the ecosystem services provided could empower and protect certain groups, especially poor and indigenous communities (Ferraro 2001, p.995; TEEB 2010, p.154; Tongson & Balasinorwala 2010), and provide a practical basis for the protection of ecosystem services (Salzman 2005; Davis 2010, p.339).

Additionally, it can be noted that there has been a variety of approaches to environmental protection that restrict the freedom of landowners in the use of their land. However, examples of this are seen in numerous jurisdictions regardless of whether an ecosystem services-based approach has been adopted (Bates 2001; Eves & Blake 2013, p.4).

19.2.6   Substantive environmental rights

Substantive environmental rights are those that entitle the right-holders to a specific standard to which the environment should be maintained. Although a substantive right to a 'clean' or 'healthy' environment on a global level can probably, at best, only be regarded as an emerging human right, many national constitutions do contain such provisions, as do certain regional human rights treaties. In practice there are only a limited number of cases in the regional human rights tribunals related directly to such rights, but in certain national jurisdictions substantive constitutional environmental rights have been used more extensively (Turner 2014, pp.25–28).

While currently these types of cases tend to relate to extreme degradation at local levels that affects individuals or communities, the development of these rights do in theory have the potential to relate to all types of ecosystem services, whether they be provisioning, regulating, cultural, or supporting services.

19.2.7   Procedural environmental rights

Procedural environmental rights are those that entitle citizens to access information, participate in decision-making, and to access justice relating to environment matters. Such rights are important to individuals and communities who may be affected by the potential degradation of ecosystem services through new and existing industrial developments. They are found in a variety of areas of law, but particularly manifest themselves within planning law and the law relating to environmental impact assessments and strategic environmental assessments. With more focus now being placed on ecosystem services within environmental impact assessments and strategic environmental assessments (Landsberg et al. 2011; Geneletti 2013), there is greater potential for procedural rights to play an important role in ecosystem services-based approaches (Blanco & Razzaque 2009).

Procedural environmental rights are well established and recognized worldwide. Having said that, the rights they confer ultimately provide no guarantee that the environment or ecosystem services they relate to will be protected in individual cases.

19.3   DEVELOPMENTS WITHIN THE FIELD OF ENVIRONMENTAL RIGHTS WITH THE POTENTIAL TO OPERATIONALIZE SUSTAINABLE ECOSYSTEM SERVICES-BASED APPROACHES

19.3.1   Limitations of the existing systems

Research analysing the field of ecosystem services-based approaches highlights the lack of a comprehensive legal regime for the protection, maintenance, and improvement of ecosystem services on national and international levels (Salzman *et al.* 2001, p.311; Bishop 2010). There are ongoing debates relating to the use of Payments for Ecosystem Services, how ecosystem services can or should be valued, and who should be responsible for protecting and restoring them. One of the core challenges is how to develop the legal obligations that would provide the responsibilities for ecosystem services to be maintained, protected, or improved.

19.3.2 Developments with the potential to address existing limitations

This section will focus on specific research that has developed a draft rights-based governance framework that could create a more comprehensive system of legal obligations for the protection of the environment, including ecosystems and the services they provide (Turner 2009; 2014). The framework is designed to lead to the direct application of human rights-based obligations beyond state actors, who normally bear the primary obligation to protect and fulfil human rights, to other actors such as international organizations and corporations (Turner 2014, p.70).

This draft framework of governance, the draft Global Environmental Right (Turner 2014, p.73), would create a duty for all decision-makers to avoid causing degradation to the environment unless certain provisos applied, and specific offsets or insurance were purchased. It would therefore have the effect of creating a market for offsets, thus producing an ongoing flow of public and private finance into projects for the protection and enhancement of ecosystem services. In this way it would provide a mechanism to potentially achieve net 'ecological impact neutrality' (McGillivray 2012, p.418; Turner 2014, p.85).

The framework envisages processes of assessment that would assess all aspects of the environment and as such would include ecosystem services (Art. 15). This would be a logical extension of the inclusion of ecosystem services within environmental impact assessments, which already takes place to a certain extent (Landsberg *et al.* 2011; Geneletti 2013). The assessment would detail any aspects of the environment that would or could be negatively impacted through the activity/policy concerned. The assessment would detail the measures that would be necessary to achieve 'no net loss' or 'ecological impact neutrality'. This would create a legal obligation upon the decision-maker to purchase direct environmental compensatory offsets (from a registered supplier) for those activities that *would* cause degradation to the environment and also environmental insurance (from a registered supplier) for any aspects of the environment that *could* cause degradation to the environment.

The framework provides the potential for the registered suppliers of environmental insurance to be required to invest a proportion of the premiums into ongoing environmental and ecosystem enhancement projects, credits of which could be surrendered in the event of a claim. Therefore, a mechanism would exist for the potential creation of Payments for Ecosystem Services schemes that could assist the most vulnerable communities and the most vulnerable ecosystems. Additionally, the registered suppliers of environmental insurance could be required to provide emergency relief services in the event of environmental disasters.

In terms of the institutional arrangements that would be required to ensure the fairness and smooth running of such a system, the draft Global Environmental Right contains provisions to ensure that the purchasers of the environmental offsets or the environmental insurance provide 'environmental accounts' which would be submitted and made public alongside their financial accounts (Arts 4, 5, 7, 8, 10, and 11). In addition, that information would need to be submitted to a World Environmental Organization that would have the ultimate authority over the quality of the offsets or insurance (Art. 13). Similarly, the registered suppliers of environmental offsets or insurance would be required to account to the World Environmental Organization for the offsets that they had provided; the oversight and authority over their registration would also be a matter for the World Environmental Organization (Art. 13). These provisions are consistent with calls for an international framework within which both states and businesses would be required to submit accounts relating to their use of natural capital and ecosystem services (Barbier 2013, p.133).

While at this time the draft Global Environmental Right is aspirational in nature, it is included here to contribute to the debate relating to the potential governance options relating to the protection of ecosystem services. It draws from the scholarship of a wide range of academics who have sought or suggested reforms to the existing systems of international environmental governance. It can, of course, be argued that the extensive nature of the reforms suggested within the draft Global Environmental Right would take a long time to implement and that piecemeal developments are far easier to achieve, but equally it can be argued that the international community should be pulling together existing expertise to iterate a long-term, coherent strategy that will lead to the accomplishment of sustainable development goals.

19.4   CONCLUSION

This chapter has shown that there are issues that can be seen both as human rights issues and ecosystem services issues simultaneously. It has also helped to illustrate the differing approaches within the fields of law, economics, and ecology, each of which has its own language and range of tools for analysis.

While the interface between human rights-based approaches and ecosystem services-based approaches to environmental protection has not been routinely articulated within the literature, it is clear that there are practical relationships between the two approaches. First, there are the ways that human rights can be used to strengthen the advocacy for ecosystem services-based approaches in decision and policy making (this would correspond with core element 4 put forward in Chapter 2). Second, there are the ways that ecosystem services-based arguments can strengthen claims relating to the impacts upon specific human rights in decision-making processes. As such, further analysis of these dynamics and the full realization of the potential of these relationships would arguably serve to strengthen both approaches. To achieve this there is undoubtedly a need for greater integration between the work of ecologists, lawyers, and economists in this area.

Finally, this chapter has observed that there is a lack of a governance system on an international level (or even at national levels) that responds adequately to the issue of ecosystem services protection. It is possible that recent research developing the basis of a new rights-based system of governance could play a significant part in addressing that absence. (This also would correspond with core element 4 put forward in Chapter 2.)

References

Achterman, G. L. & Mauger, R. (2010). The state and regional role in developing ecosystem service markets. *Duke Environmental Law and Policy Forum* **20**, 291–337.

Anton, D. K. & Shelton, D. L. (2011). *Environmental Protection and Human Rights*. Cambridge University Press, Cambridge.

Barbier, E. B. (2013). Wealth accounting, ecological capital and ecosystem services. *Environmental & Development Economics* **18**, 133–161.

Bates, G. (2001). A duty of care for the protection of biodiversity on land. Consultancy Report, Report for the Productivity Commission, Ausinfo, Canberra.

Bell, S., McGillivray, D., & Pedersen, O. W. (2013) *Environmental Law*, 8th edition. Oxford University Press, Oxford.

Bishop, J. (2010). *TEEB: The Economics of Ecosystems and Biodiversity Report for Business (Executive Summary)*. Earthscan, New York.

Blanco, E. & Razzaque, J. (2009). Ecosystem services and human well-being in a globalized world: assessing the role of law. *Human Rights Quarterly* **31**, 692–720.

Botkin, D. B. & Keller, E. A. (2003). *Environmental Science*. Wiley, Hoboken, NJ.

Boyd, D. R. (2012). *The Environmental Rights Revolution: A Global Study of Constitutions, Human Rights and the Environment*. UBC Press, Vancouver.

Costanza, R., d'Arge, R., de Groot, R., *et al.* (1997). The value of the world's ecosystem services and natural capital. *Nature* **387**, 253–260.

Daily, G. C., Alexander, S. E., Ehrlich, P. R., *et al.* (1997). Ecosystem services: benefits supplied to human societies by natural ecosystems. *Issues in Ecology* **2**, 1–18.

Davis, A. I. (2010). Ecosystems services and the value of land. *Duke Environmental Law and Policy Forum* **20**, 339–384.

Eves, C. & Blake, A. (2013). Assessing the long-term viability of leasehold rural land in Queensland. 19th Annual Pacific Rim Real Estate Conference, Melbourne. Available at: www.prres.net/papers/Eves_Assessing_The_Long_Term_Viability_Of_Leasehold_Rural_Land_In_Queensland.pdf (accessed 19 May 2014)

Fabra, A. & Arnal, E. (2002). Review of jurisprudence on human rights and the environment in Latin America. Joint UNEP–OHCHR Expert Seminar on Human Rights and the Environment – Background paper No. 6. Available at: www2.ohchr.org/english/issues/environment/environ/bp6.htm (accessed 2 August 2013).

Ferraro, P. J. (2001). Global habitat protection: limitations of development interventions and a role for conservation performance payments. *Conservation Biology* **15**(4), 990–1000.

Gathii, J. & Hirokawa, K. H. (2012). Curtailing ecosystem expropriation: ecosystem services as a basis to reconsider export driven agriculture in economies highly dependent on agricultural exports. *Virginia Environmental Law Review* **30**(1), 1–27.

Geneletti, D. (2013). Ecosystem services in environmental impact assessment and strategic environmental assessment. *Environmental Impact Assessments Review* **40**, 1–2.

Haines-Young, R. & Potschin, M. (2010). Proposal for a Common International Classification of Ecosystem Goods and Services (CICES) for Integrated Environmental and Economic Accounting. Department of Economic and Social Affairs Statistics Division. United Nations. ESA/STAT/AC.217UN CEEA/5/7/BK. Available at: www.nottingham.ac.uk/cem/pdf/UNCEEA-5-7-BK1.pdf (accessed 19 December 2013).

Landsberg, F., Ozment, S., Stickler, M., *et al.* (2011). WRI : ecosystem services review for impact assessment (Working Paper). World Resources Institute.

Lele, S. (2009). Watershed services of tropical forests: from hydrology to economic valuation to integrated analysis. *Current Opinion in Environmental Sustainability* **1**(2), 148–155.

Lugo, E. (2008). Ecosystem services, the Millennium Ecosystem Assessment, and the conceptual differences between benefits provided by ecosystems and benefits provided by people. *Journal of Land Use and Environmental Law* **23**, 1.

McAfee, K. (2012). The contradictory logic of global ecosystem service markets. *Development and Change* **43**(1), 105.

McGillivray, D. (2012). Compensating biodiversity loss: the EU Commission's approach to compensation under Article 6 of the Habitats Directive. *Journal of Environmental Law* **24**(3), 417–450.

Mertens, K., Cliquet, A., & Vanheusden, B. (2012). Ecosystem services: what's in it for a lawyer? *European Energy and Environmental Law Review* **21**(1), 31–40.

Millennium Ecosystem Assessment (2005). *Ecosystems and Human Wellbeing: Biodiversity Synthesis*. World Resources Institute, Washington, DC.

Ojea, E., Martin-Ortega, J., & Chiabai, A. (2012). Defining and classifying ecosystem services for economic valuation: the case of forest water services. *Environmental Science & Policy* **19**–20, 1–15

Razzaque, J. (2002). Human rights and the environment: the national experience in South Asia and Africa. Joint UNEP–OHCHR Expert Seminar on Human Rights and the Environment – Background paper No.4. Available at: www2.ohchr.org/english/issues/environment/environ/bp4.htm (accessed 2 August 2013)

Salzman, J. (2005). Creating markets for ecosystem services: notes from the field. *New York University Law Review* **80**, 870–961.

Salzman, J., Thompson, B. H., & Daily, G. C. (2001). Protecting ecosystem services: science, economics and law. *Stanford Environmental Law Review* **20**, 309–332.

Social and Economic Rights Action Center and the Center for Economic and Social Rights *v.* Nigeria, Communication 155/96 (African Commission of Human and Peoples' Rights, 27 October 2001).

TEEB. (2010). *The Economics of Ecosystems and Biodiversity for Local and Regional Policy Makers*. TEEB, London.

ten Brink P., Russi, D., Farmer, A., *et al.* (2013). *The Economics of Ecosystems and Biodiversity for Water and Wetlands. Executive Summary*. TEEB, London.

Tongson, E. & Balasinorwala, T. (2010). Payment for Ecosystem Services, Sibuyan Island, Philippines. Available at: www.teebweb.org/wp-content/uploads/2013/01/Payment-for-Ecosystem-Services-Sibuyan-Island-Phillipines.pdf (accessed 10 August 2013).

Turner, S. J. (2009). *A Substantive Environmental Right*. Wolters Kluwer, Leiden.

Turner, S. J. (2014). *A Global Environmental Right*. Routledge, Abingdon.

20 Water ecosystem services

Moving forward

Julia Martin-Ortega, Robert C. Ferrier, and Iain J. Gordon

Understanding water ecosystem services requires both an elucidation of the interrelationships between hydrology, landscapes and ecology, and a contextualization of how water influences human livelihoods and wellbeing and how ecosystems themselves are affected by human activities. Ecosystem services-based approaches, as defined in this book, aim to understand these complex relationships to support more efficient and sustainable decision-making. Society needs to recognize the requirement to balance and manage the benefits derived from water resources, rather than simply managing the resource itself (United Nations Environmental Programme 2009). These benefits come from the realization of a whole range of provisioning, regulating, and cultural services provided by ecosystems. This way of interpreting water systems and water resources represents a change from traditional sectoral control policies and approaches to delivering an integrated view of natural resource management.

In this book the editors and authors took up the challenge, ten years after the publication of the Millennium Ecosystem Assessment (2005), of reflecting on what has been achieved, what lessons have been learnt, and how to improve the application of ecosystem services-based approaches for managing water ecosystems in the future. By proposing a structured definition of ecosystem services-based approaches (in Martin-Ortega et al.) and exploring the forefront of their application at the conceptual level and through a series of national and regional case studies from across the world, the authors have completed a comprehensive vision of the current knowledge and challenges of applying ecosystem services-based approaches to address water challenges.

In this concluding chapter we reflect upon the key messages that have emerged from the discussions contained in this book and on the way forward. We organize this discussion around the four parts in which the book is structured.

20.1 HOW CAN ECOSYSTEM SERVICES-BASED APPROACHES HELP ADDRESS MAJOR GLOBAL CHALLENGES?

Capon et al., Febria et al., and Salman and Martinez explore how ecosystem services-based approaches have been, and can be,

applied to address three of the critical global challenges that humanity currently face, namely climate change, biodiversity loss, and meeting the growing population's food and energy demands (notably in the developing world).

In general, the authors find that ecosystem services-based approaches can help address these challenges, for example by identifying and prioritizing climate change adaptation options, by focusing attention on highly valued and/or vulnerable ecosystem services, and by assessing whether risks associated with climate change are due to services' supply and/or demand. Similarly, ecosystem services-based approaches can help in finding new strategies to sustain agricultural growth while preserving other services, by enforcing an analysis of trade-offs between provisioning services (such as food and energy) and other regulating and cultural ecosystem services.

The usefulness of applying ecosystem services-based approaches to managing biodiversity seems less certain. Managing biodiversity under an ecosystem services-based approach requires, according to Febria et al., a social decision about whether biodiversity should be seen as a final service, or as supporting ecosystem function as an intermediate service, or some combination of the two. Plus, critically, more knowledge is required on the biophysical processes that underpin freshwater biodiversity, ecosystem function, and ecosystem services.

20.2 IS THE NOTION OF ECOSYSTEM SERVICES USEFUL FOR WATER MANAGEMENT AND NATURE CONSERVATION?

The management of catchments or watersheds requires the recognition that land and water are inextricably intertwined and that every land use decision is a waters decision. It requires a vision for planning and management that embraces all aspects of the ecosystem and the linkages between them, and that recognizes that decisions for water must not be taken in isolation or independently of the people who depend on those ecosystems

(Ferrier & Jenkins 2010). Ecosystem services-based approaches as defined in this book provide a good basis for this vision.

Niasse and Cherlet and Blackstock *et al.* reflect on how the notion of ecosystem services is useful for Integrated Water Resources Management generally, and in the context of the European Water Framework Directive specifically. Niasse and Cherlet conclude that Integrated Water Resources Management can benefit from an ecosystem services-based approach to better implement its environmental sustainability pillar, as the approach: helps to communicate the importance of the multiple values of services provided by ecosystems; sets in place inclusive consultative platforms involving the poor as custodians of ecosystems; and bridges water and ecosystem management. The authors suggest that including an ecosystem services component in national Integrated Water Resources Management strategies and planning can help countries to better understand and factor in their natural capital. However, they warn about the risk of the approach being overly anthropocentric and short-sighted and advocate for a combined approach.

Similarly, Blackstock *et al.* find crucial elements by which ecosystem services-based approaches can help the Water Framework Directive deliver wider policy imperatives of sustainability, integration, and subsidiarity, and live up to its original ambition of meeting good ecological status in European water bodies. However, the authors also warn about the risk of focusing on the most easily monitored immediate benefits to society, ignoring the less visible or less immediately relevant factors within the systems.

In relation to biodiversity conservation, Leisher, similarly to Febria *et al.*, discusses the weak correlation that some provisioning and regulating ecosystem services have with biodiversity, and reflects on the implications that this has for conservation initiatives. He suggests that agencies and organizations should use ecosystem services-based approaches primarily as an information and advocacy tool, but take a more cautious approach when using them for conservation design (this relates to Febria *et al.*'s conclusion that ecosystem services-based approaches can serve as a complement, not a replacement, for conservation-based ones). For payments for watershed services initiatives, Leisher suggests that investments should be limited to where water treatment costs can be reduced substantially by better watershed management, and focus on the worst pollution sources first.

20.3 INTEGRATED BIOPHYSICAL AND SOCIO-ECONOMIC ASSESSMENTS OF ECOSYSTEM SERVICES

The third and largest section of this book provides examples of assessments of ecosystem services through a number of case studies from across the world illustrating how ecosystem

services-based approaches as defined in this book (Martin-Ortega *et al.*) are put into practice. In general, these show the latest advances in the integration of the biophysical quantification of ecosystem service delivery with economic valuation techniques.

By reviewing the UK National Ecosystems Assessment, Schaafsma *et al.* highlight how ecosystem services-based assessments at the national level may help reveal the importance of water and ecosystems to human wellbeing, but warn about how a full assessment will require the development of new knowledge. Similarly, and also at the national level but this time focusing on a particular type of ecosystem, Kang *et al.* have tested a 'rapid' spatially explicit ecosystem services analysis of coastal areas in South Korea. This type of analysis shows the potential for identifying synergies and trade-offs among multiple ecosystem services and identifying hot-spots that can be targeted for conservation, but requires care in verifying outcomes with local-scale data.

At the river basin level, Crossman *et al.* present a comprehensive assessment of a range of ecosystem services provided by the Murray-Darling Basin in Australia. Using a characteristic ecosystem services-based approach, the authors show how monetized benefits are in the same order of magnitude as management costs and reflect on the usefulness of this kind of assessment for policy making. Another Australian case, in the Southeast Queensland region (Maynard *et al.*), uses a method of value weights as an alternative to monetary valuation. The authors emphasize the need for a framework to consistently conduct ecosystem services assessments across stakeholders, so these can feed into Integrated Catchment Management.

Villa *et al.* and Mulligan *et al.*, address the issue of how to integrate biophysical and economic modelling to support decision-making. Mulligan *et al.*, by exploring the Rio Daule in Western Ecuador, show how tools for the assessment of ecosystem service and management interventions can be coupled with tools for the optimization of investments in water funds, spatially and across multiple objectives. However, they conclude that tools are often ahead of the available data for ecosystem services assessment. Villa *et al.* use the Ankeniheny–Zahamena Forest Corridor in eastern Madagascar as a case study for testing an integrated approach based on the use of artificial intelligence. This approach places the emphasis on beneficiaries, looking at four dimensions of ecosystem services value: input productivity, economic value, sustainability of supply, and quality of supply. The authors conclude that looking at this broader suit of dimensions, beyond strictly monetary values, can better support multi-criteria decision-making.

20.4 BROADENING THE PERSPECTIVE

Part IV of this book broadens the perspective on water ecosystem services beyond its typical boundaries. Houdet *et al.*, Church *et al.*, and Bril *et al.* discuss aspects of ecosystem services-based approaches that have only started to gain momentum: the notion of ecosystem services in the business world, cultural ecosystem services, and the role of community partnership in implementing ecosystem services-based approaches. Turner and Corral Verdugo *et al.* provide thought-provoking contributions to unexplored areas: the interface between human rights and ecosystem services and the psychological dimensions of ecosystem services-based approaches.

Houdet *et al.* suggest a set of guidelines to properly engage the business community in the conservation and management of ecosystem services, which include the need for promoting precise mapping of water ecological infrastructure and their uses, sharing information among businesses and stakeholders, and the promotion of water ecosystem services stewardship and water footprint guidelines. By reviewing the famous New York City Catskill watershed management initiative and comparing it with other initiatives in Europe, Bril *et al.* reflect on the keys to successful application of ecosystem services to water management. These keys to success are underpinned by the participation, information-sharing, and communication between stakeholders and communities.

Church *et al.* find that the understanding of cultural ecosystem services is still hampered by lack of consensus regarding definitions as well as overly focusing on recreation/tourism and aesthetics, to the detriment of other service categories such as knowledge systems, sense of place, and spiritual and religious values. Despite these difficulties, cultural ecosystem services have helped expand the approach beyond tangible benefits provided by ecosystems. But Corral Verdugo *et al.* warn about the fact that critical aspects of human wellbeing tend to be ignored, even when looking at non-monetary assessments. They explore the psychological aspects associated with human wellbeing (pleasure, spirituality, happiness, mental restoration) and their role in the realization of values from water ecosystem services. Moreover, the authors broaden the perspective of sustainable decision-making beyond the higher-level spheres (e.g. catchment management or integrated water resources management), to the human level, and advocate that acknowledging the psychological dimension of ecosystem services can help promote water conservation behaviours at the individuals' level.

Finally, Turner reflects on how accepted contemporary understanding of the rights to life, health, food, and water place obligations upon states to protect certain ecosystem services, establishing a link between human rights legislation and ecosystem services-based approaches. As a result, the author argues that these approaches can support each other in advocacy and policy recommendations. Turner concludes that there is currently a lack of an international system of governance for the protection of ecosystem services, but that research indicates that such a system is possible in the medium to long term.

20.5 THE CHALLENGE AHEAD

As discussed by Martin-Ortega *et al.*, and demonstrated in the other chapters, the concept of ecosystem services has inspired collaboration across scientists from different disciplines. It has also awakened the interest of policy makers and conservation agencies and organizations, and, more recently, it appears to be appealing to the business world. However, popularization of the concept has not only resulted in lack of clarity about the meaning of ecosystem services, but also in a risk of 'over appropriation' of the term and confusion over what ecosystem services-related approaches and frameworks might entail. The confusion around terminology has, in many circumstances, proved to be a barrier to the more formalized adoption of ecosystem services analysis, particularly within operational contexts. Increasing clarity around terminology and the adoption of agreed core elements is needed for the development of a 'shared language' and improved delivery of both outputs and outcomes.

There is also a growing concern about the gap between the conceptualization and the endorsement of the ecosystem services 'rhetoric' and the actual use of ecosystem services-based approaches in natural resources management practice (Nahlik *et al.* 2012), although some of the cases presented in the book have shown specific ways to take this forward in relation to integrated catchment management and water fund investments.

There are also risks associated with the enthusiastic expansion of ecosystem services-related concepts and ecosystem services-based approaches. These range from an over-use of the terminology and part (but not all) of its core elements, which devalue the concept, to risks associated with the promotion of an excessively anthropocentric view of the natural world. Mechanisms to monitor the effectiveness, on the one hand, and, on the other hand, the effects of applying ecosystem services-based approaches in the governance of natural resources management are needed.

Regarding effectiveness, throughout this book the authors have pointed out a number of current limitations to the application of ecosystem services-based approaches. Some common themes emerge in this respect, notably concerns about the need for more and better data to feed into the new integrated assessment tools. In general, environmental resource information is becoming increasingly available and integrated and such 'big data' provide a useful platform for the integration of ecosystem services-based approaches to resource management. Along with improved approaches to the collection, curation, and management of information, there is a concomitant development of new software approaches through which to analyse and process such information. Adopting an ecosystems services-based approach, though, brings additional difficulties. Coordinating data collection, harmonization, and sharing among multiple political jurisdictions still remains a major challenge. Moreover, the availability of categorical evidence of the effects of interventions (e.g. river restoration) in terms of final ecosystem services and how these translate into goods and benefits that are perceived by the public remains a critical challenge. For that, there is only one way forward: a sustained and ever-increasing dedication of efforts to integrate knowledge from the natural sciences realm with that of the social sciences. Although this has been said many times, it is still not happening to the extent that is needed. Increasingly complex problems require such integration of knowledge from across disciplinary boundaries, confronting differences and seeking common solutions. Building interdisciplinary capacity requires a shift away from traditional thinking and a more integrated approach, including social learning, systems-based approaches, and sustainability science. A critical aspect is the need for research funds and time to accommodate the research processes required to breach disciplinary boundaries.

Regarding the other challenge, the effects (positive and negative) of employing ecosystem services-based approaches in the governance of natural resources management is as critical and as urgent as the former, but has somehow received much less attention. Moreover, Waylen and Young (2014) find that initiatives to improve knowledge to assist in environmental policy (including ecosystem services assessment such as the Millennium Ecosystem Assessment (2003, 2005) and the UK National Ecosystem Assessment (2011)), are not accompanied by an evaluation of the (short- and long-term) implications of this knowledge use. Authors such as Gomez-Baggethun *et al.* (2010) and Peterson *et al.* (2010) have expressed concerns that mainstreaming ecosystem services may result in applications that diverge from the original pedagogic purpose of the concept and might move into the commodification of nature for trade in potential markets (Corbera & Pascual 2012). In addition, and as discussed by Martin-Ortega *et al.* in this book, excessive, uncritical faith in the potential of management approaches based on some form of an ecosystem services framework is likely to result in disillusion if solutions prove to be unsatisfactory. If core element 3 (transdisciplinary) of ccosystem services based approaches, as defined in this book, was properly applied in its pure form, some of these concerns may appease. Several chapters of this book have emphasized how ecosystem services-based approaches should and can engender public and political support during assessment and planning processes. Stakeholder participation, public engagement, and cross-section communication seem critical aspects in this process and, if taken seriously, could contribute to ensuring that ecosystem services remain a tool and not the aim.

As witnessed by the pollution of waterways, the depletion of aquifers, and the degradation of wetlands, the pressures on the planet's water resources are already significant. With growing human populations, aspirations of increased economic growth, and climate change, these pressures are likely to increase for the foreseeable future. If our water-based ecosystems are to continue to provide the services that support human life and spiritual and material wellbeing to the current and future generations, society will have to start taking seriously the values (in their different dimensions) of the services ecosystems produce. In this book we set out to provide a comprehensive and up-to-date vision of the potential of ecosystem services-based approaches. As demonstrated in the chapters of this book, ecosystem services-based approaches are not a panacea, and they need to be implemented critically and monitored. To date, ecosystem services-based approaches have shown the great virtue of stimulated dialogue about how natural resources are valued and used; now it is important to make sure this dialogue remains meaningful, critical, and purpose-driven.

ACKNOWLEDGEMENTS

This book is part of the UNESCO and the James Hutton Institute Global Dialogue on Water Ecosystem Services (www.hutton.ac.uk/research/themes/managing-catchments-and-coasts/ecosystem-services/unesco-global-dialogue (last accessed 12 December 2015). The editorial effort has been supported by the Scottish Government, through the Rural Affairs and the Environment Portfolio Strategic Research Programme 2011–2016. The editors are extremely grateful to all the authors participating in this book, for their invaluable contributions and their significant efforts to conform to the editorial requirements. We are also very grateful to Emma Kiddle and Zoe Pruce from Cambridge University Press for their continued support and guidance, as well as to four independent reviewers. Thanks also to Kirsty Holstead for early support. Finally, we would like to express our deepest gratitude to Carol Kyle, without whose priceless assistance this book would have not been accomplished.

References

Corbera, E. & Pascual, U. (2012). Ecosystem services: heed social goals. *Science* **335**(10), 355–356.

Ferrier, R. C. & Jenkins, A. (2010). *Handbook of Catchment Management.* Wiley: Chichester.

Gomez-Baggethun, E., De Groot, R. S., Lomas, P. L., *et al.* (2010). The history of ecosystem services in economic theory and practice: from early notions to markets and payment schemes. *Ecological Economics* **69**(6), 1209–1218.

Millennium Ecosystem Assessment (2003). *Ecosystems and Human Wellbeing: A Framework for Assessment.* Island Press, Washington, DC.

Millennium Ecosystem Assessment (2005). *Ecosystems and Human Wellbeing: General Synthesis.* Island Press, Washington, DC.

Nahlik, A. M., Kentula, M. E., Fennessy, M. S., *et al.* (2012). Where is the consensus? A proposed foundation for moving ecosystem service concepts into practice. *Ecological Economics* **77**, 27–35.

Peterson, M. J., Hall, D. M., Feldpausch-Parker, A. M., *et al.* (2010). Obscuring ecosystem function with application of the ecosystem services concept. *Conservation Biology* **24**(1), 113–119.

UK National Ecosystem Assessment (2011). *The UK National Ecosystem Assessment: Synthesis of the Key Findings.* UNEP-WCMC, Cambridge.

United Nations Environment Programme (2009). *Water Security and Ecosystem Service: The Critical Connection.* UNEP, Nairobi.

Waylen, K. A. & Young, J. (2014). Expectations and experiences of diverse forms of knowledge use: the case of the UK National Ecosystem Assessment. *Environment and Planning C: Government and Policy* **32**, 229–246.

Index

adaptation options addressing climate change risks to water
 ecosystem services, 20
 See also climate change
Africa, 39
ARIES (ARtificial Intelligence for Ecosystem Services), 111

benefit-sharing mechanisms, 100
 See also payments for ecosystem services
biodiversity management, 33
biodiversity loss, 26
biofuels, 36

climate change, 17, 111
 See also adaptation options addressing climate change risks to water
 ecosystem services; risk to water ecosystem services under a
 changing climate
coastal ecosystem services, 119
conservation, 27
Convention on Biological Diversity, 4, 50, 66
core elements, 1–4, 8–9
cultural ecosystem services, 148
 See also scenario analysis scope of cultural ecosystem services

disproportionality, 59
 See also Water Framework Directive

ecosystem approach, 4
ecosystem services, 3
ecosystem services-based approaches, 8
environmental rights, 164

final ecosystem services, 7, 33
 See also goods

Global Environmental Right, 167
 See also human rights
goods, 7
 See also final ecosystem services

human rights, 163
 right to food, 165
 right to health, 165
 right to life, 164

right to property, 166
right to water, 165
 See also Global Environmental Right

Integrated Catchment Management, 90
Integrated Valuation of Environmental Services and Tradeoffs, 121
 See also InVEST
Integrated Water Resources Management, 49
InVEST, 100, 121

Millennium Ecosystem Assessment, 3
Murray-Darling Basin scenario, 84

payment for ecosystem services, 3, 67, 143
 See also benefit-sharing mechanisms; upstream thinking; water funds
procedural environmental rights, 166
 See also human rights

risks to water ecosystem services under a changing climate, 17
 See also climate change

scenario analysis, 77
 See also Murray-Darling Basin scenario; UK National Ecosystem
 Assessment scenario analysis
scenarios for interventions at the Rio Daule Ecuador. *See* scenario analysis
scope of cultural ecosystem services, 149
substantive environmental rights, 166
 See also human rights

UK National Ecosystem Assessment, 73

UNESCO's Ecohydrology initiative, 11
UNESCO's International Hydrological Programme, 11
Upstream Thinking, 143
 See also payment for ecosystem services

water conservation, 156
Water Framework Directive, 57
 See also disproportionality
Water funds, 99
 See also payments for ecosystem services
WaterWorld, 100
Wealth Accounting and the Valuation of Ecosystem Services, 110